BODIES IN THE BOG

KARIN SANDERS

BODIES IN THE BOG
and the Archaeological Imagination

The University of Chicago Press | Chicago and London

KARIN SANDERS is professor of Scandinavian at the University of California, Berkeley.

The University of Chicago Press, Chicago 60637
The University of Chicago Press, Ltd., London
© 2009 by The University of Chicago
All rights reserved. Published 2009
Printed in the United States of America

18 17 16 15 14 13 12 11 10 09   1 2 3 4 5

ISBN-13: 978-0-226-73404-0   (cloth)
ISBN-10: 0-226-73404-8   (cloth)

Library of Congress Cataloging-in-Publication Data
Sanders, Karin, 1952–
    Bodies in the bog and the archaeological imagination / Karin Sanders.
        p. cm.
    Includes bibliographical references and index.
    ISBN-13: 978-0-226-73404-0 (cloth : alk paper)
    ISBN-10: 0-226-73404-8 (cloth : alk paper)
    1. Bog bodies—Europe.   2. Human remains (Archaeology)—Europe.   3. Water-
saturated sites (Archaeology)—Europe.   4. Antiquities, Prehistorical—Europe.
5. Europe—Antiquities.   I. Title.
    GN803.S24 2009
    936—dc22
                                                                            2009010355

♾ The paper used in this publication meets the minimum requirements of the American
National Standard for Information Sciences—Permanence of Paper for Printed Library
Materials, ANSI Z39.48-1992.

For Kenny

[CONTENTS]

[**PREFACE AND ACKNOWLEDGMENTS**]

The seed for this study was planted decades ago. In 1965 the Danish archae-ologist P. V. Glob published *Mosefolket* (*The Bog People*) and I remember find-ing a copy of the book in my primary school library outside Copenhagen. The thrill of reading about the uniquely preserved human beings in the peat bogs of northwestern Europe and the powerful effect of the black and white photo-graphs of faces, feet, and hands from these ancient yet seemingly contempo-rary people made me ponder whether I too should become an archaeologist. The realness of the bodies combined with an aura of mystery—even if it was of a rather melancholic and sometimes frightening kind—took me on many imaginary time travels.

Eventually my interests in literature and the arts won out over my fas-cination with archaeology, and I started digging out the past as a literary scholar. But the lure of archaeology and my interest in bog bodies have lin-gered through the years. The Irish poet Seamus Heaney's bog poems, from *Wintering Out* (1972) and *North* (1975), reopened the door to archaeology for me and provided a bridge between the world of words and that of ma-terial remains. Heaney's poems on Tollund Man and other bog bodies made Glob's book present and familiar again, and caused me to revisit *The Bog People* (this time in its 1969 English translation). I discovered that Heaney and Glob were not the only ones who had searched these ancient faces and bodies for arsenals of archaeological tropes to be used in literature or visual art. The roster of international responses to bog bodies—including Sigmund Freud, Carl Jung, Heinrich Himmler, Joseph Beuys, Williams Carlos Williams, Geoffrey Grigson, Hugo Claus, Serge Vandercam, Margaret Atwood, Wallace Stegner, Anne Michaels, Désirée Tonnaer, and many more—is so long and impressive that alone it makes the case that a study of bog bodies outside of the discipline of archaeology is warranted.

I soon realized that the discovery of a bog body is not the culmination of its journey through time, but rather a point of departure from which a new journey begins. To follow this journey I have chosen a variety of written texts (poems, novels, dramas, scientific studies) and visual representations (paintings, sculpture, photography, museum displays, facial reconstructions), which, each in its way, situate bog bodies in relation to the archaeological imagination. This book then is concerned with the textual and visual corpus surrounding the mummified bodies. It is *not* an archaeological examination but rather an attempt to show how we see, read, and experience these remarkable remains as human time capsules capable of connecting past and present in ways different from other archaeological artifacts. It is about how various discourses (archaeological, literary, political, poetical) imagine bog bodies and make stories about their past and present lives. And it is about how visual imagination has been able to use bog bodies in art.

When Glob published *The Bog People* it was a response to public demand but also a means to set the record straight in terms of numerous misconceptions that had filled the pages of the news media since the 1950 discovery of the bog body named Tollund Man and the equally spectacular 1952 unearthing of the so-called Grauballe Man. Questions needed to be addressed regarding ethics in the display of human corpses, queries as to their humanness, concerns about their disrespectful commercialization, misidentification of and disregard for the depth of time they represented, and their place in the canon of the nation's prehistory.

The Bog People has since become a classic. It is still in print in English and, more important, it continues to this day to be the source book for artistic expression on bog bodies. At one point Glob tells us of a bog body discovered in 1942 by peat cutters in Denmark that was found "doubled up like a question-mark."[1] It is easy to imagine this corporeal "question-mark" as a particularly fitting visualization of the many intricate problems and methodological challenges that spring from the corpus of bog bodies. As forensic archaeological objects, they are obviously subject to the particular methodology and history of the archaeological discipline *proper,* but in the kind of analysis I will propose here, the material object itself cannot be understood through just one prism or one particular methodological approach. I suggest, rather, that we see bog bodies as unique go-betweens on many fronts, straddling not only the binaries of time and space, past and present, text and image, and ethics and aesthetics, but also the disciplinary boundaries between archaeology, history, literary studies, and art history.

Throughout this book I center on how the material body is negotiated when it "ruptures" time and space. I show how, when the past is seen in the form of archaeological material, the value given to it is inevitably wedded to its ability to serve as authentic testimony. To *see* something, or to *touch* something, is presented as authentic precisely because the senses seem to allow for transparent and direct access to knowledge about the object at hand. The authentic in archaeology is, after all, a promise of closeness to what is historically far removed. All of the examples discussed in this book, then, have at least one common denominator: they all in one way or another negotiate the pressure of the bog bodies' material presence. This pressure, which is also a pressure from the past, gives urgency to the reality of the physical body—to the poetic and artistic mind, to the hand that writes or sculpts or paints. The "stubborn" materiality of the body's corporeal existence is often vividly confronted with the material resistance offered by each medium's own particularity, be it canvas, clay, wood, bronze, or words. Since bog bodies are in some sense born as scientific specimens, museum pieces, and preserved artifacts so different from what they once were, they are estranged from us even as they mirror us. It is precisely this combined identification and estrangement that gives bog bodies (and other mummified human remains, for that matter) a valence that is different than that of other archaeological objects. If we project human traits onto material artifacts that already exhibit—or seem to inhabit—those same traits, our projections are caught in the paradox of their very existence. Bog bodies are both inanimate objects and human beings.

I argue that bog bodies gain mutable identities in their complicated and sometimes boisterous afterlife. My intention is to consider fundamental questions about culture and humanity—heritage, origin, nationality, genetics, ethnicity, and gender—by way of bog bodies. I show that while the archaeological imagination situates and actualizes the bodies' humanness and historicity by use of display strategies, photography, rhetorical gestures, and so forth, their situatedness and actualization stipulate sensitivity to and negotiation of the fuzzy boundaries that the bodies have to navigate in their travels through time and space: their imagined morphing from being persons to becoming things, and conversely from being things to becoming persons.

In this process, as we shall see, almost all bog bodies discussed here regain personal-pronoun attributes. In a kind of transposing of the body to its linguistic equivalent, "it" becomes once more "he" or "she." Behind this drive to humanize is then a drive to see humanity as both singular and universal, to see the bog bodies both as gendered individuals and as human beings in abstract terms. No one has expressed this with more acumen than Seamus Heaney when he brings the material past up close and makes it personal as he describes the kind of reanimation, but also loss of identity, that happens on the level of personal pronouns:

> Once upon a time, these heads and limbs existed in order to express and embody the needs and impulses of an individual human life. They were the vehicles of different biographies and they compelled singular attention, they proclaimed "I am I." Even when they were first dead, at the moment of sacrifice or atrocity, their bodies and their limbs manifested biography and conserved vestiges of personal identity: they were corpses. But when a corpse becomes a bog body, the personal identity drops away; the bog body does not proclaim "I am I"; instead it says something like "I am it" or "I am you." Like the work of art, the bog body asks to be contemplated; it eludes the biographical and enters the realm of the aesthetic.[2]

The plethora of material on bog bodies in the "realm of the aesthetic"—the volume of which far exceeded my expectations when I first embarked on this project—takes such diverse shapes and forms that it has been a challenge to house it on the pages of one study. I have therefore had to make necessary selections. Not all poems on bog bodies are included; not all novels, nor all visual art or museum displays. Whenever possible I have chosen works that are representative of each chapter's specific concerns: the ethics of display, the uncanny, erotica, politics, and so forth. At the same time I have strived to hold onto the multifaceted corpus of texts and have allowed each chapter to unfold the different peculiarities of the material.

Ultimately, as should be apparent when we come to the end of this study, in spite of the heterogeneity in bog body representations we will find two fundamental tenets at play throughout the material. One concerns time and the ruptures and reverberations that come with untimely reappearances of dead bodies; the other concerns the ethics of aestheticizing human remains from an ancient past. The bodies represent our humanity—and inhumanity. Reconnection with the past is frequently intended to *stabilize the present*. Yet bog bodies, uncanny and liminal, more often than not refuse to be constant; indeed, they *destabilize*. "We stand face-to-face with our ancestors," claimed Glob. Other archaeologists have called the bodies "silent witnesses"[3] or claimed that they "open a window to the past, through which we see things we would otherwise seldom see."[4] To be face-to-face with silent witnesses who can open windows

to the past is no small matter. As we shall see, these witnesses are far from silent; they speak volumes across time. In fact bog bodies, I have discovered, present a particularly rich opportunity to investigate a familiar unfamiliarity with
human beings from the past. They allow us to see a "spectacle of human activity," to borrow a phrase from Marc Bloch, full of twists and turns—a complex
ethical and aesthetic "quagmire" from which we can pull information about
literature, art, and archaeology that goes well beyond that of the bog bodies
themselves.[5]

In my exploration of the various forms bog bodies take in literature
and visual representation, I have found myself immersed in a sort of cross-
disciplinary traveling that resembles what Mieke Bal has outlined in her book
Traveling Concepts in the Humanities. She proposes that a cultural analysis must
find its "basis in concepts rather than methods." The object of cultural analysis,
she goes on to say, often changes in the process, so that "after returning from
our travels, the object constructed turns out to no longer be the 'thing' that so
fascinated you when you chose it. It has become a living creature, embedded
in all the questions and considerations that the mud of your travel spattered
onto it, and that surround it like a 'field.'"[6] Bal's expression, that "it has become
a living creature," has of course a particular kind of irony when we speak of bog
bodies. But the point she makes—that the object at hand becomes animated
and malleable when looked at through the concepts of cultural analysis, and
not through one particular or inherent methodology—resonates with my own
experience of traveling the muddy fields and "digging" out bog bodies.

In *The Bog People,* Glob was sensitive to the peculiar time factor in relation to writing, publishing, and doing research on archaeological bodies.
He calls his book a "long letter" written for a group of fifteen young English
schoolgirls—Veronica, Catherine, Elizabeth, Prudence, etc.—who had con-
tacted him asking for more information on bog bodies, and also for his own
daughter, who was the same age. As he signs off with an elegant gesture toward
the past, the present, and a possible future, the conclusion of his preface echoes ingeniously the very themes of his book and the time it took to complete
his work with it:

> But I have all *too little time,* so that it has taken me *a long time* to finish my
> letter. However, here it is. You have all *grown older* and so perhaps are now
> all the better able to understand what I have written about these bog people
> of *two thousand years ago.* Yours Sincerely P.V. Glob, August 13th, 1964.[7]

My own study, which is indebted to Glob's archaeological imagination in
so many ways, has been in process for nearly a decade. But I find solace in a re-

mark by French philosopher Gaston Bachelard, who inspired some of the bog body art analyzed in this book: "Slow is not just fast with the brakes on. Slowness when imagined also desires its own excess."[8] I have had the joy of "slowing time" and enjoying its "excess" in the company of wonderful colleagues, students, friends and family in America and Denmark and throughout Europe. I owe a particular debt of gratitude to my Danish family for housing me and nurturing me on my many research trips.

My research was assisted by the University of California at Berkeley in the form of a Townsend Center for Humanities Initiative Award in 2002 and a Humanities Research Fellowship in 2004–05. Meg Conkey served as my archaeology mentor and responder during the Townsend fellowship semester, and her kindness and enthusiasm for cross-disciplinary research helped ease my anxiety about trespassing into a field of study hitherto unfamiliar to me. In my home department at Berkeley, Scandinavian, I work daily with the most intellectually exciting and supportive group of colleagues one could ever hope for. Mark Sandberg, Linda Rugg, and John Lindow each read my manuscript in its entirety and offered detailed feedback and encouragement—as did Carol Clover, who also generously provided funds from her Class of 1936 Chair of the Humanities to help defray permission costs. I am immensely thankful to them all.

Dialogue with a host of Berkeley graduate and undergraduate students has been crucial to my work. They have listened patiently to me and have offered inspiring comments when bog bodies inevitably crept into courses I was teaching on matters seemingly unrelated to my book project: word-image studies, national identity, material culture and literature—even courses on Hans Christian Andersen and Isak Dinesen. I especially thank Dean Krouk for his assistance in locating material on Seamus Heaney.

My mentor in Denmark, Jette Lundbo Levy, read the final manuscript and, as always, offered sage advice, sharp criticism, and warm support. I am indebted also to Finn Hauberg Mortensen—for his steady hand and his willingness to discuss my project—and to my "lunch friends" from the Royal Library in Copenhagen: Jørgen Vogelius for encouragement, enthusiasm and assistance with historical details, and Nanna Damsholdt for helping me think about ethics. The Department of Scandinavian Studies and Linguistics at my alma mater, Copenhagen University, hosted me as a visiting scholar during 2004–05; I thank Niels Finn Christiansen and Thomas Bredsdorff for arranging ideal working conditions for me. At the National Museum of Denmark the late Jørgen Jensen shared stories about P. V. Glob and bog bodies, understood the links between archaeology and poetry, and gave me confidence that my project was important and feasible. Pauline Asign at the Moesgaard Museum discussed display strategies with me, shared her curation notes, and was generous to a fault with access to the museum's archives and resources. Niels Lynnerup offered helpful forensic information on mummies; Christian Fischer and the Silkeborg Museum allowed me access to their newspaper archive on Tollund Man. Flemming Kaul guided me through the newly curated exhibit of a bog body (Huldremose Woman) at the National Museum of Denmark and allowed me to "step in the footprints" of the gods. I have benefited greatly from invitations to give lectures on bog body reception at various academic institutions in the United States, Denmark, Norway, Finland, England, and Germany. I thank my many hosts and the participants for constructive feedback and ideas.

An enormous amount of material had to be sifted and culled for useful examples of bog body representation to complete this book. I have had to rely on the help of friends and colleagues, too numerous to mention here, for suggestions of material. Many will be thanked in the notes. No one individual, however, offered more direct assistance than did Wijnand van der Sanden, whose book *Through Nature to Eternity: The Bog Bodies of Northwest Europe* still remains the most important source of knowledge about bog bodies. In Assen, Holland, he shared with me his files on bog-artists and poets—an extraordinary and rare generosity for which I am immensely thankful. Vincent van Vilsteren guided me through *The Mysterious Bog People* exhibit at the Drent Museum and explained the logic behind the curation strategy. John Prag at Manchester University took time to discuss face reconstructions with me and made me consider the extent and limits of this enterprise. Toril Moi accompanied me to the British Museum to view the Lindow Man, and prompted me to think deeper about the relationship between human and thing. I thank them all. I'm also grateful to Nancy Ruttenburg, who believed that my bog work would become a published book long before I thought it possible. Early in the process I benefited from discussions about bog bodies and archaeological imagination with Jennifer Wallace and Christine Finn, and have since enjoyed reading their work. I am especially grateful to the many artists and poets who have helped me and allowed me to use their work. In particular I wish to thank Lori Anderson Moseman, Désirée Tonnaer, Trudi van der Elsen, Catherine Harper, Silvia Kantaris, and Kathryn Vaughan for going the extra mile in sending me sometimes difficult-to-locate visual and written materials.

At the University of Chicago Press I have been in the best of hands from day one. Susan Bielstein has been an enthusiastic and encouraging editor, Anthony W. Burton and Christopher Westcott have offered steady and savvy guidance, and Renaldo Migaldi's sensitive eye has rescued many sentences from being either too vague or too "inventive." I am extremely grateful to them all. Small portions of the various chapters have been published in shorter versions in journals and anthologies. A segment of chapter 4 found an early form in Danish in "Fra 'sølen' og 'pølen': et moseligs fortælling i material og litteraturhistorien," published in *Kampen om litteraturhistorien,* and a longer form in "Portal through Time: Queen Gunhild," published in *Scandinavian Studies,* 2009. Other early published versions of the research material include "Bad Bog Babes" in *Passage* #50, "Bodies in Process: Bog Bodies in Contemporary Art and Poetry" in *Edda 4/04. Scandinavian Journal of Literary Research,* and "Imagining Origin: Bog Bodies in the Discourse of Archaeological Ambivalence," in *Culture and Media Studies: European Perspectives,* 2009. No chapter, however, has previously been published in full. I am grateful to the various editors for allowing me to use materials that first appeared in their journals or anthologies.

Finally, I am deeply beholden to my family. My son Daniel, who was graduated from both high school and college while I worked on this book, continues to lift my spirit with his remarkable talent for living in the present and his contagious capacity for spreading confidence and optimism. And to his father, my husband Kenny, "first reader," friend and unfailing supporter, my gratitude exceeds all measure. He has read and commented on all versions of my work. Without his generous spirit, infinite patience, and strength of mind this book would never have seen the light of day.

:::

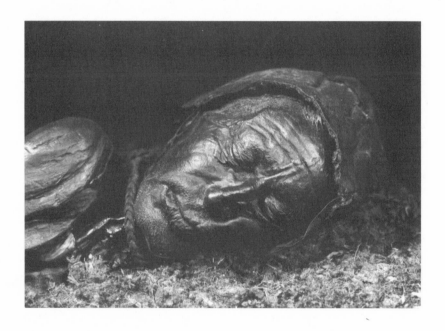

Fig. 0.1. The face of Tollund Man is an "accidental masterpiece"—a human being frozen in time, but also a corporeal time-traveler between past and present. Photograph © Silkeborg Museum, Denmark.

Remarkable Remains

Let me start with a face. In May 1950, an approximately two-thousand-year-old-body of a man was found in a bog by peat diggers on the main peninsula of Denmark and soon became the star amongst the so-called "bog people." He was called Tollund Man after the place where he was found. The emotive power of his countenance, the ease with which we identify with it, the delicate minutiae of his wrinkles, the traces of time so present in his day-old beard stubble and the peaceful expression on his face—at odds with the disturbing rope around his neck—have served as a screen upon which poets, visual artists, archaeologists, museumgoers, and even Internet browsers have projected all sorts of imaginings. The attraction of Tollund Man rests in the distinctiveness and striking individuality of his face: it is camera-friendly, offering a snapshot of what looks like a distant human relative. Caught by the mechanical lens, he is bound by nature but produced by culture. He functions as a gateway to a past in which we can imagine ourselves because he brings us to it—*face-to-face*. He is an accidental masterpiece: an anachronism born from contradictions, and a master reconciler of binaries![1] In his face, death is complete with life; the absent past is tangible and present; atrocious deeds—violent death, even human sacrifice—are glossed over by congenial traces. But he also *ruptures* our sense of time, *shatters* our sense of order, and *complicates* our sense of propriety.

Seamus Heaney points to a strange untimeliness and temporal foreshortening in Tollund Man's "pursed-lipped" face. The bog man resembles someone recently dead, a family member, his Great-Uncle Hughie, "a man whom I first saw standing tall in a turf-cart, profiled and gaunt, mustached and remote, as lean-featured in life as he would be when he lay in his coffin in death."[2] Yet in spite of Heaney's filial gesture, the ancient man who lies "in peace and in profile on the turf of a Jutland bog" remains a stranger. He does not come from our time; he is a visitor from another actuality, more than two thousand years ago. He lived, as P.V. Glob notes, "in the early Iron Age, around the time of Christ" and belongs to a peculiar group of archaeological remains of human beings: men, women, and children deposited intentionally or in some cases accidentally in the raised bogs of northern Europe and uncannily preserved through the unique properties of peat.[3] The anaerobic condition of the bogs provides an environment that slows decomposition significantly. The bog water and peat in some instances contain chemicals that naturally convert bodies into leather-like envelopes of preserved skin and hair. As a result, bodies in bogs have been able to survive for centuries with perfectly preserved features, fingerprints, nails, hair, and other distinct individual traits. Most are found in parts of northern Europe and stem from what are believed to be Iron Age sacrifices or victims of punishment.[4]

After lying dormant under the peat-blanket for centuries, these bog mummies have surfaced as uncannily well-preserved remains of human history, touching and morbid reminders of the prehistoric north European past and its practices. Many are found stripped of clothes, some with cords around their necks that suggest hanging or garroting. Others have been decapitated, and still others have suffered multiple injuries before their deaths. Some are severely marked by postmortem damage. A few have dried naturally, not needing further conservation; but the majority of the still existing bog mummies have undergone various secondary preservation processes such as freeze-drying, tanning, and waxing.[5] A number of the best preserved bog bodies are displayed in museums in Denmark, Germany, Holland, England, and Ireland. But some have disappeared or have been reburied, and are known only through written accounts.

Paper Bodies

One of the first descriptions of the peculiarity of bog mummification, a short note by Charles Leigh from 1700, reads: "One thing had almost slipt me, how sometimes in mosses are found human bodies entire and uncorrupted, as in a moss near the Meales in Lancashire. These are the most remarkable in phaenomena I have observ'd in morasses, I shall not therefore swell these sheets with unnecessary recapitulations . . . . "[6] We may well wish that Charles Leigh had decided to swell the sheets and given us a more detailed recapitulation of this remarkable phenomenon. He did not. But already in 1773, a more fully recorded bog body known as Ravnsholdt Man provides us with a meticulously fleshed out description.

On June 18, 1773, a judge by the name of Hans Christian Fogh, from Ravnsholdt on the Danish island of Funen, placed an inquiry in the local newspaper *Odense Adresse-Contoirs Efterretninger.* The letter, written on June 4, described a corpse which, it seems, had been found on the same day. The corpse, it states, was found three feet ("one and a half ells") into the peat:

> As they recognized this as human, four men dug off the peat from above, when a fully preserved male body was found, which was seen by me, the undersigned [Fogh] and two others who observed the following: the body lay stretched flat on his back with both arms crossed behind the back as if they had been tied together although there was no trace of bindings on the arms. The body was entirely naked except for a sheepskin wrapped around the head. When it was removed it was clear to see that the man had a reddish beard and shortly cropped hair as if he had worn a wig. The skin, before it was touched and turned over, was intact over the entire body, except on his throat where one could see straight through to the bone; and the teeth looked as if pushed in. The body was otherwise complete and all the limbs were clearly visible except for the one foot that had been severed by the peat-spade. The body was decayed and rotten and had most likely been [in the bog] for many years since there was thick and dense peat-soil over him [. . .] One can only assume that this person has been placed there on purpose since [. . .] some small branches into which other little sticks are placed in a crisscross above him so as if to prevent him from surfacing or sitting up. Notwithstanding, the peat lay compactly above him, dense and entwined as the rest of the peat-land and without any signs that this was an old grave. From this, one can conclude that he must have lain here for many years. Thus we declare: if anyone can offer any information in this case, regarding any person that might be missing or so forth, to contact me. Ravnsholdt the 4th of June 1773. Hans Christian Fogh. County Judge and Recorder.[7]

This remarkably well preserved bog body survived only in words; the physical remains disappeared and were most likely reburied in an unknown grave. Yet the sensibility and accuracy displayed in Fogh's detailed observations, the restraint in interpretation, and the detective-like assembling of clues makes his description convincingly archaeological in nature. In Fogh's account we find many of the fine points which we now know to be characteristic attributes of bog bodies. At the same time, it is the missing information—the identity and individuality of the person—that pulls readers into the seductive caesura of what might have happened. We want to know the life story of the man. Who was he? When did he live? How and why did he die? The well preserved body tells an incomplete story. Did his cropped hair, as suggested in Fogh's description, mean that he had worn a wig? Was he placed deliberately in the bog? Why was he seemingly prevented from floating out of the bog? Was it the fear of ghosts that prompted his executioner or executioners to pin him down?

These questions are literally written on the body of Ravnsholdt Man in that he is known in bog body literature as a so-called *paper body,* preserved by the embalming power of words. The Dutch archaeologist Wijnand van der Sanden, who coined the expression, describes paper bodies as bog bodies known exclusively from newspaper reports, books, or letters. These descriptions, he points out, are not reliable sources of information; indeed, some might be invented to boost the roster of bog body finds.[8] Archaeologists agree, however, that the Ravnsholdt Man paper body not only appears to be authentic and intact, but offers a blueprint for many bog body descriptions to come.

Bog people have left no written words to explain their practices, no narratives to help us understand them—they are, to borrow a phrase, "shadowy figure[s] lost in the long corridors of deep time."[9] We are beholden in some measure to the words and interpretations of their literate contemporaries in the South, and in particular to the earliest source of information about the practice of placing humans in the bog: the Roman historian Cornelius Tacitus's *Germania* from circa 96 AD. At a time when people in the North were still lodged in preliterary culture, the South had fully active historians capable of describing, as Tacitus does, the practices of a people for whom human sacrifice in holy groves in the ancient North was apparently common: "The mode of execution varies according to the offence. Traitors and deserters are hanged on trees; cowards, shirkers, and sodomites are pressed down under a wicker hurdle into the slimy mud of the bog. This distinction in the punishment is based on the idea that offenders against the state should be made a public example of, whereas deeds of shame should be buried out of men's sight."[10]

As the first ethnographic probing of the north European region and its religious and cultural practices, Tacitus's description remains an important source of information. From the early nineteenth century until the present day, his

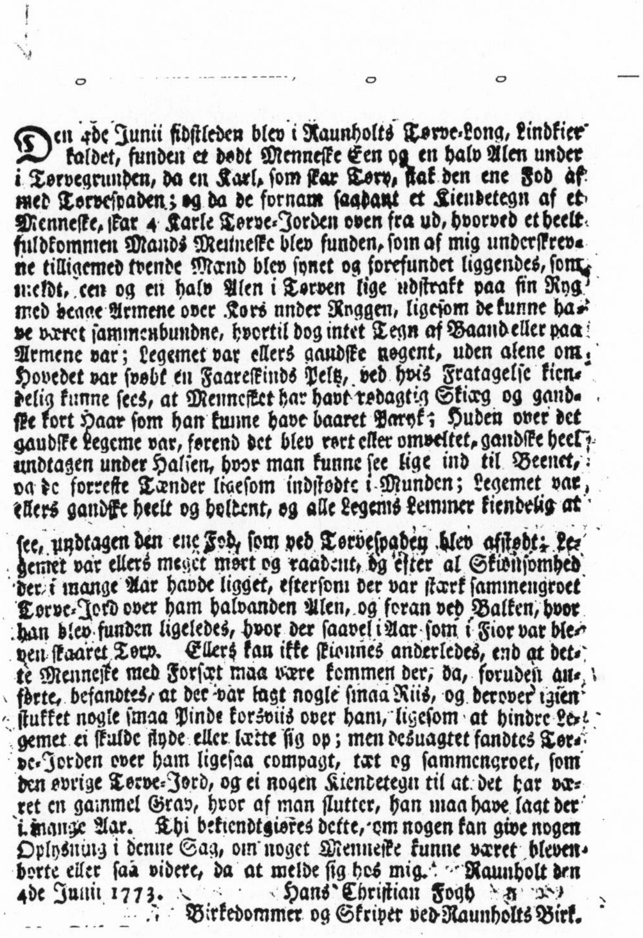

Den 4de Junii sidstleden blev i Raunholts Torve-Mose, Lindkier kaldet, funden et dødt Menneske. Een og en halv Alen under i Torvegrunden, da en Karl, som skar Torv, stak den ene Fod af med Torvespaden; og da de fornam saadant et Kiendetegn af et Menneske, skar 4 Karle Torve-Jorden oven fra ud, hvorved et heelt fuldkommen Mands Menneske blev funden, som af mig underskrevne tilligemed tvende Mænd blev synet og forefundet liggendes, som meldt, een og en halv Alen i Torven lige udstrakt paa sin Ryg med begge Armene over Kors under Ryggen, ligesom de kunne have været sammenbundne, hvortil dog intet Tegn af Baand eller paa Armene var; Legemet var ellers gandske nøgent, uden alene om Hovedet var svøbt en Faareskinds Pelt, ved hvis Frataqelse kiendelig kunne sees, at Mennesket har havt rødagtig Skiæg og gandske kort Haar som han kunne have baaret Paryk; Huden over det gandske Legeme var, førend det blev rørt eller omveltet, gandske heel, undtagen under Halsen, hvor man kunne see lige ind til Beenet, og de forreste Tænder ligesom indstødte i Munden; Legemet var ellers gandske heel og holdent, og alle Legems Lemmer kiendelig at see, undtagen den ene Fod, som ved Torvespaden blev afstødt. Legemet var ellers meget mørt og raadent, og efter al Skiønsomhed der i mange Aar havde ligget, eftersom der var stærk sammengroet Torve-Jord over ham halvanden Alen, og foran ved Balken, hvor han blev funden ligeledes, hvor der saavel i Aar som i Fior var blevven skaaret Torv. Ellers kan ikke skiønnes anderledes, end at dette Menneske med Forsæt maa være kommen der, da, foruden anførte, befandtes, at der var lagt nogle smaa Riis, og derover igien stuffet nogle smaa Pinde korsviis over ham, ligesom at hindre Legemet ei skulde flyde eller lætte sig op; men desuagtet fandtes Torve-Jorden over ham ligesaa compagt, tæt og sammengroet, som den øvrige Torve-Jord, og ei nogen Kiendetegn til at det har været en gammel Grav, hvor af man slutter, han maa have laat der i mange Aar. Thi bekiendtgiøres dette, om nogen kan give nogen Oplysning i denne Sag, om noget Menneske kunne været blevven borte eller saa videre, da at melde sig hos mig. Raunholt den 4de Junii 1773. Hans Christian Fogh Birkedommer og Skriver ved Raunholts Birk.

Fig. 0.2. The Ravnsholdt Man "paper body" was reported in 1773 by a Danish newspaper in the hope that someone could identify the remarkable remains of a man found in a bog. The body no longer exists, except in written form. © National Museum of Denmark.

descriptions have continued to be evoked by most archaeologists interpreting bog bodies.[11] His claim, for example, that bog deaths were ordered for sexual transgressions such as adultery, and his description of bog bodies as *corpores infames,* "infamous bodies," has resonated with a great deal of verve throughout the fictional and nonfictional description of them, as has his observation regarding the punishment of women for infidelity: "A guilty wife is summarily punished by her husband. He cuts off her hair, strips her naked, and in the presence of kinsmen turns her out of his house."[12]

In *The Bog People,* Glob grants that although "scholars may be correct in saying that Tacitus' book on the Germani is not an historical document so much as an unreliable record of oral tradition, nevertheless a long series of archaeological finds agree with what he tells us."[13] Building on Tacitus, Glob proposes that bog bodies constituted the remains of a practice of fertility sacrifice in which humans were offered up to the goddess Nerthus, Mother Earth— the product of a strange mix of religions, and a female deity surrounded by mystique. During the Iron Ages, Glob explains, bogs were sacred places for religious worship, spiritual depths into which votives of various kinds, including humans, were placed. They were gateways to another world, channels of communication to the spiritual powers of the hereafter.[14] Once a year, Nerthus's image was carried around the fields to ensure fertility, and during this time no weapons were to be carried; peace and happiness had to rule. To conclude the ritual, the goddess was returned to the bog land and humans were sacrificed in her honor.[15] At the end of the Late Roman Iron Age these deposits in the bogs seem to have ebbed out. The Danish archaeologist, Flemming Kaul, gives this explanation:

> There is much to suggest that the [bog sacrifice] rituals did not completely change as such, but that their execution, including offerings of various kinds, was moved from the wetlands to dry land. Written sources and certain iconographic evidence suggest that the sacrifice of human beings did not stop; they simply no longer ended up in the bogs. [...] One explanation could be that the sacrificial acts were now institutionalized under more central leadership, closely related to a new or stronger aristocracy. Sacrificial acts and rituals were no longer performed in the bogs and wetlands, but were moved into or close to the property of the magnate, perhaps in specially appointed places or houses. The social condition and political relations changed, and in this dispensation there was no longer room for *the sacred bog.*[16]

Today, most archaeological interpretations have centered on three possibilities: the bodies are remains of people who have been placed in the bogs as transgressors to be punished, as sacrifices to the gods, or as a combination of the two. It is likely, therefore, that the bodies are products of a variety of practices, sacral and magical, judicial and political.[17] But in resurfacing, they have become uncanny products of the bog and as a result their unique materiality and peculiar temporality provide us with a story that is not just about the practice of ritual sacrifice or about archaeology proper. Rather, as I hope to show, it is a story about the rich and varied afterlife they have led, and about the rare opportunity they offer for understanding the presence of the past across disciplinary lines and through the presence of a human body.[18] To fully understand these challenges, let me briefly look at the spatial and temporal peculiarities that the bog and its bodies present.

Entropic Bogs and Liminal Bodies

There is something fundamentally contradictory about bogs. They are solid *and* soft, firm *and* malleable, wet *and* dry; they are deep, dark, and dangerous; but they are also mysterious, alluring, and seductive. Neither water nor land, bogs are liminal spaces, thresholds between surfaces and depths, ambiguous sites of origin. As landscapes of nocturnal obscurity they bring about spatial and temporal disorientation: here, one can easily get lost; here, time is eerily suspended; here, things can disappear as if gulped down by strange forces and then reappear as if "frozen in time." Bogs are unhomely homes for whomever is placed in them; they are loci of paralysis, but also of explosive volatility. Steeped in such contradictory powers, they present fuzzy morphologies, effortlessly co-opted by historical, cultural, and psychological anxieties. In sum, bogs are "deeply" uncanny.

Like haunted houses, bogs represent—to make use of an expression from the American art historian Anthony Vidler—exactly "that mingling of mental projection and spatial characteristics associated with the uncanny."[19] In both haunted houses and in bogs, different as they may be, we can locate "a mental state of projection that precisely elides the boundaries of the real and the unreal in order to provoke a disturbing ambiguity, a slippage between waking and dreaming."[20] While the bog and its properties have gradually invaded, distorted, and covered the dead, the bodies have managed to survive under the surface of the landscape and—when or if they are uncovered—testify to the fact that the past can be *corporeally* preserved and rediscoverable. This corporeal imperishability makes them natural objects for archaeological research, but their "immortality" also provides ideal tropes for all sorts of uncanny fabrications of the human past.

Tacitus saw bogs as ill-omened places and described them as *paludimus foeda,* "foul bogs"; in *Beowulf* we learn that omniscient forces are harbored there.[21] In folklore, bogs and wetlands contain seductive elves recognizable only by their hollow backs. They trick the eye and mind, and (by way of so-called "elf-shots") seduce innocent men into a certain death.[22] Like their cousins, the sirens, who by way of seductive singing lure men into a watery grave, they are connected to deep moistness.[23] The mist and fog that rise from

Fig. 0.3. The bog lake close to where Tollund Man was found is a serene surface that conceals the human remains that may one day break through from its depths. Photograph © Karin Sanders.

the bogs have been seen in popular imagination as something primeval but also uncertain, indeterminate, and unlimited; the vapors are known in popular imagination as "misty maidens," "white ladies," and "brew of the bog woman." Less poetically, the depths have been imagined to be latrines full of human waste. And the uncouth creatures that lurk in them have been called "bogan, boggart, bocan, bodach, bogey-beasts, boogle boos, bogies, bogles, bogy, bugan, buggane, bugs, bug-a-boos, boggle-boos and bog bears."[24]

If bogs are slippery, the bodies in them are doubly so. In fact, there is a critical slippage in our use of the nomenclature *bodies,* which constitutes an evasion of the normal ontological shift we are supposed to acknowledge when death turns a *body* into a *corpse* or a *cadaver.* Socrates proclaims that when a man dies, "the visible part of him, the body, which lies in the visible world and which we call a corpse [. . .] is naturally subject to dissolution and decomposition."[25] He grants, however, that this natural ontological shift is *suspended in mummification.* The body, in short, has shifted from being an external vessel for the soul to being a presence of an absence.

Mummifications of human beings always interrupt what we consider "normal" and "natural" (either you are dead or you are not), yet not all mummifications are the same. While Egyptian mummies, the object of Socrates's reflections, are intentional artifices with names and narratives and history written all over them, bog bodies are less knowable because they seem to lack this intentionality. The question is: does it matter if mummies are not man-made, but nature-made? Does it matter that they are found not by design, but by accident? Is there a significant difference? I think so.

First: bog bodies disrupt the conventional archaeological three-step process—*excavation, classification,* and *interpretation*—because the first step, *excavation,* is almost always missing in bog body archaeology. In fact, as I will argue throughout this study, it is in part because bog bodies are *not* folded into the "proper" threefold process, but are always accidentally discovered, that they are so easily entangled with the modern and mistaken for modern people (like Seamus Heaney's Great-Uncle Hughie); a fate that distinguishes them from other "properly" excavated ancient human remains (such as Egyptian mummies). Yet in spite of the apparent ease with which we assign identities and life stories, all bog bodies are inevitably caught in the grip of anonymity. It is for all intents and purposes *a lack of a place in the proper,* including a lack of proper name, which throws them up for grabs and permits the conflict between the familiar and the unfamiliar to be played out. The names of bog bodies are with very few exceptions *place names* (we have already met Tollund Man, but also Grauballe Man, Lindow Man, Yde Girl, Windeby Girl, Weerdinge Couple, and many others will be discussed in this study), and as such, they are assigned to a taxonomic site in the archives that only allows them to move halfway into commonality and community with the still living. Using place names as proper names is not a trivial matter; it points to the precarious balance bog bodies occupy on a seesaw between being perceived as humans and as inanimate objects. With this we find, to use British literary scholar Nicholas Royle's words, "a disturbance of the very idea of personal or private property including the properness of proper names, one's so-called 'own' name, but also the proper names of others."[26] It is this strange anonymity that allows not only for projections but also for reflections upon the temporal "nature" of human life in general.

Second: the strange familiarity of bog bodies; the tug-of-war involved in assigning proper names and identities; the uncertainty, liminality, and threshold existences they embody; and the familiar unfamiliarity they visualize is tied not only to the uncanny per se, but to a particular aspect of the uncanny: the precarious and often treacherous vacillation between being human and becoming an inanimate object. That is to say, bog bodies in multiple ways negotiate the liminality that comes with having to travel between their material reality as archaeological artifacts (mummies) and the temporality that comes with their humanness. In this sense bog bodies become a kind of "archaeological uncanny" that perplexes as much as fascinates the living. This perplexity can be compared with the appearance or return of other culturally and historically marginal figures ("others") and as such, to borrow again from Vidler, should be "interpreted through a theory of the uncanny that destabilizes traditional notions of center and periphery" and with it "the spatial forms of the national."[27] Bog bodies become not only emblems of anamnesis, or corporeal time capsules in which the presence of the past is quite literally palpable. They also become anachronistic manifestations which not only challenge a sense of time's stability in favor of a sense of time's elasticity, but also give beholders of the physical remains pause to consider the appropriateness or inappropriateness of their public display. Are they persons or things?

Third: in spite of the appearance of being "nature's nature," bogs can be seen as strangely *un*-natural in that their unique properties disturb the regular historical and chronological layer-upon-layer-ness of time; subvert the proper archaeological "order of things," and allow preserved bodies to emerge, *un*-naturally and out of time, in order to disrupt the logic of "normal" historical processes.[28] The entropy (randomness and disorder) that bog bodies represent, the accidental ways in which they are found, and the emblematic ways in which they have been used make a strictly chronological (historical) reading difficult and makes the bogs particularly well suited to test the viability of the reconsiderations of chronological models that many critical theories have called for in past decades.[29] Bog bodies, in other words, seem perfectly fitted to complicate a neatly layered sense of history as constituted by specific moments in time; they ask us instead to consider a series of overlapping moments out of which various equally overlapping histories can be culled. In short: the *present* into which the *past* surfaces—in form of bog bodies— begs us to consider a temporal model that has an entropic *past-present* at its core.

To illustrate the challenge that the temporality of bog bodies presents, let me compare two sets of dates. The first we can conflate into *one period of time* approximately two thousand years back; we can call it the time of origin without the availability of any precisely ascertainable dates. The second is a long row of exact dates: 1645, 1700, 1773, 1797, 1818, 1835, 1871, 1879, 1897, 1938, 1950, 1952, 1984, and so on, each date denoting a specific moment in time in which an individual bog body has surfaced. The logic of the first time period is tied to a particular practice, already described, of placing humans in bogs. The logic of the second, or rather the *lack* of logic, represents the arbitrary moments when bog bodies have ruptured from their hidden graves. Each in this second series of moments can be seen as a bursting bubble in a landscape of time. Every burst produces reverberations, some few and brief, others numerous and long-lasting. The ones I examine in this study are those that have resonated in some significant form or manner and have produced secondary traces outside of archaeology proper: in literature, art, museum display, and so forth.

I argue throughout this study that this double temporality in bog bodies, the tension between the first and second moments, is particularly poignant because the archaeological object we examine is human and thus is already and always marked by time in the way all humans are. The fact that the bodies have been suspended for centuries inside the realm of bogs disrupts normal temporality, and with it also the conventional sense of linear and chronological time. The natural order of things, including the expected decay and corruption of the dead, has been unnaturally subverted, and the reappearance from obscurity of a dead body inevitably leads to a sense of instability. This instability comes about not only because the corpse produces goose bumps, shivers, disgust, and horror, or because of its uncanny nature and potential monstrous doubleness, but because the very physiognomy of violence connected to a dead body is fraught with a particular kind of ambivalence. It places the perplexed observer on a seesaw between anxiety and reassurance; a both unsettling and comforting position of changeability (life inevitably becomes death) and instability (the return of the dead). As the French philosopher Georges Bataille has pointed out:

> We perceive the transition from the living state to the corpse, that is, to the tormenting object that the corpse of one man is for another. For each man regards it with awe, the corpse is the image of his own destiny. It bears witness to a violence that destroys not only one man alone but all men in the end. The taboo which lays hold on the others at the sight of a corpse is the distance they put between themselves and violence, by which they cut themselves off from violence.[30]

Although human sacrifices in the Iron Age were most likely meant to recreate order, the boggy wetlands in which they were placed did not provide peaceful graves on which epitaphs secured and anchored the remains with identity and name. To stabilize these unruly reappearances, as this study will show, our imagination paradoxically often relishes the anachronisms of present-day encounters.

Emerging from this multilayered historicity, bog bodies can be seen as corporeal contact zones between things that historically have been separated.[31] This can be illustrated spatially if we allow ourselves to imagine a three-dimensional cartography that charts the places of the various bog body discoveries by using their names, which are almost always also place names: Haraldskjær Woman, Weerdinge Couple, Yde Girl, Tollund Man, Grauballe Man, Lindow Man and so on. If we compile data (much in line with the Italian literary historian Franco Moretti's distance-reading model[32]) of the volume, frequency, and time span in which they have been used in artistic production (anything from poems to paintings and so on), some "mountaintops" will be barely noticeable, others will be high and steep, and still others will be low yet wide-ranging. Some will cross over national borders and become part of a shared iconography amongst artists and writers internationally (the fact that Tollund Man emerged at a fortuitous moment in historical time when technology—the camera lens and museum preservation, most obviously—stood ready not only to capture his countenance for "eternity" and to transmit his image across borders from a small nation to an astonished *international* audience goes to the core of the spatio-temporal logic I outline here); others will stay relatively close to the place of origin and become part of a national identity, perhaps even serving as its "voice." With some exceptions, the bog bodies discussed in this study tend to be those that protrude from this illustrative map.

The next question is: if bogs and the bodies in them destabilize our sense of natural and historical order, even destabilize a sense of national space—if, in short, they are foreigners from a past country, to spin off David Lowenthal's well-known book title, *The Past is a Foreign Country*—how do we account for our fascination with these strange strangers? What is the relationship between the bog landscapes and the identity questions attached to the bodies in them? What is the relationship between the "natural" landscape and the "national" landscape? In *Landscape and Memory,* the British historian Simon Schama has argued that "once rooted, the irresistible cycle of vegetation, where death merely composted the process of rebirth, seemed to promise true national im-mortality." Such national identity, indeed national immortality, is drawn from a landscape that is beholden to mental constructions so that "before it can ever be a repose for the senses, landscape is the work of the mind. Its scenery is built up as much from strata of memory, as layers of rock."[33] Human agency is detectable in the most "natural" of places, even if only in form of a semantic gesture—a framing and naming, or, as Schama says it: "The wilderness, after all, does not locate itself, does not name itself."[34] Bog landscapes, then, wild as they may seem, are not natural, given that "it is our shaping perception that makes the difference between raw matter and landscape."[35] They come to con-note the sort of depth we associate with "roots" in the meaning of ancestry, pedigree, family tree, and national identity. Bog bodies fascinate, we can sur-mise, because they are both "rooted" in and "uprooted" from a sense of na-tional identity; they are both familiar and strange.

Yet even more importantly, as bog bodies connect us to the past they are subject to the kind of "dialectic of remembering and forgetting" that defines memory in the French historian Pierre Nora's optic: "unconscious of the distor-tion to which it is subject, vulnerable in various ways to appropriation and ma-nipulation, and capable of lying dormant for long periods only to be suddenly reawakened."[36] History, on the other hand, is an intellectual and discursive reconstruction, and as such, Nora implies, it is cold and lifeless in contrast to the warm, vivid, and vital memory. Placed into this dynamic, bog bodies are essentially co-opted from history by a more mnemonically oriented archaeological imagination and made "warm," vivid, and vital again. Under the "bark" of history's tree—or deep in the bog, as it were—mnemonic properties in the form of blood and sacrifice and human stories seep out and emerge as part of national and ethnic claims for identity and right of place, even as they force us to question our assumptions about history, nationhood, and representation.[37]

But bogs also connote a different sort of depth. While Schama bemoans the fact that historians cannot *touch* the past directly, but can do so only through texts and images "safely caught in the bell jar of academic convention; look but don't touch," the French philosopher Gaston Bachelard, in *Earth and the Reveries of Will,* sees bogs as eminently *touchable* sites of conception. In his psycho-phenomenological optic (which is of great importance for some of the bog artists I shall discuss later), bogs are paradoxical places that connote both a dead end and a point of origin, gestation, regeneration, and rebirth.[38] Bachelard's philosophical claim—that the empirical world is never senseless and arbitrary—makes the bog an active creative agent in what he calls "the material imagination." There are, he writes, "countless texts characterized by contradictory attitudes, positive and negative, towards swamps, mud, or the dark, wet earth," and therefore we "must concede that the moist earth strikes a nerve in the material imagination. The experience of it resonates with our own inner experiences and repressed reveries, and restimulates ancient associations, associations as old for the individual as they are for the human species."[39] The bog's paradoxical powers provide not only a temporal reassembling of the past as pieces of a puzzle, but also the soil for archaeological (academic!) activity at large, in which we can dig down and submerge ourselves while we rewrite, reconstruct, and re-present the material past.

The logic of this spatio-temporal definition suggests that wetlands become creative and fertile in the widest sense of the word when they touch the human body, directly or metaphorically. In the material unconscious, with its "nocturnal forces and subterranean powers," we find a nostalgic birthplace sutured to natural material elements.[40] But the dirt we dig in, particularly when it is not beholden to a sentiment of chronology or to an understanding of time as being layered consecutively, allows for surprising discoveries and fascinating ruptures.[41] These discoveries and ruptures are submitted to and organized by what I call the archaeological imagination. It is time, therefore, to reflect on the meaning or meanings of this nomenclature.

Archaeological Imaginations

It has become commonplace in recent years to add *imagination* after the terms for fields of study, such as the *historical* imagination, the *ethnographical* imagination, the *geographical* imagination, the *architectural* imagination, and so forth.[42] While each field has its own emphasis and particular temporal or spatial objects and structures to examine, the addition of *imagination* has become a way to unleash the objects under investigation from the "correctness" or "tyranny" of predetermined disciplinary methodologies. Imagination, as the word connotes, is primarily associated with mental and psychological processes and also with the freedom to "abstract" from material realities. It is noteworthy that the grammatical arrangement is ordered so that *imagination* occupies the central place as the noun to which qualifying adjectives (*historical, ethnographical, geographical, architectural,* and so on) *do* something. This is important because *imagination,* if we allow ourselves to think optimistically, can come to serve as a kind of collective reservoir of thoughts—an interdisciplinary landscape in which concepts can travel relatively freely between fields of study. This also means that there is no such thing as *the* historical, *the* ethnological, or *the* geographical imagination, but that each of necessity comes in multiple forms.

It follows that the *archaeological* imagination, too, takes various shapes and finds theoretical homes in many well-established paradigms. If archaeology at its core is about scientific excavation and the study of humanity's past through material evidence, the archaeological imagination can be seen as an extension of this definition. As such, it seems to be a rather elastic term that can take on many different kinds of definitions. First and foremost it is concerned with the relationship between depth (past) and surface (present). But, as the British archaeologist Clive Gamble writes, "archaeological imagination" is also "about excitement" and "about intellectual curiosity and finding ways to turn that curiosity into knowledge about people in the past."[43] Another British archaeologist, Julian Thomas, sees the archaeological imagination as an essentially quotidian exercise: "In everyday life, human beings grasp elements of the material world, and constitute them as evidence for past human practice. The 'scientific' enterprise of archaeology is based upon this prescientific way of being attuned to the world, which I will call 'the archaeological imagination.'"[44] The "deep" science of archaeological digging through strata of time is expanded in Thomas's definition to an all-embracing detective model of everyday life, a model with interesting literary resonances. Most humans are able to cull and connect clues and traces left in their mundane world (muddy footprints, crumbs from a cake tin), and as such we are all more or less attuned to the archaeological imagination.

But is the archaeological imagination more than (just) a way to "deep-read?" Another British archaeologist, Christopher Tilley, proposes that "material forms, unlike words, are not just communicating meaning but actively *doing* something in the world as mediators of activity in the world [. . .] in specific material contexts which are historically determined."[45] Although we may disagree with Tilley's implication that words do not "do" something in the world (speech act theorists would certainly disagree) his understanding of material metaphors accentuates that which cannot be translated from the physical world to language. The resistance from the material and from the soil, then, becomes intimately tied to the essence or ontology of the archaeological imaginary. The archaeological imaginary, Tilley implies, is a way of insisting on the materiality of objects and artifacts and on a lack of representational intent—and it is precisely this residual element in his proposal that I wish to pursue here. What interests me particularly is the way in which the stubborn *thereness* of archaeological objects as material testimony seems to leave an element of something both unavoidable and mysterious, something that begs for imaginary fill-ins and fill-outs. They function as conduits between the material (their *thingness*) and language. Or, said differently, in archaeology the very materiality of the artifact appears to simultaneously offer itself up for interpretation and refuse to be *known*.

But there are other ways to think about the archaeological imagination. A well-known model that is inherently archaeological in structure and logic, and that seems particularly suitable in the discussion of bog bodies, is to be found in the Russian literary critic and philosopher Mikhael Bakhtin's definition of *chronotope*. A chronotope is the "intrinsic connectedness of temporal and spatial relationships that are artistically expressed in literature."[46] Although Bakhtin stresses that the chronotope is *almost* a metaphor and *specific* to literature, his description seems to suggest that other cultural articulations could be seen through the same lens. When, for example, he tells us that time "thickens, takes on flesh, becomes artistically visible," and that "space becomes charged and responsive to the movements of time, plot and history," it is not difficult to visualize bog bodies as chronotopic materializations.[47] In them time thickens and takes on flesh rather manifestly, not just in its existence as matter but also and more in tune with Bahktin's intention, as a literary element. Indeed, as Bahktin goes on: "The image of man is always intrinsically chronotopic." The human movement through time and space is fraught with chronotopicity.[48] The Canadian literary scholar Anthony Purdy has also made the observation that the idea of chronotopicity lends itself particularly well to examinations of bog bodies as "'the place where the knots of narratives are tied and untied,' where '[t]ime becomes, in effect, palpable and visible.'" He argues that the bog body chronotopes' "primary function is to materialize a past in the present, to serve as a vehicle for personal and cultural memory," and he proposes to call them *mnemotopes*—that is, "a chronotopic motif that manifests the presence of the past, the consciousness or unconscious memory traces of a more or less distant period in the life of a culture or an individual."[49]

Archaeology, of course, has been used as a metaphor in a number of other analytical practices and theories, including those of Sigmund Freud, Walter Benjamin, Michel Foucault, and Martin Heidegger to mention the most obvious. Yet whether we use a Freudian optic, looking at the dark continent of (cultural and personal) repression; a Benjaminian optic, questioning authenticity and aura; a Foucaultian optic, homing in on material and temporal ruptures, and particularly on the notion that discursive formations are not fixed "things" but can be seen as vertically defined correspondences between different (and ongoing) temporalities; or a Heideggerian concept of materiality and time, bog bodies make us acutely aware of interpersonal relations and inter-human codes of conduct.[50] No matter which theoretical model of archaeology one subscribes to—and I am obliged to say that this study will not subscribe to any one archaeological reading in particular, but will instead allow the research material on bog bodies to inform the questions of analysis—there is an underlying consensus which says that the archaeological object "tells" us about itself, but in our own language. Or, said differently, the metaphorical model of interaction and communication between people provides a model for our interaction with and understanding of artifacts.[51] Because bog bodies represent both people and archaeological artifacts, the dynamic between those two things (which is also a dynamic between life and death, time and space) gives them great potential as corporeal time travelers. Consequently, as the following chapters will show, they are often called to the witness stand in the "court of history" and/or the "court of memory" to testify about the past and its (in) human practices.

As seen, a simple definition of the archaeological imagination is difficult to present, but I agree with the British literary scholar Jennifer Wallace when she proposes that the archaeological imagination requires "a poetics of depth," that it can be seen as an ability to read "between the lines" and "to see things in a new way, to turn the imagination 'inside outwards' in order to ponder what is hidden below the earth rather than what is visible above."[52] Thus the term "archaeological imagination" describes not only the discursive strategy of archaeological writing proper, but also embraces a more general sense of digging the depths of the past—one in which, to cite a phrase from Clive Gamble, "We also need to carry the mute body of prehistory with us rather than step over its corpse on the road to civilization and the lure of texts."[53]

Bog/Book

Since Glob's *The Bog People*—ironically, *bog* in Danish means "book"—plays such a key role in the history of bog body reception and in the archaeological imagination that is attached to bog bodies, allow me to look briefly at this "bog/book." The magnetism of Glob's text lies not only in the sensationalism of its subject but in its narrative style, its mixture of scientific-archaeological discourse and mythological-poetical narration, and not least its use of photography (by Glob himself and by Lennart Larsen). Deeply influenced by historical imaging but also aware of contemporary public fascination, Glob summons forth specters of the past, which speak not only to popular imagination but to the strategies of archaeological imagination. The Danish archaeologist Kristian Kristiansen has pointed out that there is a "slightly piquant and macabre appeal" in Glob's book, which feeds part of the public's craving for archaeological entertainment: "Archaeology has long ago abandoned its nationalistic commitment and has become entertainment for the rising middle classes."[54] While that may be so, Glob's rhetorical approach and the inclusion of rich black-and-white photographs also illustrate how he places the reader in a bifurcated position as imaginary traveler to the past *and* rational observer of the present.[55]

As the indisputable ur-text, Glob's book became a site from which fiction writers, poets, and visual artists would cull material for representing and reconfiguring in words and images the interstices of time and matter in the antiquity of human remains. Glob's verbal craftsmanship demonstrated that he had a sharp sense of the power of writing and was willing to provide an archaeological narrative that slipped out of the confines of science and flirted with literary tropes and genres. With Glob as guide, readers were placed both as imaginary travelers to the past and as rational observers of the present. Glob's narrative strategy appeared to succeed in presenting archaeological material without compromising scientific data. But more importantly, as Anthony Purdy has pointed out, "within a few short years, the tropes and topoi of a recognizable poetics were well established, all of them originating in Glob's book, the founding text of a narrative genre characterized by the conflicting constructions of discourses running the gamut from forensic to lyric."[56]

Glob, in other words, was profoundly influenced by the potentiality of interaction between fact and fiction. Drawing on the historian Hayden White, the British archaeologist Ian Hodder has suggested that archaeological narratives need storytelling to situate "particular propositions within a larger argument."[57] While literary-poetical imagination speaks for itself because we always presume that (most of) what is given to us in a fictional form is indeed *imagined,* archaeological imagination needs explanation.[58] Hodder's key points—that the narrative interpretation which takes place in archaeological texts has agency and audience—place focus on the archaeologist as practitioner of a specific form of communication: telling the reader about the meanings of material things from the past. Glob would have been unfamiliar with such theoretical considerations, but he was no doubt familiar with the idea of his countryman Martin A. Hansen that there exists an unavoidable alliance between historical materiality and creative imagination. "Archaeologists dig out fairy tales," Hansen exclaimed.[59] Glob certainly made use of rhetorical ploys associated with fiction literature, and seemed to agree with Hansen that archaeologists are facilitators who not only unlock magical stories hidden in remnants from the past, but also inevitably become active agents in creating these "fairy tales" as penholders who dig, archaeologists as storytellers.[60]

The past, of course, is not just fiction or fairy tale—and to imagine the archaeologist as someone who excavates fairy tales is *not* to say that archaeological writing denies the presence of physical facts. Nor is it to say that those facts cannot be described in language stripped of metaphors or other linguistic or artistic strategies, which are presumably auxiliary to archaeology. Clearly they can, and obviously they are. It *is* to say, however, that most archaeological texts, at least when it comes to bog bodies (and I suspect that it also goes for many other archaeological accounts), draw extensively on rhetorical strategies ranging from plot to metaphor, and that most of these descriptions are indebted to the artifice and maneuvers of literary and poetical genres—none more so than Glob's bog body book.

This is best described by the opening paragraph of the first chapter of *The Bog People.* Notice the lyrical tone.

> An early spring day—8 May, 1950. Evening was gathering over Tollund Fen in Bjaeldskov Dal. Momentarily the sun burst in, bright and yet subdued, through a gate in blue thunder-clouds in the west, bringing everything mysteriously to life. The evening stillness was only broken, now and again, by the grating love-call of the snipe. The dead man, too, deep down in the umber-brown peat, seemed to have come alive. He lay on his damp bed as though asleep, resting on his side, the head inclined a little forward, arms and legs bent. His face wore a gentle expression—the eyes lightly closed, the lips softly pursed, as if in silent prayer. It was as though the dead man's soul had for a moment returned from another world, through the gate of the western sky.[61]

Evening stillness, damp beds, grating love calls, gentle expressions, and lips softly pursed in prayer make us think not of archaeological prudence and restraint, but of literary attentiveness to the slightest details. It is *not,* of course, as Glob wants us to imagine, through the "western sky" that the soul returns to the dead body. It is, rather and quite literally, Glob's own Promethean pen and imagination that serve as ignition here.

As Glob penetrates the mystery of the dead man's face, he both submits it to a scientific desire to *see* and *examine* and protects it against sacrilegious attacks on the peace of the grave. This is to say that by combining science with aesthetics, Glob solves the dilemma of wanting to know, dig into, examine, and inevitably destroy part of the body while also wanting to preserve and retain the magic of its existence. His parsing of Tollund Man's face, for example, pulls us away from the abstract past and into the immediate accessibility of face-to-face encounters. It is the enthused archaeologist who gives a verbalized spark to the bog man's enigmatic smile, character, and "amiable" personality; but equally important, it is the prudence of a scientist's pen which pulls us out of these poetic musings and blends poetic imaginings with scientific observations. Elsewhere in his bog/book Glob makes an inventory of Tollund Man's body, and labels and describes its parts in a manner congruent with the scientific instructions of his profession, yet at the same time as he describes the autopsy and vividly spells out the arrangement and content of the corpse's interior organs, he also sutures the dissected body through poetic and photographic glossing. Such a glossing—almost a re-embalming—in its own way preserves the body for the reader and spectator while implying an erotic connection between the body and the fertility goddess to which it appears to have been sacrificed—an erotic potential that was not lost on the many poets, writers, and artists who were inspired by Glob's narrative.

Glob was aware of the misuse of archaeology by his own generation of archaeologists, and implicitly aware that archaeologists had to "excavate" their "own core concepts" and "unquestioned background assumptions" (to paraphrase American archaeologist Margaret Conkey), which structure the ways in which they select and interpret material and data.[62] That is, archaeology is not just about materiality but also about mental attitudes, and this also goes to the core of how bog bodies were understood and interpreted both within and outside archaeology. Although Glob may not have known French historian Marc Bloch's unfinished study *The Historian's Craft,* he seems to have been intuitively in sync with the questions raised in it. Bloch proposes that the drive and genesis behind engagement with archaeological matters rests in the entertainment value and aesthetic pleasure we get when we are drawn into the "unquestionable fascination of history." He goes on: "The spectacle of human activity which forms its particular object is, more than any other, designed to seduce the imagination—above all when, thanks to the remoteness in time and space, it is adorned with the subtle enchantment of the unfamiliar."[63] Such subtle enchantment of the unfamiliar is seductive, Bloch suggests, precisely because "the spectacle of human activity" is also so very familiar. The archaeological imagination and fascination which wraps the remains of bog people is tied, I think, to the magnetic pull of this unfamiliar familiarity. Bloch, for his part, offers no apology for the pleasures of historical pursuits and warns that we should "guard against stripping our science of its share of poetry." Yet his outline of the vortex of time and space offered by history (and archaeology) is not exclusively intended to whirl us into aesthetic gratification. It is also meant to reinforce ethical questions about humans and humanity in time.

There is no better place to find the "unquestionable fascination" with bog bodies than in *The Bog People*. It can be seen as a sort of hybrid *archaeo-literature*, loaded with all the appurtenances (corpses, violent deaths, erotic possibilities, myths) that facilitate cross-fertilization between disciplines dedicated to fact (archaeology, history), and those committed to fiction (literature, poetry, art). It not only illuminates how narrative style and metaphor play out in archaeological interpretation, but also highlights the significance of authorial voice and author saturation.[64] The slippage between a scientific and literary-poetic pen also corresponds in a number of ways—albeit on a different level—with the experience we have of anamnesis and anachronism in the material bodies themselves. More importantly, as already suggested, Seamus Heaney is most famously beholden to the soil provided by Glob's *The Bog People*. In fact, many of Heaney's most famous bog body poems are not written on observations of actual bog bodies, but instead use Glob's book as source.[65] Heaney explains:

> The head of the Tollund Man and the body of the Grauballe Man have a double force, a riddling power: on the one hand, they invite us to reverie and daydream, while on the other hand, they can tempt the intellect to its most strenuous exertions. And it has always seemed to me that this phenomenal potency derives from the fact that these bodies erase the boundary-line between culture and nature, between art and life, between vision and eyesight, as it were. The head of the Tollund Man, the body of the Grauballe Man, the head of the girl from Windeby Fen, even the terrible head of the decapitated girl from the fen at Roum (*all of which I first met in the photographs in Glob's book*) these can now be classed as *objets,* if you like, because they seem to belong with heads made of clay or bronze or marble heads that we find in the art museum.[66]

Heaney, in other words, digs into the field of Glob's archaeological prose (his "bog"/book), which he sees as analogous to his own poetry; it contains "the charms of poetry itself" and is indeed, Heaney enthuses, "a requiem grounded in research."[67] This requiem consists of different archaeological layers of representation: a dead corpse accidentally unearthed, captured by the eye of a camera, put into writing by the archaeologist, treated and refashioned into a museum object, enclosed in a glass frame, and then finally "penned" by the metaphorical spade of a poet into poetry.[68]

Before we look into the complexities of poetry and other fictional narratives, let us step back for a moment and consider the "eye of the camera" and Heaney's admission that the photographs in Glob's book constituted his first encounter with Tollund Man, Grauballe Man, and Windeby Girl—all of whom would play significant roles in his bog poetry. As the following chapters will show, not only Heaney but also a host of other visual and verbal artists are beholden to the power of the camera. Photography, in other words, is sine qua non for our investigation. With that in mind, let us enter "nature's own darkroom."

CHAPTER 1

Nature's Own Darkroom

On April 28, 1952, a peat digger in Eastern Denmark found the dead body of a man buried in the layer of peat known as "dog flesh." Even before he was fully released from the soil, this new archaeological sensation was given the name of Grauballe Man, so called after the nearest village. He was a full-grown man with long curly red hair, eyebrows, day-old beard stubble, and a wrinkled forehead. His body was unusually well preserved. The skeletal parts were gone, but the limbs were intact and the muscles were clearly defined. The body lay in a fetal position, with the torso raised from the rest of the body and the head bent slightly backwards. During the night after the discovery, one of the curious onlookers who had gathered at the location of the find stepped on Grauballe Man's head, partly squashing the face and thereby adding to his appearance as a "forceps baby," as Seamus Heaney would later call it.

Twice Photographed

In the first photographs of Grauballe Man in situ, his head is clearly visible, with his upper torso still struggling to emerge from the tight embrace of the peat-grave. He is caught between

Fig. 1.1. Caught by the camera at the moment of discovery in 1952, Grauballe Man is both an ancient man and a newborn: a "forceps baby," in the words of Seamus Heaney. Photograph © Moesgaard Museum, Denmark.

realms: lingering, resting, and hesitating before the final contraction releases him from the darkroom of the bog and into the light and availability of the camera. Spawned from and caught up in the paradoxes of bog and photography respectively, the bog man himself forms a third paradox in the odd shape of an "ancient man-baby." He is both newborn and prehistoric. The bog has "killed" him and now "gives birth" to him in front of our very eyes—for which the camera and photographer are proxy. He is, writes Glob, "*naked as the day he was born.*"[1]

But before the bog man is rescued and embalmed by the photographic lens, he has already been "shot" and mummified by the bog acids. In this sense, he has been "shot" twice. In the first "shot" the agency involved can only be called *nature* and *time* in a rather abstract way. In the second, the agent can be named, someone has literally snapped the shutter, and the photographs in question can be signed and dated. In the first, the body has in some sense become a representation of itself without losing its original material corporeality that is its core and trademark. In the second, this corporeality is flattened and displaced so that in spite of the old assumptions of a photograph's ability to reproduce, as German cultural critic Siegfried Kracauer put it, "nature with a fidelity 'equal to nature itself,'" its material corporeality is lost.[2] Yet the structural similarities are hard to overlook.

In fact, it is tempting to see bogs as nature's own darkroom with powers similar to those of photography.[3] If we think of the advancement in technology from the earliest daguerreotypes to the digital cameras of today's world, we could place the bog as a kind of ur-camera or as a pre-photographic natural darkroom that acted as "nature's own pencil" and a "mirror of memory" before man-made inventions lay claim to such magic capabilities. Taking their time, literally around two thousand years, the mnemonic properties in these bog-laboratories of northern Europe have produced nature's own equivalent of photo-sculpture. The inventions, advanced chemical processes, and technological capabilities of the nineteenth century and the artificial memories of the twentieth and twenty-first centuries have substituted the preserving acid of the bog with a more up-to-date magic. But both bog and photography can, each in their way, be seen as laboratories for *prosthetic memories.* A prosthesis is a material replacement of something missing, of something that *has been* but is now lost. Its function is to be an (artificial) extension and to mimic the functionality of that which has been amputated. If we take prosthetics metaphorically, as we must when we add memory to the term, cultural history offers a host of prosthetic memories, many of which resonate directly with the materiality and metaphoricity of the bog realm—photography being the most obvious example. Dutch historian of psychology Douwe Draaisma extends the term back in time to Plato's wax tablet, which is described in *Theaetetus* as a membrane finely tuned between surface and depth: the wax tablet must be neither too hard nor too soft or the imprint—the memory—will be lost.[4] Correspondingly, the bog must consist of just the right properties to serve as a mnemonic, to hold the material traces.[5] In the development of prosthetic memories over time from wax tablet via various pre-photographic machines—such as those known to produce "physionotraces"—to photography proper, and eventually to computer memory, the bog offers its own kind of prosthesis that shares the metaphoric and alchemic associations we have with such memory tools.

Because bog bodies are *made in nature,* it may seem counterintuitive to exercise the nomenclature of prosthetics—the *artificial* extension or replacement of something original and missing—about that which is very much *natural* and very much *in existence.* Yet as with other prosthetic devices, the extension and replacement that bog bodies offer is seemingly liberated from the circumstances of time and space—no matter how "natural" their temporal-spatial reality seems to be. The analogy to French film critic André Bazin's well-known thesis on the ontology of the photographic image is almost too obvious. He hypothesizes that "the photographic image is the object itself, the object freed from the conditions of time and space that govern it. No matter how fuzzy, distorted, or discolored, no matter how lacking in documentary value the image may be, it shares, by virtue of the very process of its becoming, the being of the model of which it is the reproduction; it is the model."[6] As a natural geological laboratory a bog is similarly able to capture images of the objects and humans placed in it, and like photography, it stops and freezes and transports images through time, so that the end result, as with the birth photo of Grauballe Man, becomes an uncanny copy of reality—a "natural" artificiality connected to the earth, to use Roland Barthes's expression, by an "umbilical cord that the photographer gives life."[7]

Authenticity or Simulacrum

While the birth image of Grauballe Man as "forceps baby" helps illustrate the analogous ontology of photography and bog, Tollund Man allows us to consider questions of authenticity and inauthenticity, original and copy. He too is "twice photographed"; he too adds to the analogy between bog realm and photography, albeit with a different twist. In *The Bog People,* Glob gives us a detailed description of Tollund Man's remains in toto (with stomach content of the last meal and so forth), only to surprise the reader (and even most present-day museum visitors) when he concludes that in the end only Tollund Man's "splendid head" was selected for a year's treatment with formaldehyde, acetic acid, alcohol, paraffin, and heated wax, and was eventually displayed at Silkeborg Museum. The rest of the body was "left to dehydrate and offered to various scientific institutions for research."[8] Glob goes on:

> In the process of conservation, the proportions of the head and the features of the face were happily completely retained, but the head as a whole had shrunk by about 12 per cent. In spite of this it has emerged as the best preserved head of an early man to have come down to us so far. The majestic head astonishes the beholder and rivets his attention. Standing in front of the glass case in which it is displayed, he finds himself face to face with an Iron Age man. *Dark in hue, the head is still full of life and more beautiful than the best portrait by the world's greatest artists, since it is the man himself we see.*[9]

Fig. 1.2. The preserved head of Tollund Man looks like a decapitated bust. Photograph
© Silkeborg Museum, Denmark.

Glob's claim is rather bold, but the elegant and evocative assertion that it is
"the man himself we see" is easily disputed. To borrow a phrase from French
writer André Malraux's study of "museums without walls" (*The Voices of Silence*), "each exhibit is a representation of something, differing from the thing
itself, this specific difference being its *raison d'être*."[10] What Glob wishes to do,
however, is to deem the body authentic even if it has been reduced to a "splendid head," just as he has authenticated the entire body in his words and in the
photographs. Nonetheless, authenticity seems to have been in some peril when
only the head underwent the lengthy preservation process and was displayed
in the first public exhibit. Glob tells us that the torso and limbs were disposed
of, partly because it was thought inappropriate and too morbid to display a
dead corpse, and partly due to lack of adequate preservation techniques.

Unlike the multitude of bog bodies, Tollund Man was not in need of a *face*
reconstruction; he was in need of a *body* reconstruction. Yet not until the mid-
1980s, with help from the preservative powers of photography, was the splendid head reunited with the body. The original remains of Tollund Man's body

Figs. 1.3 and 1.4. Before and after—real or fake? Tollund Man's body was not preserved, but has been reconstructed from photographic archives. Photographs © Silkeborg Museum, Denmark.

were poorly kept and scattered for years in various locations, which meant that the original photographs had to serve as prosthetic memory aids. In 1987 the complete bog man was finally exhibited.[11]

Tollund Man's museum body in its present display is, in other words, a curious hybrid made by means of bog preservation, photography preservation, and museum technology. It consists of an amalgam of original and copy, in which the authenticity of the head has been fused with the photo/copy reconstruction of the body. The result teeters precariously close to becoming an example of unadulterated *simulacrum* in which the representation replaces the reality it represents—or, said differently, the copy takes on the reality of the original.[12] If this causes the displayed bog body to become pure constructedness, *Schein,* or illusion, the display of Tollund Man and the synchronized artificiality and authenticity he embodies can also be seen as a fascinating and instructive instance of how "raw" authenticity (and with it the curiosity-effect) has been pried out of the "natural" and reinstated as hybrid. Ironically, as a consequence, the most "natural" and authentic access we have to Tollund Man today is through the original photographs. This brings questions about authenticity and inauthenticity to the fore. Lionel Trilling, for one, has asserted that the fact that "the word [authenticity] has become part of the moral slang of our day points to the peculiar nature of our fallen condition, our anxiety over the credibility of existence and of individual existences."[13] Authenticity, then, is often seen as "merely not being inauthentic" or as something that "involves a degree of rough concreteness or of extremity."[14] Bog bodies inevitably and continually renegotiate these questions as they travel back and forth or linger in a realm between authenticity (realness, humanness) and simulacrum (fakery, thingness). And Tollund Man's story, in particular, stresses the complicated interrelation of photography, authenticity, and simulacrum.

So is he (it!) the real thing? Walter Benjamin has taught us that the "presence of the original is the prerequisite to the concept of authenticity," and Roland Barthes has argued that in photography "the power of authentication exceeds the power of representation."[15] But as we have seen, such authentication—so closely sutured to bog body photography (and with it a sense of *certainty* about represented "pasts")—has been uncannily tested in Tollund Man's remains.[16] If time is already stopped in bog bodies and stopped again in the photographic process, then photographed bog bodies have undergone a kind of double-fossilization that in turn mythologizes both them and their past.[17] Furthermore, if death and photography are somehow enmeshed as the shutter holds life captive, it would seem that bog bodies when photographed undergo not only a kind of enlivenment, but one that is fraught with the propriety (or lack thereof) in aestheticizing dead bodies.

Postcard Aesthetics

Let me offer an example of this. Two photo postcards of Grauballe Man purchased in a museum shop have decorated my office during the years it has taken to complete this book. One shows his hand, the other shows his feet. Each postcard is deliberately and cunningly cropped from what seems to have been much larger black and white photographs.[18] In the postcard of the hand, the wrist is turned upward so that the clearly ascertainable veins face the viewer. The hands have long, fine fingers with manicured but dirty fingernails. Even the crescents (lunules) at the bases of the nails are plainly visible. The lens has captured the wrinkled skin as a microscopically detailed surface. The feet are sturdier, but they too are uncannily "real" with slightly swollen ankles and finely chiseled veins. In Glob's words: "Like his fingerprints, the lines on the soles of the Grauballe man's feet were as sharp *as when they were formed in the embryo,* more than one and a half thousand years ago."[19] The intense sharpness of the postcard photographs and the focused light also reveals what looks to be imprints into the skin made from the fabric of peat in which the man rested for centuries. The cropping of the photographs makes them "easy on the eye" and highlights an effect of seeing which seems intended to solicit emotive responses. Adjectives like fragile, delicate, exposed, and vulnerable spring to mind, and it is easy to imagine, for example, that if one came into near contact with the bog-hand, one would want to enfold it in a protective handshake.

Similarly, museum postcards with photographs of Tollund Man's face, as Anthony Purdy has pointed out, are fraught with Walter Benjamin's notion that "the human countenance was the last site of resistance in early photography for cult value against the encroachment of exhibition value." Purdy reminds us of Benjamin's comment that "The cult of remembrance of loved ones, absent or dead, offers a last refuge of the cult value of the picture. For the last time the aura emanates from the early photographs in the fleeting expression of the human face. This is what constitutes their melancholy, incomparable beauty."[20] And he concludes that although Tollund Man has become a tourist attraction, he "retains his extraordinary capacity to move."[21] My museum postcards' power to move (in this case the power of Grauballe Man's hands and feet) resembles, I would argue, that of Tollund Man's face. We are implicitly stirred by the individuality of the lines and patterns. We may not see the hands and feet as individual characters or personalities, such as we do faces—but as the photo postcards suggest, when the photos are deliberately and strategically cropped to underscore their humanness and individuality, they take on "face" nonetheless. Indeed, when medical students are asked to perform their first autopsies, they are often protected from the discomfort of cutting into a dead

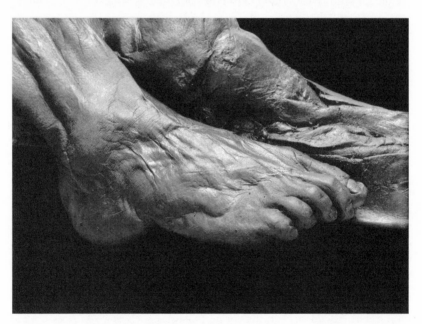

Figs. 1.5 and 1.6. Picture postcards of Grauballe Man sold in the gift shop of Moesgaard Museum. Photographs © Moesgaard Museum, Denmark.

human being by having not only the cadaver's face but also its hands and feet covered over.

Yet what is at issue here is not the bog bodies' actual faces, hands, or feet, but rather the particularly rich relation that photography as a medium has with them. Given that death has been a key player in the practice and theory of the photographic medium since its inauguration, bog bodies are well suited to keep company with photography in more ways than one. Almost immediately after its invention photography was theorized as a kind of open casket, a memorial apparatus in which we can see, chemically preserved, that which is no longer in existence yet mystically guarded against time's "relentless melt."[22] It is worth repeating here what has become a well-known refrain in photography theory: namely, that the duplicity of photographs, the simultaneous act of making the dead alive and the living dead, has a particular and paradoxical poignancy when the object photographed is in fact already a dead body. The "funereal immobility," as Barthes has it, paradoxically makes "the corpse [...] alive, as *corpse*."[23] Barthes, in his early analysis in "The Photographic Message," also points out how the "utopian character of denotation" makes photography appear as naïve and innocent, "a kind of Edenic state of the image."[24] This resonates with the bog's equally naïve archaeological photo album. And Barthes's later phenomenological reading in *Camera Lucida,* in which he favors the uncoded, mnemonic aspects of photography (punctum) over the "rational intermediary of an ethical and political culture" (studium) also seems to cast light on the astonishment and wonder that the nameless bog people evoke when we look at them in photographs. Thus both the ontology of photography and the emotiveness of its effects (its *affects*) have a bearing on bogs and on the bodies preserved in them.

With this said, it appears that photography, at least since the discovery of Tollund Man and Grauballe Man, has become the bog bodies' visual medium par excellence, and I would go so far as to claim that without photography many of the artistic representations discussed in this book would never have seen the light of day.[25] As we shall see later, most poets, authors, and visual artists working with bog bodies since the middle of the twentieth century are beholden to the transformative powers of the photographic lens. From the archives of bog photography, the archaeological gaze in all its forms can unearth numerous images of humanity and its perils.

Glossing Trauma

It seems evident by now that when modernity's cameras meet the faces, hands, and feet of a human being from an ancient past, a litigious tension between ethics and aesthetics takes place. To what degree, we need to ask, do the museum postcards constitute a "guilty" pleasure? To what degree is looking at pleasantly cropped close-ups of the bog bodies' hands, feet, and faces an exercise in substituting trauma with nostalgia—or vice versa? And to what degree is this tension rooted in or released from a peculiar kind of synergy between the ontology of the photographic medium and the realm of the bog? Does the paradox of bog and photography, each in its way—combined and in sync—make it possible to see a bog body photograph as lingering between the atrocious and the sublime? Can we see photography as part of this "competing way" of treating bodies?

While photography has allowed archaeologists new and interesting ways to explore the material world,[26] the fact that bog bodies are human remains asks us to consider the trauma (Greek *traumat-, trauma:* wound) that is shown in the mutilated bodies—a trauma that was inflicted by someone and experienced by someone else, even if it was part of a voluntary sacrificial act—and the projections of nostalgia (Greek *nostos:* return home; *algia:* longing) which are often applied to bog body photography, particularly in the form of bog landscapes. Even the most nostalgic of bog landscapes, when coupled with the trauma represented in the bodies that have come from them, press the question of complicity. How can we enjoy looking at, and buy, postcards of the cropped hands, feet, or faces of dead people? French writer Georges Bataille famously created a kind of protective shield when he proposed that the dead body had become a (tabooed) thing, but it is in American writer Susan Sontag's photo-optic, which homes in on the question of atrocity, that we find the material with which to dig deeper into a bog body photo-ethics. In other words, the distance created with taboos by Bataille has shifted for Sontag to a distance of a different kind, one that concerns itself with the relationship between ethical and aesthetical demands. For her the torment is double: it is represented in photography, and it is part of structuring the relationship between the various agencies involved in the photographic process and reception. In photography, she claims, we look for authentic death testimony, which means that in "photography of atrocity, people want the weight of witnessing without the taint of artistry, which is equated with insincerity or contrivance."[27] The "weight of witnessing" without the "taint of artistry" means that the onlooker will not accept, or will see as artificial, unseemly, and even unethical, any kind of tampering with the raw power of the photographic lens such as intentional—intrusive—cutting and glossing, as occurs in my photo postcard.

Yet the question remains of whether artistry is experienced as a taint in the case of photographed bog bodies. I would argue that it is not; which has to do with bog bodies' particular place in time (and representation). Although bog body photographs prompt ethical questions, they do not suggest a sense of guilt in viewing. Why? As a colleague observed when I asked about her response to photographs of bog people compared to photographs of other dead bodies: "When I know that the person who took the photograph is so far removed from the dead, it creates a buffer, and I feel relatively safe and unscathed by the image; the photographer after all is directing his lens at a body already placed in a museum."[28] This reaction bears out Sontag's observation that there is (or can be) a difference in how we experience, for example, a painted representation of the flaying of Marsyas (although Sontag confesses that she finds it hard to look at Titian's painting of this grisly motif) and how we experience a photograph of a similar real event. There is "shame as well as shock in looking at the close-up of a real horror," she says; only "legitimate" viewers of such photographic images, like surgeons or others who can learn from them, are exempted from being like the "rest of us," voyeurs.[29] Photographs transform reality. And photographic art—even if we imagine that it is imbued with more authenticity and immediacy than other art forms and therefore is closer to traumatic events—can make that which is horrible or excruciating become beautiful and pleasurable. But as Sontag insists that while "transforming is what art does, [. . .] photography that bears witness to the calamitous and the reprehensible is much criticized if it seems 'aesthetic'; that is, too much like art."[30] Photography's promise to give evidence, to document without deceit, is (or has been seen by some as being) at odds with its ability to aestheticize. This is particularly relevant for photographed bog bodies in that the onlooker of bog body photographs appears to escape the embarrassment of "indecent" objectification in a number of ways. But this does not answer Sontag's ethical qualms vis-à-vis photographs of atrocity.

In order to address these questions, it is necessary to return to the analogy between photography and the bog. If we agree that chemical processes in the bog have already started the process of "fixing" or "photographing," then the bog bodies are, as I have proposed above, caught in a *double-take* which creates remoteness. It is the long exposure time of the first "shot" that provides just enough remoteness so that we can experience both "the weight of witnessing" and aesthetic "artistry"—but without the sense and experience of undesirable artificiality that such artistry can entail. Distance in time has lifted the pressing and disturbing actuality that recent photographs conjure up (and I count as "recent" the practice of photographing death and mutilation that has persisted from the beginning of the invention of the camera until today). The two thousand years, give or take, between the subjects' death and our viewing of their remains creates a temporal buffer zone. But this buffer zone does not cancel out the fact that bog bodies are human subjects, since it is precisely the fact that they *are* human that makes them so vitally present in photography. American literary scholar W.J.T. Mitchell's observation about the "taking" and "use" of human subjects in photography is useful here:

> The "taking" of human subjects by a photographer (or a writer) is a concrete social encounter, often between a damaged, victimized, and powerless individual and a relatively privileged observer, often acting as the "eye of power," the agent of some social, political, or journalistic institution. The "use" of this person as instrumental subject matter in a code of photographic messages is exactly what links the political aim with the ethical, creating exchanges and resistances at the level of value that do not concern the photographer alone, but which reflect back on the writer's (relatively invisible) relation to the subject as well and on the exchanges between writer and photographer.[31]

So, how can bog bodies, if we regard them as human subjects, escape being "taken" or "used?" If, when we look at photographs of atrocities closer to our time, we are burdened with guilt, how are we freed from this guilt (if indeed we are) when we look at these prehistoric bodies?

Again *time* and *distance* seems to play a part. First, unlike images of contemporary violence which evoke guilt in the viewer, bog bodies—even if they are horrid and even if they put on view atrocious practices or evil deeds—are often experienced as being so far removed from modern experience that they can be quarantined safely under the nomenclature "archaeological remains." Thus the onlooker can avoid the experience of being caught in a "shameful" position as unseemly voyeur of tormented bodies. Second, if we stay within a vocabulary of blame and guilt, the viewers are not "blameworthy" of shooting the victim alongside the photographer, since the victim is already "shot" while latently resting in the age-old darkroom of the bog. Third, since many of the photographed bodies are already museum displays—that is to say already "embalmed" (even replicated, as was Tollund's body) in a representational form that cushions and softens any experience the onlooker may have of uncomfortable closeness with a dead body—it alleviates the sense of complicity between our eyes and the photographer's hand.[32] The result of this is that bog body photography not only invites artistic interpretation but in fact appears to gain in palatability and popularity if, contrary to Sontag's observation, it carries the "taint of artistry."[33]

As an example, let us look at an extreme close-up of the naked face of Tollund Man, so often replicated in bog body literature (fig. 1.7). The face, or part of it as seen here, can take on emotive and ethical signification and as a result

Fig. 1.7. The noticeable beard stubble on Tollund Man emblematizes the uncanny afterlife of bog bodies. Hair often appears to grow for a few days after death, as if it has a paradoxical life of its own. Photograph © Silkeborg Museum, Denmark.

reach far beyond the body's physical reality. The close-up has been cropped and a part of the face has been selected which gives a sensual hue, in which the prohibition against showing "the naked face" of the dead seems overridden by an aesthetic maneuvering that pulls us into the humanness of the man instead of removing us from it. The premise is, of course, that we not only accept the artistic license of the hand that operates the zoom lens and crops the picture, but also that we are informed about the context. If we are made to believe that we are looking at a close-up of a recent victim of disaster or war (the tsunami in Asia, the war in Iraq, or 9/11) the "taint of artistry" would no doubt weigh heavily on our eyes.

The experience of being in the presence of a twice-photographed face creates a distance but not a detachment from what we observe. This distance does not hinder an experience of being pulled into a very intimate face-to-face encounter; quite the opposite is true. The *double-take,* I would argue, not only tolerates but actually facilitates a sense of intimacy and proximity between the photographed man and us. The close-up of Tollund Man makes him vulnerable, but also makes us as viewers sensitive and receptive to his naked face—a face that takes on the burden of being an emblem of (human) reality. The photograph's special claim on reality is doubled here by the bog man's own claim to both *being* and *having been* real. As such, there is a metonymic chain pointing from the photograph back in time past bog body to man; and a pronominal trajectory which points from "it" (photograph) to "it/he" (bog body) to "he" (man). This is not a claim of realism, but of reality.

If "cameras [are] clocks for seeing," as Barthes proposed, photography offers a pictorial "museum" in which the clock has stopped.[34] These stopped clocks allow us to come face-to-face with bog bodies in a way that tests not only the abject, authenticity, trauma, and nostalgia, but also how real museum-walls present the viewer with a "colossal mirror," as Georges Bataille puts it, in which "man contemplates himself finally in all his faces."[35] Another French critic, Maurice Blanchot, in line with most photo-theorists has described a "cadaverous presence" in the image; the individual in death becomes a thing, an object, an "unbearable image and figure of the unique" turned to "nothing in particular." The corpse somehow materializes "in the strangeness of its solitude as that which has disdainfully withdrawn from us." Death, in other words, "suspends the relation to place, even though the deceased rests heavily in his spot as if upon the only basis that is left for him."[36] As photographic representation is set in motion, the corpse starts resembling itself in a monumental way, as a way of being present in its absence. The dead body, photographic convention holds, represents the very essence of the photographic image, a kind of phantom, both anchored in and removed from reality. "To take a photograph," as Sontag points out, "is to participate in another person's (or thing's) mortality, vulnerability, mutability. Precisely by slicing out this moment and freezing it, all photographs testify to time's relentless melt."[37] While the very fabric of the medium, its ontology, may concern death (the large and growing body of work done on death and photography would testify to this), what interests me here is how photographs of bog people not only highlight and articulate (either directly or indirectly) ethical concerns and qualms we have or do not have when looking at photographs of dead bodies, but also continue to provide material for projections of various kinds that spill over into political and ideological considerations and articulations in poetry, literature, and visual art.

Memento Mori

Let me return to Glob's *The Bog People,* where we find one more instance in his handling of photographic illustrations worth parsing: it takes the form of a metaphysical abstraction. When, for example, under a close-up of Tollund Man's face he places a caption containing a quotation from the oldest heroic epic, *Gilgamesh,* he deliberately gives us a metaphysical key with which to understand the photograph. The caption reads: "The dead and the sleeping, how they resemble one another." In the main text he fleshes out this point of view: "Majesty and gentleness still stamp his features as they did when he was alive [. . . .] It is the dead man's lightly-closed eyes and half-closed lips, however, that gives this unique face its distinctive expression, and call compellingly to mind the words of the world's oldest epic."[38] Through this caption, and with the question of who is dead and who is alive, Glob draws on a poetical discourse to contrast—perhaps even challenge—the archaeological one. The possibility that the man in the photograph is only sleeping, the illusion brought on by the half-open eyes and mouth, functions like an ironic memento mori, reminding us both of our mortality and of a possible immortality. As an emblem of reality, the face of Tollund Man suggests a slightly different way of viewing traditional memento mori such as we know them from art history. His face is a reminder of our death, perhaps more forcefully so than most memento mori precisely because *he* (the personal pronoun is an important part of fleshing out his face) is so "up close and personal." Contrary to the traditional use of *vanitas* emblems such as the cranium, the Tollund Man is not reduced to a pure cool sign, but has maintained an individuality which mirrors our own. His full-faced presence invites an affectionate close-up. The many, often quite emotive, responses to his photographed portrait in poetry, novels, archaeological discourses, paintings, and sculpture—examples that fill this book—bear out this point.

The grotesque image of an exhumed and (presumably) executed body has with the caption turned into an emotional response of empathy. By way of the photographs—through these "living images" of a dead body—Glob evokes a past life, ended by sacrifice, revived through words, and captured on photography prints before it disappears or "ends" once more. The *Gilgamesh* quotation and caption illustrate that in Tollund Man's uncannily lifelike countenance, we face our own mortality yet are allowed the dream of immortality through the preserved "life" that emanates from the photograph. In his mummified face, the residue (or punctum, to use Barthes's term) is not "frozen in time" but kept in perpetual transit—between the living and the dead. The caption-image relationship adds to the elegant hybridism already so evident in Glob's writing. Jefferson Hunter has made an astute analysis of this and argues that

> If Glob were the kind of captioner who constantly nudged the reader toward generalized pathos, or the kind of archaeologist who ascribed preservation solely to the gratified ego of the fertility goddess, he would be censurable by Sontag's standards. But he is in fact antisentimental, both in the dozens of

photographic plates he comments on in a strictly factual way and in the care he takes, when meditating on the Tollund Man, to turn the reader's sensibility in a specific direction for limited and specific purposes. He regards the past which his photographs bring him as a subject for study, conceiving "study" in an admirably broad way, devoting equal attention to the fertility goddess and tannic acid. He also regards that past as a basis for historical judgment.[39]

I only partly agree. While Hunter rightly maintains that photography cannot be separated from the context in which it is couched, the caption in this case presses our empathy response to the fore, and the place of pride that Glob gives to the photographs of Tollund Man's features itself speaks volumes about the humanity of the man they depict. Throughout Glob's book, photographs and captions suggest that the past is both scientifically researchable and poetically present and re-presentable. There is an implied promise of other bodies or objects yet to be dug out of the soil and reanimated for our archaeological or imaginary perusals and desires.

The Danger of Projection

Glob was keenly aware of the bog as a paradox, a taker of life, and a preserver of life. And in his opening chapter on Tollund Man he implies that the unearthed body is connected to a mysterious and mythological past, even to a danger that extends to the present. According to archaeological lore, disturbing the peace of the dead seldom goes unpunished; indeed, opening doors into the undisclosed tombs of the Pharaohs in Egyptian pyramids or (as here) digging into the bog imposes grave danger on archaeologists or other "transgressors" who ultimately meet with untimely deaths.[40] The removal of the Tollund Man body, likewise, Glob informs us, was not accomplished without loss: "One of the helpers overstrained himself and collapsed with a heart attack. The bog claimed a life for a life; or as some prefer to think, the old gods took a modern man in place of the man from the past."[41]

Glob, as it were, had already faced dangers of another nature. The "dangers" had to do with hostility, public prejudice, and outspoken resistance to archaeological interpretations.[42] But there were other hazards of which Glob was clearly cognizant, namely the risks in projecting national traits onto the faces of bog finds. Both Tollund Man and Grauballe Man were routinely described in the press (and still are today in much bog body literature) as having aristocratic facial traits and honorable visages. Tollund Man's high forehead and noble features prompted archaeologists to compare him with *Queen Nifertete,* the most beautiful woman in the world, and were said to tell "of a man with certain intelligence." Glob writes that like Tollund Man, Grauballe Man "gives an impression of how this man looked on the threshold of death, many years ago. This time the effect is not one of tranquility but of pain and terror. The puckered forehead, the eyes, the mouth and the twisted posture all express it. The circumstances that led to his death were probably not the same as in the case of the Tollund man."[43] Because the years 1950 and 1952, when Tollund Man and Grauballe Man were found, still resonated with memories of Nazi archaeological fabrications, the Danish archaeologists might well have wished to tread lightly in arguing for any evidence of a Germanic-looking body from the past. Most newspaper articles on the bog bodies were remarkably free of descriptions of national or racial features. And Glob, mindful of the pitfalls in interpretations of ancient features as having racial or national characteristics, stays carefully within the parameter of a psychological, and not racial, profiling.

Not all would follow such a careful protocol. In fact on May 2, 1952, one of the main Danish newspapers, *Jyllandsposten* (recently and infamously known for its publication of the Mohammed cartoons), published an article called "Grauballe Man, a Beautiful Representative of our Ancestors," which glorifies the "Germanic" body for what it might have been: "He has a beautiful and stately shape, almost as tall as a palace guard and very solidly built; possible speculations about a certain coarseness in our Germanic ancestors from the Iron Age are knocked over by the fact that he had beautiful hands and almond shaped fingernails. His thick red hair has most likely been bound in a so-called Germanic Knot by the left ear, a Germanic fashion from the period around the birth of Christ."[44] "Silvanus" the anonymous author of the article goes on: "The slightly Mongoloid imprint on the face need not originate from birth, but might well be caused by a violent injury to the head."[45] While most avoided such direct references to racial characteristics, many observers saw signs of a "noble" humanity in bog people's faces, such as in Tollund Man's "clean and even features with the thoughtful wrinkles in the forehead."[46] Nothing, the newspapers reported, in the two men's facial features indicate that they were slaves; in fact Tollund Man's expression suggests that he most likely went to his death out of free will in an effort to save his community from starvation. He was a willing self-sacrifice. Grauballe Man too sacrificed himself, it was proposed, but his countenance tells a slightly different story; after all, his face did not possess the tranquility of Tollund Man's. In another Danish newspaper, *Nationaltidende,* we read: "Whether Grauballe Man in the last minute was seized by panic, and the ritual-priest has had to strike him down with a blow from a stick, is of course mere conjecture. But is it such a wild guess?"[47]

These readings of ancient faces and bodies are, of course, part of *both* a popular interest in giving them a story *and* a continued interest in understanding nations and identities. If we see faces as not being fixed in prescribed dichotomies—either as unnatural human beings or as glorified ideals—they may help us to examine a more complex humanity. As I will show at the end of this study, plastic face reconstructions (more so than the full-body reconstruction discussed above) need to carefully navigate the territory laid out by their history at large, both by acknowledging the conflicted record of face reconstructions and by paying attention to the associations and the ways in which faces and "face" (in the sense of responsibility proposed by Emmanuel Levinas) can be considered vis-à-vis plastic representations. I shall return to these questions, but for now let me offer just one observation on face reconstruction that has direct bearing on photography.

While photographs of bog bodies are infinitely reproducible and widely distributable surrogates that allow for close-up experiences at a safe distance from the bodies themselves, face reconstructions provide a rather different experience. They are both closer to and further from the realism of the camera lens. Like photography they replace the real thing, but unlike photography they neither zoom in on selected delicate details nor collapse into one shared ontological realm. To a certain extent the ontological challenges of face reconstructions are both straightforward and complicated. They are straightforward in the sense that the reconstructions quite literally carry out what the idiom specifies: they re-construct faces to look as they looked (or at least close to how they may have looked) before pressures of time, processes of nature, and of course actions of fellow humans altered their original appearances. But they are complicated in the sense that the one-to-one relationship between the ravished face and its re-facement—the presumption, indeed the promise, that the face is simply restored to itself—is contradicted by the artificiality of the medium, an artificiality that photography seems to escape. In photography the materiality of the medium is almost always magically absent, or at least relatively unimportant. It is never *it* that we see, as Barthes pronounced. That is, it is never the actual materiality of the flat paper that we notice, but what is recorded on the paper. Photography is always about material absence, even as we feel psychologically or emotionally pressed into an intimate face-to-face experience.

Landscape Nostalgia

Finally, what about the many bog photo-landscapes in Glob's *The Bog People*? Is there a way in which they offer another kind of "face?" The first landscape photograph (fig. 1.8) early in Glob's *The Bog People* shows a rural dirt road, carved through an idyllic Danish landscape, that leads to the "scene of the crime." It is difficult to see whether the path was shaped by a twentieth-century

Fig. 1.8. The road leading to the location of the Tollund Man find: A blossoming tree that looks to be an alder serves as a kind of stand-in for the body in the bog, a still-living double of the dead. Photograph © Moesgaard Museum, Denmark.

automobile or by an Iron Age cart. In fact, it looks as if the photograph has been cropped intentionally so as to wipe out signs of modern life. It is obviously not included in the book to authenticate the site, to provide a recognizable map to the scene of the find, or to tell us that we are entering an archaeological excavation. Rather, it serves as a photographic echo of the temporal play in the written text, as a way to fine-tune a mythopoetic attentiveness in the reader. It is a deliberate mollification of the "atrocious"—a calculated gloss over the harsh, often gruesome and shocking sight of the dead bodies. Glob establishes a particular kind of reality for the bodies in the bog, in that he sets up a landscape context that adds to the weight of mythology and heightens our sense of a presumed immortality in both bodies and bogs. As the dirt road forms a kind of transhistorical trajectory through the scenery, the site photographs favor the nostalgic over the atrocious.

The use of a tree as main figure, not only in this photograph but in the many other such landscape photographs in Glob's book, can be seen as providing a living deputy for the many wood figures that were sacrificed in the bog along with (or perhaps as surrogates for) humans. The tree becomes to the wood figures what the living human being was (or is) to the bog body. Add to this that the world tree in Norse mythology, Yggdrasil, is connected to human sacrifice (Odin's self-sacrifice was to hang from the tree), and Glob's use of the tree takes on a deep-rooted connection to a Nordic mythological past so that it points (indirectly) to the sacrificial hanging of Tollund Man before he was placed in the unseen bog at the end of the road. British archaeologist Miranda Aldhouse-Green has pointed out a similar kind of perceived synergy between wood and flesh. The flesh-like degeneration of wood, she offers, has in some communities been tied to a "sense of horror, mortality and impurity; indeed the smell of the rotting timber could be perceived as akin to that of a decaying corpse which, though undoubtedly finished as an active living thing, is still (in many traditions) regarded as still retaining part of the original life-force."[48] The original life force is reflected "in the very movement of matter and the leaching of fluids that decomposition involves." But above that, the landscape photographs suggest a nostalgic, bittersweet longing for the past as a home— not only a personal longing, but also one tied to nation and to the identity that nationhood brings with it. In short, what Glob's landscape photography presses to the fore is the fact that in bog bodies and bog landscapes we are moved by a kind of nostalgia of *presence* as well as nostalgia of *origin*.[49] And the umbilical cord that photography ties to origin and memory will, as we shall see later, play a key role in much bog body literature and art.

Photography as a medium always has something latently uncanny about it; beneath the placid surface of the photo-landscape we find not only the "bogeyman" but also a "mass" of historical and sexual repression.[50] The uncanny "flash of the past," like Walter Benjamin's "disinterred corpses of the ancestors,"[51] will come back to haunt us. The latent uncanny potential, it goes without saying, is not only fixed to *photographed* bog bodies but is implicitly tied to bog bodies in whatever cultural incarnations they embody. And who better to help us uncover this haunting subterfuge than the master of the uncanny, Sigmund Freud—and, for that matter, Carl Jung. Both men, it turns out, knew about bog bodies.

CHAPTER 2

The Archaeological Uncanny

I have already suggested that bogs are archaeologically uncanny places *and* spaces, both geographical and geological substances *and* abstract constructions; that they are material *and* mental; that they permit the spatial and temporal to collapse. I want now to suggest that their particular logic and physiognomy is (uncannily) similar to the definition of *unheimlich* used by Ernst Jentsch and elaborated by Sigmund Freud: "something one does not know one's way about in"—a *locus suspectus*.[1] In Freud's psychosexual optic, the uncanny is intimately linked to the ultimate "home," the womb, and more specifically to the experience of the "unhomely" entrance to this home, the female genitals. The unhomely then "is actually the entrance to man's old home," he writes, "the place where everyone once lived," and "the negative prefix *un*- is the indicator of repression."[2] This sexualizing of the uncanny will, as we shall see later, play a key role in the conflation of death and the erotic in much bog body art and literature. But before we delve into this, I want to look at the uncanny from a slightly different perspective: one that relates to the archaeological metaphor that Freud adopted.

In 1905, Freud confessed his great debt to archaeology: "In face of the incompleteness of my analytic results, I had no choice but to follow the example of those discoverers whose good fortune it is to bring to the light of day after their long burial the priceless though mutilated relics of antiquity. I have restored what is missing, taking the best models known to me from other analyses; but, like a conscientious archaeologist, I have not omitted to mention in each case where the authentic parts end and my constructions begin."[3] For Freud, the voyage of self-discovery became a voyage to the depths of the past, "through the deepest strata of the mind."[4] His most celebrated example is found in *Civilization and Its Discontents,* where ancient Rome is chosen as an analogy for the unconscious: "Now let us, by a flight of the imagination, suppose that Rome is not a human habitation but a psychical entity with a similarly long and copious past—an entity, that is to say, in which nothing that has once come into existence will have passed away and all the earlier phases of development continue to exist alongside the later one."[5] All palaces and castles and statues and artifacts are still there, deeply layered, as are all historical periods making time graspable as a spatial playing-field where "the observer would perhaps only have to change the direction of his glance or his position in order to call up the one view or the other."[6]

The temporal depth that Freud locates in the ancient civilization of Rome finds a northern (albeit more "uncivilized") pendant in bogs. Bogs seem perfectly fitted as spatial tropes for psycho-archaeological depth, and it is easy to imagine that bog bodies would have been of interest to Freud. Nevertheless, although Freud knew about the existence of bog bodies, he paid little attention to them and dismissed their importance as cultural traces. This dismissal is all the more surprising because bog bodies seem tailor-made for his theory of the uncanny (*das Unheimliche*). In fact there are a number of ways in which the uncanny has root not only in Freud's interest in the process of archaeological digging and in conflations of psychological and cultural strata, but also—and, for the present study, importantly—in our relationship to dead bodies. It seems appropriate, therefore, to investigate how Freud's optic bears on bog bodies.[7] This is important because Freud's archaeological interests allow us to widen our perspective from the northern European bogs to the classical archaeological soil of Southern Europe, and permit us to see how, in a comparison between the two, place and identity turn out to be full of twists and turns.[8]

The uncanny, as Freud famously argued, is often experienced in connection with "something which ought to have remained hidden but has come to light." In this sense, all "found objects" have something uncanny (unexpected, repressed) about them. But, more importantly, "anything [that has] to do with death, dead bodies, revenants, spirits and ghosts," is at "the acme of the uncanny."[9] Freud's observation is connected to an emotive ontology:

Yet in hardly any other sphere has our thinking and feeling changed so little since primitive times or the old been so well preserved, under a thin veneer, as in our relation to death. Two factors account for this lack of movement: the strength of our original emotional reactions and the uncertainty of our scientific knowledge. Biology has so far been unable to decide whether death is the necessary fate of every living creature or simply a regular, but perhaps avoidable, contingency within life itself. It is true that in textbooks on logic the statement that 'all men must die' passes for an exemplary general proposition, but it is obvious to no one; our unconscious is still as unreceptive as ever to the idea of our own mortality.[10]

The unexpected reemergence of bog bodies inevitably tests not only that which "no human really grasps"—mortality and death—but also the "strange repetitiveness" in the uncanny which, in Freud's optic, is connected to the death drive. Disbelief or doubt as to the material reality of the phenomena Freud locates as the key operatives in the uncanny multiplies, I would argue, when it comes to bog bodies—due not least to the distinctive material reality of the realm from which they emerge. Again, it is the testing of the material reality that needs our attention. Freud elaborates: "If we now go on to review the persons and things, the impressions, processes and situations that can arouse an especially strong and distinct sense of the uncanny in us, we must clearly choose an appropriate example to start. E. Jentsch singles out, as an excellent case, 'doubt as to whether an apparently animate object really is alive and, conversely, whether a lifeless object might not perhaps be animate.' In this connection he refers to the impressions made on us by waxwork figures, ingeniously constructed dolls and automata."[11]

If bog bodies owe their fame to the fact that they are bona fide remains, how do we know that they are (or were) real human beings? How do we know that they are not fakes, waxworks, or dolls (let us recall Tollund Man's reconstructed photo-body)? Is the fact that they look like us enough to assign them agency as fellow human beings? Do we see them as reassurance and stability against "time's relentless melt," like photography as discussed in the previous chapter? Or, conversely, are they intruders from another time that generate instability and volatility?

Whether human or thing, real or fake, bog bodies are caught in a paradox. On one hand, their uncanniness rests on our experience of authenticity; they are the *real thing!* On the other hand, their uncanniness rests on the possibility that they are *not real,* that they are ghostly *simulacra.* The irony is that if they are real, they are uncanny and that if they are not real, they are uncanny. Either way, they cannot help but cast suspicion on what Freud called "the material reality of the phenomena." Indeed, when they appear in some instances as if they had "died yesterday," they embody what Freud saw as "the most uncanny thing of all," namely "the idea of being buried alive by mistake." Whether they have returned from the dead; are fake bodies that trick the eye (doubles); or are caught between life and death (buried alive in a quasi-existence we all fear), bog bodies qualify on all fronts as the ultimate manifestation of uncanniness, locked in the space between the familiar and the unfamiliar. Above all, they harbor a particular kind of *temporal* uncanniness that is not only un-*heim*lich (un-homely) but also un-*zeit*lich (un-timely). This untimely unhomeliness happens to be part and parcel of their unique "material reality" as archaeological objects. They would seem to be "poster children" for Freud's theory.

Freud versus Jung

Yet, as mentioned, Freud's own well-documented archaeological curiosity did not encompass bog bodies, although he was apparently familiar with their existence through Carl Jung. Unlike Freud, Jung was fascinated by them and in *Memories, Dreams, Reflections* he tells us he believes that his interest in them helped precipitate the split between Freud and himself. It is worth citing in full.

> In Bremen [in 1909] the much-discussed incident of Freud's fainting fit occurred. It was provoked—indirectly—by my interest in the "peat-bog corpses." I knew that in certain districts of Northern Germany these so-called bog corpses were to be found. They are bodies of prehistoric men who either drowned in the marshes or were buried there. The bog water in which the bodies lie contains humic acid, which consumes the bones and simultaneously tans the skin, so that it and the hair are perfectly preserved. In essence this is a process of natural mummification, in the course of which the bodies are pressed flat by the weight of the peat . . . . Having read about these peat-bog corpses, I recalled them when we were in Bremen, but, being a bit muddled, confused them with the mummies in the lead cellars of the city. This interest of mine got on Freud's nerves. "Why are you so concerned with these corpses?" He asked me several times. He was inordinately vexed by the whole thing and during one such conversation, while we were having dinner together, he suddenly fainted. Afterward he said to me that he was convinced that all this chatter about corpses meant I had death-wishes toward him. I was more than surprised by this interpretation. I was alarmed by the intensity of his fantasies—so strong that, obviously, they could cause him to faint.[12]

What a remarkable scene! To Jung, Freud was "blind toward the paradox and ambiguity of the contents of the unconscious, and did not know that everything which arises out of the unconscious has a top and a bottom, an inside and an outside."[13] There was a "monotony of interpretation" in Freud's continual obsession with sexuality, and Jung goes on: "In his own words, he felt himself menaced by a 'black tide of mud'—he who more than anyone else had tried to let down his buckets into those black depths."[14] We find a contextual link between Freud's conceptualization of the uncanny and his reaction to bog bodies by comparing Jung's description of the fainting episode in 1909 with a letter from Freud to Jung written earlier the same year in which Freud takes issue with Jung's "spook-complex" and with an experiment Jung had performed in Freud's office: he had tried to produce a "so-called catalytic exteriorization phenomenon" by making a bookcase rattle without physically touching it. Freud saw the noise coming from the bookcase as logically explainable. In a letter to Jung, he puts on his "horn-rimmed paternal spectacles," shakes his "wise gray locks" at the "son's" fanciful conceptions of the paranormal, and then goes on to discuss his own preoccupations and superstitions regarding numbers. The point Freud wants to make is that just as the most uncanny coincidences of numbers are explainable, so is Jung's claimed paranormal ability to make objects move.[15]

But why did bog bodies not fit into Freud's archeological interest? Was it because they were too *uncivilized* and consequently unable to stand in for previous *civilizations*? Did the bog and the soil of the classic past offer two different kinds of depths? Or were bog bodies simply too close to "home," connected implicitly to the "primitive" father-son rivalry he located with Jung? The answers are complicated. Both men placed great symbolic value on archaic vestiges. But Jung's preoccupation with the past was also tied to his obsession with alchemy and ancestry—an obsession that would eventually, albeit not directly, lead to anti-Semitic sentiments. Given that archaeology had developed exceedingly fraught ties in Germany to the *Blut und Boden* ideology of the pre–World War II period, Jung's fascination and Freud's rejection of bog bodies point implicitly to a larger set of questions and interpretations of archaeological remains vis-à-vis politics and ideology. Freud's fear that a "black tide of mud" would suffocate his insights on the role of sexuality in the unconscious seems justified when Jung, in reaction to Freud's materialism and positivism, espoused a mythological "struggle between light and darkness" and eventually co-opted "the black tide of mud of occultism" into his writings on "the conscious and unconscious *historical* assumptions underlying our contemporary psychology."[16]

The fact that Jung, as he admits, was "a bit muddled" when he described bog bodies to Freud and had "confused them with the mummies in the lead cellars of the city" can, in part, explain Freud's fear of an aggressive death wish on Jung's part. After all, a corpse in a cellar in the city where they were meeting was closer to "home" than a body in a bog. In the memoirs, Jung follows the fainting scene with a description of a dream in which he enters a house ("it was 'my house'") that is separated in archaeological layers. "It was plain to me," he writes, "that the house represented a kind of image of the psyche—that is to say, of my then state of consciousness, with hitherto unconscious additions."[17] The top floor represented the conscious, the ground floor the unconscious. Descending through rooms with walls from Roman times, he finally lifts a slab in the floor, descends "down into the depths," and enters a cave. "Thick dust lay on the floor, and in the dust were scattered bones and broken pottery, like remains of a primitive culture. I discovered two human skulls, obviously very old and half disintegrated."[18] In the cave under the floor, Jung concludes, "I discovered remains of a primitive culture, that is, the world of the primitive man within myself." Once again, Freud insisted that Jung recognize and acknowledge the two skulls as symbolic representations of death wishes; the question was towards whom were those wishes directed! Compromising his own belief to the contrary, Jung decided to "lie" and, to Freud's relief, named "my wife and my sister-in-law—after all, I had to name someone whose death was worth the wishing!"[19]

The house dream revived Jung's interest in archaeology, and as he homed in on the mythological character of dreams and fantasies, he discovered the "close relationship between ancient mythology and the psychology of primitives." Unwilling to accept Freud's imposing authority, his theory on the death wish, or his interpretation of dreams as " 'a façade' behind which meanings lies hidden" (Jung believed dreams were "a part of nature" with "no intention to deceive"), the "son" eventually broke with the "father."[20] Nevertheless, as with Freud, archaeology continued to be an obsession for Jung. But Freud's privileged site, Rome, remained a slippery place for Jung because, as he writes: "I felt that I was not really up to the impression the city would have made upon me."[21] He goes on: "I always wonder about people who go to Rome as they might go, for example, to Paris or to London. Certainly Rome as well as these other cities can be enjoyed esthetically; but if you are affected to the depths of your being at every step by the spirit that broods there, if a remnant of a wall here and a column there gaze upon you with a face instantly recognized, then it becomes another matter entirely."[22] The temporal repositories in walls and columns of ancient Rome are anthropomorphized by Jung, and thus made into strange obstacles even as, or precisely because, they are imagined to be "easily recognized."

Jung did in time travel to Pompeii, where, as he moans, "the impressions very nearly exceeded my powers of receptivity." He was only able to visit the famous excavation site (the quintessential site of unhomeliness and uncanniness, where death, as Wilhelm Jensen wrote in *Gravida,* "was beginning to talk") after a healthy booster shot of careful study of classical antiquity and its psychology.[23] Even so, in Pompeii "unforeseen vistas opened, unexpected things became conscious, and questions were posed which were beyond my powers to handle."[24] What these unforeseen vistas and unexpected things consisted of, he does not reveal. But his conclusion may offer a clue: "In my old age—in 1949—I wished to repair this omission [never having been to Rome] but was *stricken with a faint* while I was buying tickets. After that, the plans for a trip to Rome were once and for all laid aside."[25]

If Freud fainted at Jung's tale of bog bodies, Jung in turn, it seems, fainted at the prospect of finally setting foot in "Freud's Rome," implicitly implying that "the spirit that broods there" in the faces of the columns and walls might be that of the (by now dead) "father." Thus, albeit in different ways, uncanny archaeological remains found a way into Freud and Jung's respective theories of the psyche, but also into their rivalry.

North versus South

Apart from his complicated relationship with Jung and their disagreements over the death drive, and so forth, Freud's reluctance to engage with bog bodies is also noteworthy for other reasons, particularly in view of the archaeological gaze he so notoriously directs toward ancient bodies similarly "returning" from the dead: those found in the remains of Pompeii. Besides Rome as a privileged psycho-archaeological site, Pompeii had a great influence on Freud, and his analysis of Wilhelm Jensen's *Gradiva* is a central text in his examination of the functions of the uncanny. In Jensen's story a young scientist falls in love with the image of a woman on a Pompeian marble bas-relief, and then transfers his desire to a real woman. Freud uses the story to show the interrelation of original and copy, as part of his interest in the *doppelgänger* motif and the uncanny.[26]

In many ways, Freud's archaeological interest dovetailed with a rather conventional set of paradigms of antiquarianism in the nineteenth century. To be an antiquarian meant in essence to be an archaeo–art historian. It meant—for the most part—cultivating an interest in art treasures from the classical past, and it meant finding artifacts that could testify aesthetically and in an implicitly moral way to the progress of civilization.[27] In contrast to this, a perceived sense of Nordic crudeness made it difficult for it to be seen as rivaling the grandeur and nobility of classic Greek-Roman or Egyptian antiquity. Long-established and hardwired perceptions of the north as being closer to its peoples' origin yet further from the southern regions' depth of culture implied that the north, as an archaeological site, was too "shallow" and climatically hostile to yield the plethora and profundity that sprang from the "deep" Grecian and Roman soil. In other words: to be closer in time to a past, which has to *define* the present, essentially meant to be less civilized. Even if rune stones and burial mounds would, in time, begin to compete for a place in the European archaeological imagination (particularly with the soaring of European national romanticism), the disturbing bog bodies, "these corpses," from the north could hardly vie with the beautiful marble figures of classic antiquity. Unlike the "civilized" Roman and Grecian past, the northern bogs appeared both shallow and bottomless and dark—or, to borrow from Gaston Bachelard's psycho-phenomenology, "the desire to dig in the earth immediately takes on a new aspect, a new duality of meaning, if the ground is muddy."[28] It certainly seemed to be so for Freud.

Nevertheless, Freud's fascination with Pompeii, and not least with the Pompeian dead, offers a direct comparison with bog bodies. Taking into account that Nordic prehistory has a relative dearth of statuary, bog bodies can be seen as a natural prehistorical substitute for, or equivalent of, other ancient quasi-statuary remains. They are eminently comparable not only to Egyptian mummies but, even more, to the famous Pompeii plaster casts. In fact, similarities and differences between the Pompeii plaster casts and bog bodies can help us understand both the peculiarity of bog body remains and the operations and variations of the uncanny.

In the 1870s the Italian archaeologist Giuseppe Fiorelli developed a technique of casting corpses from the cavities left by the decomposed victims of Vesuvius' disastrous eruption in Pompeii in 79 BC. Men, women, and children caught in the ashes and frozen in the midst of social situations (mothers protecting children; prisoners bolted to walls, lovers embracing) were magically pulled into view when plaster of Paris was poured into the hollow spaces left by their disintegrated remains. In the earth, the plaster was allowed to solidify into three-dimensional full-body death casts before it was excavated. This "raising of the dead"—and the manifest visualization and recasting of their psychical materiality—meant destroying the space they had occupied; it meant razing the authentic site.[29] But it also meant turning negative space into positive form.

Unlike the fleshiness and overt corporeality of bog mummies, Fiorelli's casts provided a kind of intermediary between the real body and the unreal, a filling-out from the inside out. Compared with bog bodies, this suggests an

Fig. 2.1. One of Giuseppe Fiorelli's casts from circa 1870, here in a display case in Pompeii, seems to rise from the dead. Photograph © Karin Sanders.

interesting binary: where there ought to have been presence, there is absence; and where there ought to have been absence, there is presence. This reversal of absence and presence in the Pompeian casts and in bog bodies can be made clearer if we use Gordon Beam's distinction between the *eerie* and the *uncanny*: "The absence of what ought to be present is *eerie*" and "the presence of what ought to absent is *uncanny*."[30] Pompeian casts, if we follow this prescription, are eerie; bog bodies are uncanny. But more importantly, if the Pompeian plaster casts are "negatives" turned "positives," showing what is no longer present, bog bodies are "positives" turned "negatives": the "things-themselves," but without aesthetic distance or representational detachment. In other words, bog bodies and Pompeian casts seem at surface to be ontologically dissimilar in that one *is* the dead body and the other *represents* the dead body. However, this distinction is not as impermeable as it might seem, particularly if we allow Socrates's claim that mummification suspends the ontological shift from body to corpse to encompass quasi-mummified traces and imprints like those from Pompeii.

This brings us back to photography. Photography, as I have already shown, has an uncanny ability to step squarely into the reality testing that Freud called for. To Freud, the symptom of a traumatic memory was a kind of mnemonic symbol, "an imprint of the original trace, a further development upon the original form. The task of the analyst is to interpret the distortion, to try to imagine the original experience from the painful imprint which it had left."[31]

Painful and traumatic imprints, not least Freud's focus on a "test of material reality" in the uncanny, finds a particularly resonant medium in bog body photography. The ontological challenge in dead bodies, and bog bodies in particular, resonates uncannily with the ontology of photography. In fact, I offer that the fluctuation between positive and negative I have located in chapter 1 is further complicated if we allow ourselves to see photography as being to bog bodies what marble statuary was to Fiorelli's plaster casts.

Leather Skin and Marble Bodies

The British archaeologist Colin Renfrew has proposed that the Pompeiian casts represent "involuntary art." This "involuntary creation of traces by the human body, which are found significant and worthy of contemplation by those subsequently recovering or detecting them," is art, he says, not because it is a result of human activity, but because "the traces are immediately suggestive of the body, being actual imprints [. . .] capable of simulating emotions in us as observers."[32] These emotions, he continues, are "more arresting" than those produced by other more traditional sculptures of the human body. The reason is that there is a slippage between reality and representation; the aura of the human person as presence (the body) and as proximity (the representation) results in dissimilar emotive reactions.

The iconographical power of Pompeian plaster casts both borrows from and distances them from the visual vocabulary of classical statuary. They are not quite sculptures, not quite bodies, but shaped as representations of what no longer exists; and as such they are easily associated with a lack and longing. Bog mummies, in contrast, are rarely a pleasant sight.[33] Only one or two can lay claim to physiognomic beauty. The rest are more or less unsightly human remains, often with distorted features; while ruinous in form, they are (again with one or two exceptions) not beautiful ruins. Unlike other uses of dead bodies—such as anatomical dissection paintings in the Renaissance or Romanticism's sentimentalized death tableaux, which each in their way offer dichotomies between interior and exterior with some sort of philosophical (the Cartesian split between body and soul, for example) or educational ramifications and intentions—bog bodies do not promise "deep" interiority because the interior here *is* the exterior and vice versa.

There is also an interesting process aspect at play here and it may be useful to remind ourselves of Michelangelo's legendary credo, which says that the sculptural body is born in clay, dies in plaster, and is immortalized in marble. This can—with some license—be adapted to a bog body that is then born in the "clay" of the living body, dies in the bog, and finally is immortalized as an envelope of leathery skin. The point of the analogy is that while the plaster casts are far from the leathery and fleshy corporeality of bog mummifications, they are not far from the *reality* of bodies, and are closer to the authenticity of actual bodies than are marble figures. Although the plaster of Paris sanitizes the sensations of disgust tied to cadavers, the sense of immediacy the casts bring to the traumatic events of the "mummified city" preserves in them a ghastly presence. Behind the "quotidian semblance," as Vidler has it, "there lurked a horror."[34] Like bog bodies, Pompeian casts are essentially liminal, and their ghastliness and implied corporeality places them at odds with the dominant aesthetics of classical ideals. Archaeology, "by revealing what should have remained invisible, had irredeemably confirmed the existence of a 'dark side' of classicism."[35] Or, as Jennifer Wallace argues, "archaeology highlights what has been lost or passed away naturally and what can only be understood elegiacally, through imagining what has disappeared."[36]

Still, it seems safe to claim that we experience a presentness and realness in bog bodies that we do not in the Pompeian cast. The dead from Pompeii and the dead from the bog project two different kinds of sensual corporeality. And this brings interesting connotations in regard to erotica. The "mummified city," it is well known, offered access to forbidden and tantalizing images of a pornographic nature.[37] But erotizing dead bodies at the height of the archaeological interest during the eighteenth and nineteenth centuries was possible because the dead (for the most part) were firmly wrapped in classical sculptural representations of youth and beauty, intimately tied to the aesthetics of distance offered by marble skin. Clothed in marble, sculptural bodies could be tantalizing without forcing close contact to the offending orifices, excretions, and smells of real bodies; they had no threatening interior.[38] In touching or looking at classical marble sculptures, erotic-archaeological encounters could froth safely without anyone losing sight of the difference between living and dead bodies, and without the experience of disgust or threat of decay and destruction that real dead bodies inevitably bring about. This kind of archaeological erotica has been well documented and was familiar to readers of eighteen and nineteenth century literature and poetry, Freud included.

In contrast, and decidedly unlike smooth classical skin, bog body skin is always darkened, weathered, wrinkled. Although it can be relatively soft and pliant when first uncovered, it almost immediately turns leathery and rigid. Whilst marble skin (and to some degree the plaster cast) spells eternal youth; bog body skin implies aging and decay. Since erotic encounters are almost always about touching (even if it is allocated to the mere dream of touching by proxy of the gaze), we can detect two interrelated models of touch which have different archaeological resonances. First, touch can be seen as part of a

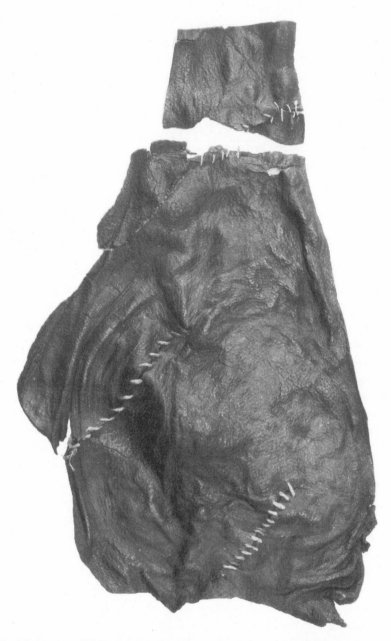

Fig. 2.2. A leathery right breast fragment from the bog body of a young girl found in 1784 in Bareler Moor near Oldenburg, Germany. Photograph by Wolfgang Kchmeier © Landes Museum für Natur und Mensch, Oldenburg, Germany.

process: the activity and gratification of excavating and digging. Second, touching and the sensual experience it brings (still including touching by proxy of the gaze) gives primacy to the "skin" of archaeological artifacts, to the pleasure of beauty, or to the displeasure in lack of beauty. We will return to these questions in more detail later in this study.

Digging for Civilization

But first, let us return to Freud and Jung. As David Lowenthal points out, "Freud treated the retention as well as the unearthing of memory traces archaeologically. Just as the burial of ancient artifacts often preserved them, and Pompeii's relics only began to decay when they were dug up and made visible, so did conscious memory wear away, leaving only what was buried and unconscious unchanged; memory fragments were 'often most powerful and most enduring when the process which left them behind was one which never entered consciousness.' "[39] It is the phrase "more powerful and enduring" when buried that is most telling here. Uncovered, repressed objects dissolve into consciousness (and presumably healing); and that which is deceased or repressed ultimately "decays" or "dissolves" when exposed through analysis. The problem is that unlike classical archaeological sites like Pompeii, bogs do not provide a stratigraphy filled with traces of previous civilizations, nor did bogs (to my knowledge) yield little figures for Freud's desk;[40] instead there is a deep dark abyss, an undefined and uncanny embodiment where the natural process of decay has stopped. In addition, and in the context of Freud's fainting spell, most important: there is no actual excavation, no "digging" for "those corpses." That is to say there is no process, no controlled therapy, but a violent occurrence of a "death wish" symbolically housed in the damp cellar of (Jung's) haunted dream and buried fratricidal desires.[41]

While it still remains puzzling that Freud did not recognize in bog bodies a rich source for his psycho-archaeological archive, we can now start to detect a pattern that might help us understand, if not explain, why these remarkable remains were excluded from his cluster of ancient relics and tropes.[42] What Freud is looking for in ancient relics is predominantly a mirror for the human psyche, but also a repository for the origin for human civilization. Freud's biographer, Peter Gay, points out that Freud, "like so many northerners [...] loved Mediterranean civilization" but also that "his antiques seemed reminders of a lost world to which he and his people, the Jews, could trace their remote roots."[43] This, of course, adds an important dimension to Freud's search for archaeological tropes. If Gay is correct that what Freud is looking for are the roots of his Jewish ancestry, then Jung's fascination with the prehistoric "bog corpses" and the inferred death wish may indirectly, and long before the more pronounced anti-Semitism of Jung, resonate with a conflict of more historical dimensions.

In bogs, to repeat, time's way of "naturally" decaying bodies is sabotaged and subverted. While this makes for a psychological uncanny that would seem to fit hand in glove with Freud's theory, it appears to have become epistemologically suspect to Freud. If Pompeii was to Freud a material manifestation of the unconscious, then the bodies in the bogs become unmanageable hyper-visualizations of this unconsciousness. While repressed "objects" excavated from infected psyches on Freud's analytical couch were kept in check by a steadfast collection of antique relics on his desk (from where they testified to the reassuring powers of a "deep" civilization and its survival), the haunting of Jung's bog "corpses" afflicted Freud deeply enough to blind him toward their potential tropic value as equally potent embodiments of (pre)historical wounds deeply buried in haunted psyches. This is ironic, of course, not least because Freud's point in *Civilization and its Discontent* is precisely that civilization is threatened when or if it falls prey to an insistence on repression of the abject, the disgusting, and the uncanny. Yet even to the father of such "deep" insights, the bodies from the bog became an impenetrable obstacle to the "proper." They ought to have been hidden, but nevertheless came into light.

CHAPTER 3

Uses and Abuses: Bog Body Politics

It is time now to investigate a range of responses in the form of fiction narratives and poems in which bog bodies serve as a "medium" (sometimes resembling ventriloquism) through which to channel a range of political and ideological agendas. Bog bodies, as we shall see, are not only perfectly fitted to serve as emblems of *psychological* repression, they are also uncanny material witnesses to the *historically* repressed; to atrocities committed in the past—even if this past is not their own. They become humanity's conscience. Let us start by looking at one terrifying moment in history.

Corpores Infames

In a speech to the Waffen-SS in 1937, Heinrich Himmler interpreted the fate of bog bodies. After having lamented the "pestilence" of widespread homosexuality in contemporary society, Himmler went on:

> Unfortunately, we do not have it as easy as our ancestors did. They only had few abnormal degenerates. Homosexuals, called Urnings, were drowned in swamps. The worthy profes-

sors who find these corpses in the bog are clearly not aware that in ninety-out-of-a-hundred cases they are faced with remains of a homosexual who was drowned in a swamp along with his clothes and everything else. That was not a punishment, but simply the termination of such an abnormal life. They had to be removed just like when we pull out nettles, stack them and burn them. It was not a question of revenge but simply that they had to be done away with. This is how our ancestors did it.[1]

Himmler evidently drew on Tacitus's hypothesis that the bodies represented transgressors, *corpores infames*. Such *corpores infames* interested the German bog body specialist, Alfred Dieck, who, spurred by the Zeitgeist in Germany during the 1930s and 1940s, interpreted Tacitus's expression as a description of cowardly men who mutilated themselves in order to escape military service; i.e., traitors to the nation. But *corpores infames* was also seen and understood to be homosexuals deposited in the bogs as punishment, and it was this interpretation that was adopted by Nazi ideologues. If, as Himmler implied, the bodies represented someone *entartet,* degenerate, they could be seen not only as testimony to a past which was able and ready to deal with "unnaturals," but also (and for political and ideological purposes, just as important) as an endorsement *from the past* of the necessary elimination *in the present* of unwanted peoples.

Bog bodies did not play a large role in the Nazi archaeological scheme, but the small parts they *did* play are important. Implicitly, they represent those who were to be destroyed: the homosexual, the gypsy, the Jew—not as "a punishment, but simply the termination of such an abnormal life." The association that was drawn by Nazi ideology between homosexuals and Jews is common knowledge; both were seen as feminine, cowardly, and lazy.[2] So, the logic holds, if the bog bodies represented such "foreigners" and "others," then they could be seen as archaeological evidence of a past that knew how to deal with "degenerates," i.e., "*entartet*" *corpores infames:* the solution was to bury them in a realm which corresponded to their own essential quality, the swamp. Tacitus's *Germania,* read through this kind of filter, became a manual for racial and sexual eugenics. Or, as Simon Schama observes: with the help of Tacitus the Nazis found something resembling a social geography. With *Germania* in one hand and an archaeological shovel in the other, they dug into the earth in an attempt to find the "birth certificate of the German race."[3] Tacitus served as guide when he wrote:

For myself, I accept the view that the peoples of Germany have never contaminated themselves by intermarriage with foreigners but remain of pure blood, distinct and unlike any other nation. One result of this is that their physical characteristics, in so far as one can generalize about such a large population, are always the same: fierce-looking blue eyes, reddish hair, and big frames—which, however, can exert their strength only by means of violent effort.[4]

In spite of their iron-red hair, bog bodies appeared to be rather unfit to provide any form of glorious image to boost German national pride from the imagined depths of the Aryan race's past. They did not offer the kind of pedigree that fit the manual for an unmixed indigenous race of ancestors, but seemed to speak about a past that had to be buried again. If *Blut und Boden* iconography was to be found in archaeological material, it certainly had to *look* different from bog bodies; and if these bodies visualized living persons from the past, such persons were not to be emulated. They were not portraits of the race and its origin, but representations of what the race had to be cleared and cleaned of.[5]

In Nazi ideology, the slime and softness of the bog, as the German sociologist Klaus Theweleit has pointed out, suggested femininity, thus compromising masculinity and consequently posing a danger to the race. Because "firmness is susceptible to erosion, it must be kept good and dry; even the merest splash of depth will soften it."[6] The failure of will, the succumbing to soft pleasures, turned men and soldiers into "things without contours." Loss of bodily boundaries threatened the steadfastness of the Aryan male. Associations with the earth, with *terra firma,* were split into two separate models for the German people and their past: one was associated with "Boden," where earth and soil provided solid ground and implied nation and firmness. The other was linked with morass, mire, slime, and dirt and connotes softness. Bog bodies, wet, soft, and uncanny, would seem to fit into this second realm.

With that said, there *are* examples of interpretations of bog bodies in Germany during the war which seemed to make the opposite claim. Wijnand van der Sanden, for example, tells of a mummified bog body of a 14-year-old girl that was found in Dröbnitz in 1939. The local newspaper, *Osnabrücker Tagesblatt,* in the January 14, 1940, issue called her "eine germanische Schönheit," a Germanic beauty.[7] While the photograph belies this idealization, the fact that the reporter saw an ideal German woman in the remains, and thus clearly differed with Himmler's 1937 *corpores-infames*–inspired diatribe, illustrates the power of projection. In case of the Dröbnitz Girl, Himmler's ideas about degenerated bog bodies seem to have been circumvented.

But this did not mean that contemporary women were exempt from becoming necessary "sacrifices" in order to safeguard the Aryan nation's purity. Tacitus's description of the punishment of women for infidelity—"A guilty wife is summarily punished by her husband. He cuts off her hair, strips her naked, and in the presence of kinsmen turns her out of his house"—might well, according to Danish historian Allan A. Lund, have inspired Himmler's proposal for a similar practice of public defamation of German women who had betrayed the nation by marrying outside their race.[8] Lund points out that the so-called Nürnberg Laws from September 1935, which prescribed the protection of German blood and German honor from contamination by intermarriage with others (Jews, Gypsies, Africans, and so on), found written justification in *Germania*.[9] The public humiliation of women, which followed, took its cue from these ancient protocols. While Himmler seems not to have known of any physical evidence for such ancient practices, in 1952 a (presumed) female bog body was in fact found in northern Germany that displayed a partly shaved head; perhaps a mark of adultery. The body was called Windeby Girl. "She" was recently determined to have been a "he," and we shall return to him/her later.

The history of archaeology, taken as a whole, shows us that the association with nationalism or national identities and the use of archaeological discoveries in nation-building contexts can easily take on accents of manipulation: the assertion of ownership of the past to support present political claims, and so forth. This is not to say that archaeological narratives are inherently deceptive or suspect, but that the archaeological stories told (and presumably verified by material remains) are not inculpable and interchangeable, each equally true.[10] The question is *how* and *when* the line has been crossed between that which is excessively politicized in order to construct and manipulate the past and that which can be carefully ascertained from available empirical data. As Walter Benjamin points out, "to articulate the past historically does not mean to learn 'how it really was,'" but it does mean, in the words of American archaeologists Philip Kohl and Clare Fawcett, that there is a fine line between "legitimate" and "questionable" research: "Some archaeological tales are not innocuous, but dangerous in that they fan the passions of ethnic pride and fuel the conflicts that today pit peoples against each other."[11] It seems easy to point to tendencies in the discipline's past in which national pride or personal biases have factored into archaeological assessments and allowed for blind spots and judgments colored by the interests of the nation and its identity rather than by the need for, or urgency of, historical "truth."[12] The relationship between archaeology and nationalism and the ways in which prehistoric artifacts frequently supply tangible aid to constructed narratives of a nation's past have been scrutinized in a number of archaeological studies. And, as another American archaeologist, Neil Asher Silberman, puts it, such archaeological narratives "cannot help but be constructed in contemporary idiom, with emphasis on each society's specific hopes and fears."[13]

With the rise of Nazi ideology, the implied national and ethnic uses of archaeology turned overtly racist.[14] Here, as American archaeologists Bernard Wailes and Amy Zoll suggest, "the equation of ethnicity with biology provided a radically pure and culturally superior Aryan antiquity to support the racial purification and cultural superiority of the Nazi present."[15] German archaeologists struck a "Faustian bargain" with the Third Reich which allowed a focus on German prehistory rather than on the traditionally preferred classical archaeology, and, as American anthropologist Bettina Arnold and German archaeologist Henning Hassmann have pointed out, "one of the contributions of prehistoric research to the Nazi propaganda machine was in the realm of symbolism and iconography." This meant that "rune research became a legitimate academic discipline with the help of party support during this time."[16] But while runes and runology temporarily lost their innocence to Nazi appropriations and to the mining of signs and artifacts for iconographic potential and political purposes, bog bodies seemed to play a different role.[17] The interesting point here, however, is that as bog bodies become images of inhumanity, dehumanized and/or inscribed with dehumanizing marks, they always return to some original position as fellow human beings, for better and for worse.

What Lies Beneath

Although bogs and the bodies in them served Himmler's extermination ideology, the crimes of the Nazis are not buried deep, but lie just under the surface of historical time. Simon Schama's analysis of landscape as inscribed with remembrance offers a parallel image of this. He tells us about his search for the graveyard of his ancestors who had perished in Polish-Lithuania. The Jewish cemetery has become almost invisible to the eye, covered by a field of dandelions so thick that the "headstones that had been lovingly cut and carved were losing any sign that human hands had wrought them. They were becoming a geological layer."[18] The dandelions that the visiting historian sweeps away like "a child making a snow-angel" are used as a symbol of cultural repression and forgetting. They must be removed to reveal that "just under the surface" of the landscape the earth is saturated with the blood of both deep *and* recent history. The blanket of dandelions, lovely and innocent looking, represents a kind of amnesia, a cover-up of the crimes of times past, a thin veneer under which a deeper and more sinister story seeps through the strata of time and space: atrocities like those committed during the Second World War—and under them, layer by layer by layer, other atrocities, ancient ones.

Placed into a similar optic, the bog's paradoxical powers provide not only a temporal reassembling of the past as a jigsaw puzzle that informs us of Iron

Age practices, but also the soil for the kind of archaeological activity that can brush away "dirt" from a more recent past.[19] As seen in the dandelion cemetery, history can be (very nearly) contemporary. Similarly, bog bodies have been asked to "bear witness" to a past that is "just under the surface." Their stories bleed through the strata of time and space so that the diachronic and the synchronic converge. Building in part on French sociologist Maurice Halbwachs' concept of time as a social construction and in part on American anthropologist Clifford Geertz' thick description, Israeli philosopher Avishai Margalit has proposed that we can think of thin and thick memory. Thick memory is shared with close relations (family, friends, peers, and so on) while thin memory relies on "some aspect of being human," but in a distant way (the human race, our humanity).[20] The archaeological imagination(s) surrounding the bog bodies, I would like to suggest, is an ongoing effort of thickening this thin relation, bringing the closeness of "family" and "nation" together within the common category of "humans." In search of our "deepest humanity," as French philosopher Paul Ricoeur reminds us, "history is reminded of its indebtedness to people of the past. And in certain circumstances—in particular when the historian is confronted with the horrible, the extreme figure of the history of victims—the relation of debt is transformed into the duty never to forget."[21] But does "the duty never to forget" extend to bog bodies? Can they be seen as *stand-ins* for other extreme figures and victims in history?

To answer this, and to allow bog bodies to "speak freely" without compromising historical, archaeological, and scientific data (that is, without doing what Himmler did), we need the elasticity of fiction. If archaeological narratives (to some degree) mimic fictional narratives, fictional narratives—or poetry for that matter—also look to archaeology for inspiration. In the history of bog body interpretations we find numerous examples of how the archaeological and literary texts bleed into each other. According to the British author Margaret Drabble, novelists have a particularly privileged access to the archaeological imaginary. "The truth is," she argues, "the novelist can conquer time and bring back the dead. The novelist, like the paleontologist and the archaeologist and the historian, is in the resurrection business. We gaze at the skulls and the bones of the long dead, and we try to give them living faces."[22] Yet resurrection of the archaeological "dead" in fiction narratives (particularly historical fiction)—but also in poetry, and herein lies the crux—must negotiate the material reality it describes. It must, in other words, overcome the resistance that the material imparts: a resistance to being "dissolved" into the world of words. In a literary narrative we accept that the information provided to us by the narrator and by the author has been made up; we do not (or should not) question whether it is real. The nature of fiction *is* that we know it to be just that: fiction. We enter a contract willfully (for the most part) in which we are supposed to understand that a world is constructed for us; we know (or should know) that it does not exist in reality. This is the condition and convention of fiction. We know (or should know) that the text we read consists of signs, marks on a page that have been assembled by the author to give us a sense of a possible world. In literary studies, then, it has always been and still is a pressing question to understand and conceptualize what happens when something factual enters the world of fiction, what happens when something "real" is fictionalized. We are accustomed, of course, to this sort of slippage from our reading of historical novels, in which story and history are constantly negotiated; one might even argue that *any* reading of fiction must hold some sense of reality effect in order to make sense, that *any* reading of fiction inevitably tests our ability to distinguish between fact and fiction. So when bog bodies are imported into fictional texts as actual objects mediated by language, we as readers must be able to slip back and forth between the worlds of fiction and fact.

Seen from a rhetorical standpoint, the use of bog bodies in fiction and poetry draws implicitly on the rhetorical device known as ekphrasis. Ekphrasis, a verbal representation of a visual representation, is often understood as a way to give voice to a silent object. But "giving" agency to a visual object in a verbal text has also been seen as problematic flirting with a potential deadly "other," as W.J.T. Mitchell has shown. Ekphrasis, he suggests, "tends to unravel [and] expose the social structure of representation as an activity and a relationship of power/knowledge/desire—representation as something done to something, with something, by someone, for someone."[23] American art historian Michael Ann Holly, too, has a particularly interesting point when in *Past Looking* she suggests that "ekphrasis, as it struggles to describe that which can no longer be seen in a variety of ways, can itself serve as an allegory for the larger project of history writing. The vanished past cannot be reclaimed, but the desire to speak images, tell pictures about it, persists."[24] Her argument is that in ekphrasis, the visual art (and we can with some right include bog bodies as art objects) that is "made to speak" is in some sense the poems', or the literary texts', past—a past that has been forgotten, destroyed, and laid in ruins, only to be dug out archaeologically and made visible through language. The essential element in ekphrasis—namely, that language can make us see something that is otherwise absent—can, as Holly proposes, be placed in relation to our writing of history and the rhetorical and aesthetic fields within which word-image dialogues are so often placed. With this, let me repeat the question posed above: does "the duty never to forget" extend to bog bodies? Can they be seen as *stand-ins* for other extreme figures and victims in history? It is time to show how fiction can use bog bodies as witnesses to a contentious past.

Escaping the Holocaust: Dehumanized

In his 1970 novel *Le roi des aulnes,* (translated first to *The Erl-King,* then to *The Ogre*) French author Michel Tournier employs bog bodies as tropes for humanity at risk.[25] Halfway through the novel, the protagonist Abel Tiffauges, a self-described myopic ogre, witnesses the autopsy of an unearthed bog body. Keenly attuned to "signs and encounters," and an incessant reader of the physical world as if it were a large semiotic library, Tiffauges is fascinated by the lecture of an archaeology professor: "Note the delicacy of the hands and feet; the fineness of the face, with its aquiline profile in spite of the broad brow; the aristocratic air reinforced by the richness of the chlamys, which looks as if it was made of gold thread, and of the objects with which the dead man was surrounded for use in the afterlife."[26] He learns from the professor that the man on the autopsy table "was one of those 'peat-bog men' periodically exhumed in Denmark and the north of Germany in such a marvelous state of preservation, because of the acidity of the soil, that the country people take them to be the victims of some recent accident or crime" when in fact they are two thousand years old. And, as the professor goes on, the "most moving aspect of our discovery" is that it is

> [...] not absolutely beyond the bounds of possibility that this last meal of a man who must have been of some importance, probably a king, eaten before a death that was horrible but freely chosen, took place at the same time—the same year, perhaps even the same day, the same hour!—as the Last Supper, the paschal meal shared by Jesus and his disciples before the Passion. Thus at the very moment when the Judeo-Mediterranean religion was springing into life in the Middle East, a similar rite may have been founding a parallel religion here that is strictly Nordic and even Germanic.[27]

Moved to tears that such Germanic depths could rival Judeo-Christian hegemony, the archaeologist trumps his own hyperbole by evoking Goethe's ballad "The Erl-King" [Erlkönig]. "It sings to our German ears, it lulls our German hearts, it is the true quintessence of the German soul." The bog body should therefore, the professor suggests, be given the name the Erl-King.

The legend on which Goethe developed his ballad may have originated in Denmark, but such nationalistic cloaking of the bog body becomes all the more complicated in the novel because of an explicit comparison of the bog man to Tiffauges. Tiffauges, however, is not only a Frenchman and a prisoner of war of the German Nazis but also a giant uncouth character with enormous hands (not quite one with the "delicacy of the hands and feet; the fineness of

the face" of the bog man) which he uses, along with seemingly limitless privilege to roam the countryside independently, to "save" children by bringing
them into the training camps of the Nazis. His innocence is compromised,
and when he finally rescues a Jewish boy, a primal scream heard throughout
the novel is at long last understood:

> He knew he was hearing for the first time in its primitive form the clamor
> suspended between life and death which was the fundamental sound of his
> whole destiny. And once again, as on the day he met the retreating French
> prisoners, but with a persuasiveness incomparably greater, it was the peaceful
> and disincarnate face of the Erl-King, wrapped in his shroud of peat, which
> presented itself to his mind as the ultimate resource, the ultimate retreat.[28]

Another bog body fragment is found close to the Erl-King: that of a head
with "a small face, emaciated, childish and sad," and, more importantly for our
purpose here, Tiffauges etches "the frail, mournful little convict's head, which
the sun was caressing for the first time after so many centuries of muddy darkness" into his own memory so that later he is able to recognize it again in the
form of the ravished little Jewish boy, an Auschwitz survivor.[29] Unhealthily
obsessed with rescuing children—he sees himself as Saint Sebastian—and fix-
ated on reading any and all things as signs and symbols, Tiffauges, blinded by
the loss of his glasses, is guided by the young Holocaust survivor, Ephraim,
away from the last spasms of war and (back) into the bog: "The deeper his
feet sank into the waterlogged swamp the more he felt the boy—so thin and
diaphanous—weighing down on him like a lump of lead." The last thing the
ogre sees before he and the boy are submerged is "a six-pointed star turning
slowly against the black sky."[30] The final visual sign in the novel, then, is an
image with cosmic, religious, political, and cultural signification. In Tournier's
optic, bog bodies are messengers from the darkness of the past, but the deliberate disappearance back into the darkness of the bog can be understood as
a survival strategy—an escape from historic atrocities—while also being preserved as a cautionary mnemonic. One day, it is implied, the Ogre and the boy
will reappear and testify to crimes against humanity from a more recent past
than the one represented by Iron Age bodies.[31]

Surviving the Holocaust: Re-humanized

Reappear they do, in the Canadian author Anne Michaels's 1966 novel *Fugitive Pieces*. If bog bodies in Nazi ideology were symbolically repressed, and if
Tournier's novel reburied them in the bog, *Fugitive Pieces* raises them again,

equally symbolically, as images of Holocaust survivors. If Nazi ideology dehumanized bog bodies and saw them as examples of the abnormal, Michaels in turn humanizes them and sees them as images of resistance. And if Tournier used them to articulate a sense of defeatism, Michaels offers a more hopeful future for humanity. By making the bog bodies into metaphorical counterimages to the interpretation used by National Socialist ideologues, she usurps the archaeological iconography Himmler had exploited and shows how the bodies can be about not forgetting the crime against humanity committed during the Holocaust. In her fictional universe, bog bodies connote survival, not sacrifice; persistence, not extermination. They are robust metaphors for memory.

The novel is saturated with archaeological imagery and pulls us into a spatial-temporal realm, back and forth between past and present, and between reality and fiction.[32] The gesture of authentication can be seen as a warranty for realness (the events of the Second World War) and as a way of putting the narrative in the same category with other assurances of authenticity: eyewitness accounts, diaries, and memoirs. But it also makes use of archaeology as a rhetorical action: historical and mnemonic testimonies are "buried," "tucked," and "concealed" and have to be "retrieved" and "recovered" through the archaeological structure of the novel. This structure is used to the full when the two protagonists and their stories are entangled and layered to form a series of chronotopic knots. Throughout the novel, bog bodies are human time capsules, hidden, dormant, yet resilient, providing a foil for the novel's two protagonists: Jakob, a Holocaust survivor, and Ben, a son of a Holocaust survivor. The novel falls into two parts: the first consists of first-person reflections written by Jakob Beer shortly before his death in a 1993 Athens car accident, and given to us posthumously. In the unfinished manuscript he remembers his family's demise and his own escape into a bog where he dug a bed and planted himself like a "turnip." "I know why we bury our dead and mark the place with stone, with the heaviest, most permanent thing we can think of," he reflects; it is to keep the dead underground. "If I can't rise, then let me sink, sink into the forest floor like a seal into wax."[33] But there is no permanent gravestone to keep Jakob in the ground; he rises and returns uncannily from the bog as an "afterbirth of earth," in a symbolic rebirth:

> Time is a blind guide.
>   Bog boy, I surfaced into the miry streets of the drowned city.... I
> squirmed from the marshy ground like Tollund man, Grauballe man, like
> the boy they uprooted in the middle of Franz Josef Street while they were
> repairing the road, six hundred cockleshell beads around his neck, a helmet
> of mud. Dripping with the prune-coloured juices of the peat-sweating bog.[34]

Cast as a bog boy, Jakob retrospectively sees himself as a historical sacrifice analogous to the Danish bog bodies. In order to recover from this traumatic past, his first-person narrative performs a kind of self-archaeologizing, in which the metaphorical entanglement with the bog bodies, the pronominal and physical identification ("I" am like "Tollund Man and Grauballe Man") and the way in which bog bodies are presented both as material and imaginary stand-ins become mnemonic links between past and present. As the past partakes in the novel's multilayered present(s), the bog bodies provide a bridge over the temporal gap, thus furnishing a mirror for the contemporary ideological and political problems that frame the story. Jacob attempts to give firm contours to his existence, and the retrospective narrative is a way to solidify his shattered identity.[35]

His rebirth from the bog, "stiff as a golem," into the arms and nurture of Athos, a Greek geologist and archaeologist, symbolizes an emergence into civilization (both ancient and modern). Athos, who had been digging in the "Polish Pompeii," Biskupin, introduces Jakob to the world of archaeology, and to an invisible world in which bog bodies reside "under a sky of mud," a place where "peat men [are] preserved as statuary"; a place of underground activity and safety where resistance fighters and fugitives can gain access if they utter "bony passwords."[36] To the archaeologist, the bog bodies are manifestations of freedom. Jakob remembers:

> Athos and I would come to share our secrets of the earth. He described the bog bodies. They had steeped for centuries, their skin tanning to dark leather, umber juices deep in the lines of palms and soles. In autumn, with the smell of snow in the dark clouds, men had been led out into the moor as sacrificial offerings. There, they were anchored with birch and stones to drown in the acidic ground. Time stopped. And that is why, Athos explained, the bog men are so serene. Asleep for centuries, they are uncovered perfectly intact; thus they outlast their killers—whose bodies have long dissolved to dust.[37]

The bog bodies have become placid freedom fighters instead of victims, statuesque and serene instead of uncanny specters of muck and mutilation. Pain and terror, fear and horror have been glossed over by the spatial-temporal blanket of the bog; and as reassurance of the ability to stay alive in the midst of a frightening reality, bog bodies are allowed to redefine time, "outlast their killers" with their unrelenting witness, and thus revise the kind of history and archaeology that Himmler and his cohorts espoused. Just as the dead bodies stubbornly rise out of the dark earth to tell *their* tales, so Jakob (and with him a nation of Holocaust victims) surfaces to tell *his* story, assaulted *and* resilient.

Jakob's "rebirth" from the bog is also a long restoration process back to the world of words since the Nazis' corruption of language had left Jakob "wild with deafness. My peat-clogged ears. / So hungry. I screamed into the silence the only phrase I knew in more than one language, I screamed it in Polish and German and Yiddish, thumping my fist on my own chest: dirty Jew, dirty Jew, dirty Jew."[38] The power of words, Anne Michaels deftly implies, is related not only to the act of remembrance *in* language (the writing of Jakob's memoirs) but also to the observation that "the German language annihilated metaphor, turning humans into objects, [as] physicists turned matter into energy."[39] The violence against human bodies seeps into language, concretely substituting humans for *Figuren:* "Nazi policy was beyond racism, it was anti-matter, for Jews were not considered human. An old trick of language, used often in the course of history. Non-Aryans were never to be referred to as human, but as 'figuren,' 'stücke'—'dolls,' 'wood,' 'merchandise,' 'rags.' Humans were not being gassed, only 'figuren,' so ethics weren't being violated."[40] In short, the reduction of humans to linguistic figures and/or to literal *Figuren* makes extermination possible: "When the [Nazi] soldier realized that only death has the power to turn 'man' into 'figuren,' his difficulty was solved."[41] Censured and unable to sing about their fate, the concentration camp prisoners use the bog as a "figure" of resistance: "The Nazis didn't allow prisoners to sing anything except Nazi marching songs while they cut the peat, so it was real rebellion to invent a song of their own." The song was called "Moorsoldaten" (Peat Bog Soldiers) and was "the first song ever written in a concentration camp, in Borgermoor [. . . .] It spread to all the camps."[42]

The "I" who remembers himself as a bog boy, as we can see, is not only caught metaphorically in a temporal void; he is also caught between places and languages. Athos, the archaeologist, eventually brings the boy back to Greece and into exile on the island of Idhra where he feeds history to the fugitive, literally nursing him back to health by making him "digest" a volume of Pliny's *Natural History.* Pliny's text sustains them not only as a book on history but quite literally as a cookbook: history made tangible, edible—a manual for survival.

Jakob's tongue, nonetheless, remains an "orphan," particularly when confronted with the new language, English, that he tries to adopt. English keeps on being foreign, "an alphabet without memory," and makes silence the only alternative: "My life could not be stored in any language but only in silence."[43] This inability to communicate is tied most manifestly to the non-loss of his beloved sister Bella and to the fact he has *not* seen her disappear; he did *not* see her dead body as he did those of his other family members. She is neither fully remembered nor forgotten.

If you cannot *see* loss, Michaels suggests, you cannot articulate it either.

Therefore bog bodies, in their manifest visual *proof* of loss, become the mate-rial evidence that can help articulate the pain of loss and thus help reclaim the past—not only for Jakob but also for his mentor Athos, who works obsessively on a book called *Bearing False Witness,* a record of how the Nazi Ahnenerbe archaeologists fabricated the past to fit their ideology. Athos, in a word, repre-sents deep ancient civilization against the inhumanity of shallow Nazi ideol-ogy. While the Nazis violently destroy the Polish Pompeii, Biskupin, and with it the bog out of which Jakob rose, all to eliminate "proof of an advanced cul-ture that wasn't German," the Greek archaeologist refuses to lose faith in the authenticity of depth, and laboriously educates Jakob about origins and about the "power we give to stones to hold human time."[44] Although Jakob longs "to cleanse his mouth of memory," Athos reminds him that it is "your future you are remembering."[45] Stones and bones become part of a resilient poetics of life, and he urges Jakob to "try to be buried in ground that will remember you."[46] This is a process Jakob has already gone through, of course, in the cycle of be-ing buried in the bog and excavated—returning from the dead as a bog body to testify about the horrors of the Holocaust.

To enable the boy to return to a normal life Athos lays out a cartography of place and provides him with a plan of survival—a kind of poetics—rooted in the "power of language to restore." The stories of bog bodies are intermixed with stories from the Polish synagogues and their sanctuaries in caves; with Biskupin, the excavation site where Jakob emerged as a bog boy; and ulti-mately with archaeology at large via the associations brought on by Athos's Greek nationality and métier as geologist-archaeologist. Yet if archaeology is seen as a positive, history is seen as treacherous, as a "Totenbuch, The book of the Dead, kept by the administrators of the camps." It is a "poisoned well, seeping into the groundwater [. . . .] This is the duplicity of history: an idea recorded will become an idea resurrected. Out of fertile ground, the compost of history."[47] While history is corrupt and amoral, memory is virtuous: "Mem-ory is the Memorbucher [*sic*], the names of those to be mourned, read aloud in the synagogue. / History and memory share events; that is they share time and space. *Every moment is two moments* . . . . "[48] It becomes clear that archaeology belongs to the mnemonic and not the historic.

This evocation of history and memory in Michaels's novel is reminiscent of the theoretical distinction made by Pierre Nora when he, by way of an organic metaphor, says that the tree is memory and the bark is history.[49] History, in other words, is the outside to memory. History is the official story, memory the unofficial, and both are folded into interdependency—described by Nora as a besiegement, penetration, and petrifaction of memory by history; a "push and pull" which produces "moments of history torn away from the movement of history, then returned; no longer quite life, not yet death, like shells on the shore when the sea of living memory has receded."[50] Every moment is indeed two moments. History can repeat itself, and does, but memory—as Michaels suggests—will pull into focus the hidden truth. It will rise and return from the bog.

If language plays a key role in Jakob's regaining of identity, the significance of photography and its mnemonic link to language and names provides a bridge to the second part of the novel and its focus on photographic witnessing. Jakob remembers that in "Birkenau, a woman carried the faces of her husband and daughter, torn from a photograph, under her tongue so their images wouldn't be taken from her. If only everything could fit under the tongue."[51] To fit a photograph under the tongue, to hold faces in her mouth, to digest the names of loved ones is, for the woman in the concentration camp, a way to hold on to the existence of what has been lost.

Photography is even more essential in the novel's second part, where the other key narrator, Ben, turns to bog bodies to help forge a link to a father who has been marked by the Holocaust:

> I had discovered the perfectly preserved bog people in *National Geographic,* and derived a fascinated comfort from their preservation. These were not like the bodies in the photos my father showed me. I drew the aromatic earth over my shoulders, the peaceful spongy blanket of peat. I see now that my fascination wasn't archaeology or even forensics; it was biography. The faces that stared at me across the centuries, with creases in their cheeks like my mother's when she fell asleep on the couch, were the faces of people without names. They stared and waited, mute. It was my responsibility to imagine who they might be.[52]

The indentations in the skin of the Tollund Man and Grauballe Man represent a kind of permanent photo-historical shadow, the markings of intimacy and proximity, like the creases made by a couch on his mother's face while she is sleeping. Ben's imagining of who they were—those mute and nameless victims of history—are fused in metaphor with his recollections and sensations of his own life and past.

While Jakob has literally emerged from the peat bog as an archaeological artifact representing past and present crimes to humanity, Ben's "bog" has morphed into a psychological map marked by the silent but insistent grief of his parents. During the war Ben's father was forced to unearth the dead bodies of people killed by the Nazis: "They dug the bodies out of the ground. They put their bare hands not only into death, not only into the syrups and bacteria of the body, but into emotions, beliefs, confessions. One man's memories then another's, thousands whose lives it was their duty to imagine. . . ."[53] The father repeatedly shows his son photographs of the victims of World War II atrocities because he cannot speak of his own losses—his two firstborn children—and so instead collects and obsessively scrutinizes photographic evidence and testimony of the horror he has experienced. While Jakob has come face-to-face with the trauma and loss of his entire family, Ben, born a generation later, finds that he is "born into absence." The "bog" of his childhood is "rotted out by grief." He reflects: "History has left a space already fetid with undergrowth, worms chewing soil abandoned by roots. Rain had made the lowest parts swampy, the green melancholia of bog with its swaying carpet of pollen."[54] Yet he too, like Jakob a generation earlier, seeks the bog for comfort; he too needs the "the aromatic earth over my shoulders, the peaceful spongy blanket of peat," and finally places himself in a wood one night to purge himself, albeit unsuccessfully, of the fears he has inherited.

Ben's absorption is neither with archaeology nor forensics, but with a kind of archaeo-*auto*biography in which he digs into the past of (the newly dead) Jakob in order to find answers and tools for self-excavation. When, for example, he searches the old house on Idhra to recover Jakob's unfinished memoir (which constitutes the first part of the novel), the meticulous and simultaneous unearthing of both his own and Jakob's past is described as that of "an archaeologist examining one square inch at a time."[55] The empty house is still marked by Jakob's nearness: "It's a strange relationship we have with objects that belonged to the dead; in the knit of atoms, their touch is left behind. Every room emanated absence yet was drenched with your presence."[56] But the presence of the departed rests not only in the form of tangible objects. The visible imprints of the bodies of Jakob and his young wife (who also died in the 1993 traffic accident) in the cushions, which were left behind shortly before their deaths, makes the departed couple sensually present. The indentations are diaphanous marks resembling those described in the bog body photographs: thin marks, different from other material objects, but nevertheless a *fact*.[57]

The marks made on human bodies by the cruelty of men are, we learn, nearly synonymous with photographic prints accidentally left behind by random lightning: "Sometimes, when lightning passes through objects and through human tissue, it imprints the object onto a hand, an arm, a belly—leaving a permanent shadow, a skin photograph."[58] In fact, for Ben, photography takes the form of a particular kind of memory. "We think of photographs as the captured past. But some photographs are like DNA. In them you can read your whole future."[59] His two siblings, dead by the hands of the Nazis, whose existence he gets to know only through the symbolic DNA of a hidden photograph, represent the future that Ben foresees. Their names, Hannah and Paul, are inscribed on the back of the photographs, but as Ben discovers, to have a name and to have your picture taken do not guarantee survival. Therefore, his own name is a non-name, because his parents hoped "that if they did not name me, the angel of death might pass by. Ben, not from Benjamin, but merely 'ben'—the Hebrew word for son."[60] Michaels, in other words, uses photography not only as a "material proof" in which private memories are kept, but also as one in which is stored the DNA of an extended family's genetics—its roots. It is a network of roots, and an example of the thickening of thin memory.

In the end, the presence of the past in *Fugitive Pieces* is not only a condemnation of the past but also a redemption of the past through the experience of physical love. Ben's continued archaeological digging into Jakob's life and his quest to absorb the life of Jakob is about love and desire. This is reflected finally in Ben's sexual affair with a young woman, Petra (a name that clearly harks back to the first inhabitant in the house, Athos, and his love of stones). As a result, archaeology is the "quest to discover another's psyche, to absorb another's motives as deeply as your own, a lover's quest."[61] To both of the novel's narrators, Ben and Jakob, the way back to life travels through the body, through the sensual and physical magic of lovemaking. In lovemaking broken (bog) bodies are somehow restored.[62]

"Undiluted Danish"

By now we have seen how bog bodies serve as tropes of atrocity *and* humanity connected to the Holocaust. But, as the next fictional examples will show, they also serve to question national identities and rootedness. "Dig anywhere [. . .] and below the picnic plastics of the present you ran into the age of iron, and below that the age of bronze, and below that the age of polished stone, and below that the age of chipped stone, and still below that the age of antler and bone. All of it, straight down from the fourth millennium B.C., was *undiluted Danish*," we read in American author Wallace Stegner's 1976 novel *The Spectator Bird*.[63] To Stegner, known best for his fiction of the American West, obvious displays of national identity in an old European country sound a note of caution and concern. And as the novel unfolds, we realize we are called on to be progressively more suspicious of any kind of deep *un*mixed past.

Utilizing its archaeological potential, the novel is structured as a flux between past and present. In the novel's present, the early 1970s, the narrator and protagonist, Joe Allston—an aging, arthritic literary agent and self-described "museum exhibit of deterioration"—is struggling to come to terms with his own mortality. The plot's mnemonic artifact comes in the form of a postcard from a Danish Countess Astrid, who has played a key role during Joe and his wife Ruth's visit to Denmark two decades earlier, in 1954. The journey to Denmark is meant to function both as a personal excavation process for Joe, who sought to solve the puzzle over his Danish-born mother's decision to leave her home country for America at age sixteen, *and* as a healing process after the loss of his only son in a drowning accident (or suicide!). "He [the son] was my only descendant, as she [the mother] was my only ancestor." The lack of both ancestors and descedants leaves Joe raw with bereavement and painfully cognizant of his implicit rootlessness. Envious of "other people's archeology," he wants to "belong to something," and in Denmark he imagines he has found a place that is "undiluted Danish."

The search for roots and for identity turns out to involve the disclosure of a human-origin experiment, circuitously linked to a bog body. But Joe becomes wary and apprehensive and fails to reconnect either to his Danish roots or to the suggestions in Danish nationalistic literature (here represented by Nobel prize winner Johannes V. Jensen) that "the Scandinavians invented everything, first sex, then fire, then tools, then shelter, then agriculture, then bronze and gold, then iron, until the human race put into gear by all that Nordic ingenuity, could be trusted to go forward on its own"; and Joe finally finds that "there was something rotten in that state, as elsewhere, and that the Danes like the rest of the world are attracted to evil, are involved in it, even dutiful toward it."[64]

To explain what is "rotten" in Denmark, the novel matches up the old world with the new world: while Americans come from a "civilization without attics," the "attic" of Denmark is full of ominous "accumulations." Countess Astrid, we discover, was married to a distant cousin who brought shame on her, not only by abandoning her for a younger woman but by collaborating with German occupation forces during World War II. Furthermore, she is related to the author Karen Blixen (known to English-language readers as Isak Dinesen), whose husband was likewise accused of collaboration with the Germans during World War I, and who also abandoned his wife for another woman. Throughout the novel these aristocratic Scandinavians are deeply compromised.

Blixen, accordingly, is imbued with characteristics culled from her own oeuvre: she is a story-telling witch who, along with other family members, has sold her soul to Lucifer. The same goes for the old count, Astrid's father. In Blixen's words, the old count was "the Doctor Faustus of genetics," and his son, the present Count Rødding—Astrid's estranged brother—has "inherited his father's gifts." In fact, as the fictionalized Blixen muses, Joe too might finally have found his roots and may well be one of her own relatives: "It takes only a little twisting to make your return [to Denmark] take on possibilities and become part of a Gothic tale."[65] Indeed, *The Spectator Bird* does turn positively gothic when

Figs. 3.1 and 3.2. Seen side by side, Tollund Man's countenance (above) seems to be *pulled* forward in time while Karen Blixen's (opposite) is *pushed* back. Two thousand years of separation melt away, bridged by different yet equally uncanny properties of fixation: bog and photography. Photograph of Tollund Man © Silkeborg Museum, Denmark. Photograph of Karen Blixen © Peter Beard / Art + Commerce 1961.

it offers a correlation between the private story and a larger, deeper history by comparing Blixen's face to that of a bog body. Astrid's estranged brother, it turns out, keeps a bog body in his small museum. Joe's diary recalls:

> His prize exhibit was under a big bell jar on a table in the middle of the room. When it was first dug up, and the air hit it, it had begun to crumble, and Rødding had rushed it to Copenhagen to get the museum there to put it under glass. Deteriorated or not, it was recognizably human. It lay curled on its side with its knees drawn up, small, a shrunken man with a bent nose and high cheekbones. An odd, cocky, Robin Hood sort of leather hat was on its head. Rawhide cords bound its hands and feet, a rawhide strangling-cord was twisted into the neck under the ear. Its eyes were closed. On its mouth was what must have been the grimace it made when the cord was tightened, but it looked like a whimsical, knowing smile.[66]

To Stegner the mummified bog body opens a door to the perversity of excessive genetic interests. Count Rødding "for a joke claimed [the bog body] as his ancestor," and Joe Allston muses that "damned it if he didn't look like it,

with the same little smirking smile. Shrink him and dry him out, and he could have been the relative that he claimed to be. Maybe, in fact, he was. That was what a real past could do for you." Yet if the count looks like the bog body, Karen Blixen, with her "leather-skinned face," has an even more uncanny resemblance to the mummy. In a snapshot taken by Joe's wife Ruth, the famous Danish author is seen as

> tiny, shrunken, her eyes as alive as snakes: as surely a witch as any old woman in one of her tales. In her hand she was holding a rune stone she had dug up only a few minutes before we arrived, and on her face was a look of glee, a smugness of secret knowledge, as if the murky work she visited every night on her broomstick had just sent her, in the cryptic markings on the stone,

a daylight message that only she and her wizard and warlock friends could read. Sure enough, she looked like Rødding, and even more like the mummy of his. The same smile.[67]

The resemblance between Blixen and the bog man ("it was a little spooky to have that lovely, subtle Danish writer looking back at me out from the snapshot with the same knowing, Old World smile [of the bog mummy]") becomes positively spine-chilling when we gradually understand that the count's interest in archaeology is less than innocent, that Blixen's words of warning about the potentiality of selling your soul to Lucifer in order to make stories (a well-known Blixen axiom) points to a predicament in the "rotten state" connected not only to archaeology, but also to genetics—or rather, eugenics.

Indeed, Stegner's story turns out to be ever more worrying the deeper our narrator digs into the secrets of Countess Astrid's family. Her Faustian father has "cultivated" his land and everything on it according to scientific principles of genetics. What's more, in the words of his son, who has followed eagerly in the father's footsteps, the old count has "*made* things, new things. He improved what he found. People talk about Mendel. My father looked through windows that [Gregor] Mendel [the "father of eugenics"] didn't even know were there."[68] The small archaeological museum holds artifacts that trace the Danes back to 4000 BC: "We're one of the rare examples of selective breeding of humans over a long period," Count Rødding boasts. "There's no evidence of any immigration or invasions [. . .] My tribe." Aware that since Hitler the concept of breeding races for purity has acquired "racist and fascist connotations," the young count tries to convince Joe that "*if it could be done scientifically,*" selective breeding would prevent the kind of "mongrelization" that has taken place in America. By now cured of his heritage envy, Joe embraces hybridism—"as an American I have to stand up for hybrid vigor"—while the count, gesturing to the bog body in the museum glass case, insists that "America will be ten thousand years developing an American type as pure as that fellow there with the string around his neck."[69]

This chilling conversation is folded into the novel's general unease with ancestry. And when the novel finally unearths the ultimate eugenic experiment, performed over many decades by the two counts, father and son, we are pulled into a perverse and frightening world. The counts have taken their experiments to the utmost. Worried about the increasing barrenness in the family (including Karen Blixen's), and after corresponding with "anthropologists who had studied primitive tribes, also the Mormons in Utah," the old count has pondered "the effects of inbreeding and the effects of polygamy" and regarded "polygamy among human beings as like ordinary stock breeding [. . .] one bull to a flock of females." Using himself as part of the genetic/eugenic experiment, he has impregnated one of his own illegitimate daughters on the island, who in turn has been further impregnated by the younger count, thus producing a human stock of "pure" descedants. Once the old count was seen as "Denmark's contribution to the world mind," but in the end, "hounded" by the public after the revelation that he "was testing Mendel with human subjects," he commits suicide.[70]

In the end, Joe Allston is freed from the burden of heritage so that he, and with him the reader, can draw a sigh of relief that he was born "accidentally" as a "hybrid" in America and not bred "deliberately" as an "undiluted" Dane amidst an origin of uncanny bog bodies and "purely" perverted aristocracy.

Democratic Roots

While Stegner pitches the new American world against the old European world, Danish author Ebbe Kløvedal Reich uses Tollund Man to juxtapose an ancient Roman world with the century-old building of a Danish conscious-ness. His authorship as a whole insists on the importance of national identity as being deep-rooted, and not constructed or "imagined," as Benedict Ander-son would have it.[71] Although Kløvedal Reich's early writings were embedded in the anti-imperialistic (anti–Vietnam War) movement, he soon turned to national history, most notably nineteenth-century national romanticism, to find characters for his popular-historical novels in an effort to protect what he saw as an inherently democratic Danish resistance to constant pressures from foreign hegemonic powers.

In his massive 1977 novel *Fæ og Frænde. Syvenhalv nats fortællinger om vejene til Rom og Danmark* (Farm and Family: Seven and-a-Half Stories about the Roads to Rome and Denmark), the historical aim is didactic—we are asked to learn from the past—and we are made privy to a contentious but shared Roman and Danish past which is used to create a mirror for contemporary ideological and political problems in Denmark in the 1970s. Set up as a grand narrative of how the Danes came to be, Kløvedal Reich's novel mixes fact and fiction into a combined historical-mythical time and place conceived as a direct response to the inclusion of Denmark in the European Union. Interspersed with black-and-white archaeological photographs and with quotations from the works of Pliny, Tacitus, Caesar, and other ancient historians, Kløvedal Reich weaves a chronicle constructed in part through an imagined life story of Tollund Man.

As the novel "digs out" a vanished past, we are asked to remember what has been forgotten ("nothing is as memorized as forgetting") and to remember when Danes became Danes. Ironically, the "first writing on Danish history" is not located in the mnemonics of the north but in the south, in inscriptions at the Augustus temple in Ancyra (now Ankara, Turkey), which tell of the initial encounter of the Romans with "the land of the Cimbri," and in Rome, where a priestess by the name of Birgit informs the Roman historian Pliny about the bog's unique properties as a site for remembrance and as a storage place of ritual artifacts and bodies and ideas.[72] In a land high up toward the northern ocean, sphagnum—the ancient bog vegetation also known as Dog's Flesh [Hundekød] (Glob describes it as a "reddish peat-stratum which was formed in Danish bogs precisely in the Iron Age," so named by peasants because of its "colour and its relatively poor quality as fuel"[73])—is given life and voice and is capable of making a pact with the inhabitants: "The old sinewy and dried out spots in the bog were given to the humans as peat, out of which they could build things and cook food. In return the humans threw various kinds of interesting bodies and other sacrifices out in the new bog, which Dog's Flesh could then practice its art of immortality upon."[74] Because of the presence of humus (here called "modvand," a neologism which literally translates as "counter-water"), Dog's Flesh has the ability to hold on to the flesh and skin of a body and "dissolve death and bones." The priestess tells Pliny how what the Romans call sphagnum—the substance they allow "the roses to grow in"—is part of something much larger, a material and mental concept: "one great idea, spread out in all the little moss plants of which it is made: the idea to submerge and immortalize everything in fresh water."[75] Like sponge, it starts as "a sucking plant" which germinates until it is full-grown and able to absorb "both water and trees and animal and humans," and as a tweaking of the Christian burial ritual—"Of water you have come, to water you shall return, and of water you will be resurrected"—the bog connotes (pre-Christian) death, preservation, *and* resurrection.[76]

Two thousand years later, the traces of this "great idea" are found in the shape of the Tollund Man when two peat diggers, ignorant of the pact made centuries earlier between Dog's Flesh and the inhabitants of the land, accidentally unearth the ancient man in 1950. The bog body's face is relaxed and dreaming: "He dreams a pleasant dream, and his lips are pursed in a little smile, which tells of peace, but also of a knowledge that had doused the ability to feel deep and prolonged happiness. It is a dream. And he looks like he is smiling only while dreaming."[77]

The novel gives Tollund Man a fictional proper name, Geppu; makes him "a foreigner," a Celt; and assigns him an age, thirty-three years (notice again the continuous reference to him as Christlike), and a face "without wrinkles and a golden glow from the sun."[78] Once a prince expecting to be elected druid, Geppu is an outlaw on the run because he has committed a ritual sin by mixing fresh pig's blood together with the old dried-up blood in the ceremonial cauldrons. He walks to the north to gain support to win back his lost position, but eventually gets into a fight with an old hunter named Odtor (seemingly a conflation of Odin and Thor) over two ritual cauldrons (based on the so-called Gundestrup cauldron now housed in the National Museum in Copenhagen). As they each run off with one of the cauldrons, they soon find that both are needed to maintain the ritual function. Sections of each cauldron are eventually mixed—female and male parts together—and given as a gift centuries later to Augustus' army when it, as we know from the inscription on the wall of the temple in Ancyra, first encounters the land of the Cimbri.

In the end, tired of being on the run, Geppu the Tollund Man willingly succumbs to his fate at the hands of his competitor:

> Odtor squeezed [the rope] and took off Geppu's clothes as well as the
> bag and dragged him to the place just before the lake became open water,
> where Dog's Flesh stood fair and young, and it was too wet and spongy for
> anything else to grow. Odtor took Geppu's left arm and leg and swung him
> around a few times./ A slurp was heard and a ripple went through Dog's
> Flesh. Then Geppu was gone .... [79]

While this is neither a ritual sacrifice nor a punishment for transgressions (*corpores infames*), Tollund Man's death, clothed in the story of Geppu, is intimately tied to rituals of the land, to transgressions of sacred laws and fateful decisions. Yet Tollund Man is not quite Danish in Kløvedal Reich's novel. He is a foreigner, a Celt, related only implicitly to the making of the Danes, moored to the landscape (both metaphorically and, after his death, literally) but *not* manifestly to the origin of the people. To be Danish here is, rather, located in the conscious decision to *call* oneself Danish.

Kløvedal Reich's epic depiction of the making of the Danes is partly a romantic vision inspired by Tacitus's view of the peoples of northern Europe as those who favored debate over submission to powers, and who practiced openness in relations between the sexes (certainly a different reading of Tacitus than the one performed by Himmler and his cohorts).[80] It is also partly an imagined scenario of how the Danes became Danes *then,* or perhaps more importantly, how they can remain Danes *now* (that is, since the referendum to join the European Union). The novel ends with an implied birth certificate for Danish democracy (the words of Kløvedal Reich's nineteenth-century idol, the Danish national romanticist N.F.S. Grundtvig—"few have too much and fewer too little"—are clearly implied). The novel is a conscript for a nation built "just with people," outside any empire, be it ancient Roman or contemporary

European. The imaginary Danish prehistory, auspiciously filtered through nineteenth-century national romanticism, repeats the political or ideological position of the European movement, with traces of a time in which the past is reimagined to find out how the nation could or should see itself as an independent place *now*. The threat of the European Union to a small nation's sovereignty parallels other times in history in which that country has been tested.[81]

More importantly, as the novel comes to a close, Tollund Man reappears as a voice from the bog; he is not quite dead. Rescue comes from the south in the form of one of the Roman twins, Remus, who has set out to the north to find a new home. Remus is nearly caught, like Geppu, in the dog's flesh. He hears the bog man's voice emanating silently from the depths, converses with him, and then finally releases him from his decade-long limbo by proclaiming that he is not responsible for the mass emigration to the north; that he has only played a minor role; and that it is Rome, the Goddess, and fate that has reshuffled people in Europe. Finally released from his limbo, the bog man's "light brown body twitched. He had peace."[82] His release is the beginning of a new national identity; it was at this time, we read, that more and more people started to call themselves Danes. That this claim to identity is not an entirely uncomplicated gesture is underscored by a mythological figure, Nornegæst (imagined to live as long as his candle will burn, a Norse Meleager), who has forewarned Geppu early on that "although you are dead, you shall see the whole hideous history with your own eyes in two-thousand years."[83] The unearthing finally becomes a resurrection and turns Tollund Man from being a witness *from* the past to being a witness *of* the present and "the whole hideous history" in between.

Therapeutic Anamnesis

So far we have seen how German ideologues and French, American, and Danish authors have used bog bodies to question national or ethnic identities. It is time now to turn to the Irish poet Seamus Heaney, whose work on bog bodies has drawn more attention than that of any other poet (or novelist for that matter). He, too, uses the bog as a place of national identity—in this case Irish national identity.[84] And he, too, calls bog bodies to the witness stand to testify not only in a courtroom of *politics,* but also in one of *poetics.* To him, the pen is a spade and language is the layered soil that the pen digs into, and from which the poetic imagination culls words to speak about present calamities. Although he acknowledges that "in one sense the efficacy of poetry is *nil*—no lyric has ever stopped a tank," poetry's power is both unlimited and paradoxical: "It is like writing in the sand in the face of which accusers and accused are left speechless and renewed."[85] Accusers and accused are left "speechless" when confronted with atrocities, yet feel "renewed" by the ruptures of bog bodies because they offer historical and poetic depth to present-day conflicts, and to the poet's pen.

Heaney carefully connects archaeology and national identity as he laboriously "digs" out a space for Ireland in a poetics that, to borrow an expression from Helen Vendler, becomes an "anthropology of the present." His aim is to mend a conflicted present in Ireland by understanding the depth of the bog, by seeing the marks on those who live on it as having been "kinned by hieroglyphic / peat." The peat bog represents a strangely familiar face from the past, a face that affirms the poet's love for his country and intrigues his sense of ritual and process. As in "Kinship":

> I love this turf-face,
> its black incisions,
> the cooped secrets
> of process and ritual;[86]

The Irish bog is to Heaney a place of endless depth: "the wet center is bottomless," he writes in "Bogland." But he maintains a sharp eye for the paradox of bogs as airless, stagnant spaces of preservation that hold to their own sinister and paradoxical logic: they kill and they preserve. In "Kinship" he pays tribute to the pliability of its properties by way of an endless list of synonyms: "quagmire," "swampland," "morass," "slime kingdoms," "domains of the cold-blooded," "*bog* / meaning soft," "pupil of amber," "Earth-pantry," "bone-vault," "sun-bank," "embalmer," "insatiable bride," "sword-swallower," "casket," "midden," "floe of history," "nesting ground," and "outback of my mind." The materiality of the bog ebbs into the "outback" of the poet's mind, and from there the bog and its objects are rearticulated.[87]

Heaney's love of the face of the bog landscape becomes manifest when Tollund Man lends his features as a stenographic symbol of the past, which must be read and deciphered to serve both as a cultural and as a personal identification.

> Out there in Jutland
> In the old man-killing parishes
> I will feel lost,
> Unhappy and at home.[88]

In the bog man's face, as already mentioned, Heaney sees not only the saints whose stories filled his Catholic childhood, and "whose bodies stayed undecomposed and fragrant in death" with "a certain Christ-like resignation, which gave him the aura of a redeemer"; he also recognizes something far less divine and more familiar: his uncle Hughie. In this way the personal identification is expanded not only to a concern for Ireland as nation, but also to a crisis of identity for the poet:

> The Ulster conflict, the violence that began in the late sixties, was already in full swing in the early seventies and the fact that both sides were ready to die, in a more or less religious, self-sacrificing way for the preservation of their land, linked in very suggestively with Glob's theory that the Tollund Man and many other bog bodies were part of an ancient rite involving human sacrifice, enacted to bring new life to the ancient earth, year after year.
>
> Since then, other scholars have suggested that the sacrifices were in response to some crisis, such as famine, or perhaps attacks of some other sort. And this too corresponds to what was happening in my own case, since the poems I wrote after reading Glob's book were in fact something similar: *responses to a crisis inside myself prompted by events outside.*[89]

The crisis with which Heaney finds himself "face to face" is connected to the "barbaric attitudes" and slaughter in the Irish civil war, and his poetic tribute to the bog sacrifices is an attempt to restore "culture to itself."[90] In this process, photography once again plays an essential role. Heaney writes:

> And the unforgettable photographs of these victims blended in my mind with photographs of atrocities, past and present, in the long rites of Irish political and religious struggles. When I wrote this poem, I had a completely new sensation, one of fear. It was a vow to go on a pilgrimage and I felt as it came to me—and again it came quickly—that unless I was deeply in earnest about what I was saying, I was simply invoking dangers for myself.[91]

Years before Heaney finally visited Tollund, Glob's "marvelous story" had prepared him for a journey into the hybrid landscape of history and story. Heaney's seismographic use of bog body images in his poetics is indebted to Glob for providing the prose-soil, but Heaney also sees deeper linguistic material, even a geological affinity with Glob's language: "The boggy landscape that Glob wrote about [. . .] and the fact that in the speech of our local district in Ulster we used the word *moss*—so close to *mose*—rather than the word bog, that made me feel even closer to the marvelous Danish story which Professor Glob had to tell."[92]

It is evident that the bog serves as a memory bank here, and Heaney acknowledges that "since memory was the faculty that supplied me with the first quickening of my own poetry, I had a tentative unrealized need to make a congruence between memory and bogland and, for the want of a better word, national consciousness." Thus, "the atrocious and the beautiful often partake of one another's reality." This allows Heaney to project into the bog bodies from Denmark and other nations the brutality of human violence and national sacrifices in Ireland. In the Danish bog bodies he finds objects of "contemplation and the violated remains of human flesh and bones," and through the bog poems he voices and makes visual what he, with William Butler Yeats, calls "befitting emblems of adversity" in relation to the many murders of the Irish civil war, now and in the past.[93] As seen in the second half of his Tollund Man poem he links the sacrifice of the Tollund Man to that of the massacre in Ireland of four young brothers:

> Tell-tale skin and teeth
> Flecking the sleepers
> Of four young brothers, trailed
> For miles along the lines.[94]

The link between Tollund Man's sacrifice and "the tradition of Irish political martyrdom" makes Heaney see not only "an archaic barbarous rite," but also an "archetypal pattern" which extends from the past to the present.[95]

If for Heaney the bodies of the past exist within a time-space configuration connected to language and history, then the visual object, the dead bodies of the bog people, function structurally as the past of the poem's present, and therefore become analogous with a larger historical and ideological temporal figuration. Yet, ironically, for Heaney the attention to aesthetics also brings the ancient bodies into conventional (fine arts) classifications as limbs and objects comparable to art representations in bronze or marble. In sum, what Glob grants Heaney (or what Heaney culls from Glob) is to see the historical body as an artifact, and as a "fiction" different from the authentic archaeological "thing itself." The actual object has already undergone a transposition through Glob's archaeological narrative (which, as we have already seen, borrows extensively from literary tropes), but through Heaney's poems they are placed in a representational minefield in which the very core of what representation is, or is not, is at issue. On all levels and in spite of the promise of authenticity, the bog body—"this man" in this "place" at this "time"—is far from the reality of *who* or *what* the human being, the person, behind Tollund Man may or may not have been. "He" or "it" is part of a complicated representational play.

With that said, we may want to repeat the observation made by Michael Ann Holly that ekphrasis "struggles to describe that which can no longer be seen in a variety of ways, [and therefore] can itself serve as an allegory for the larger project of history writing."[96] It seems evident that Heaney's ekphrastic poems, as they express a kind of solidarity with the sacrificed bog people, bring into play temporal and spatial problems in ways that become consequential[97] —not as essential categories, but rather as poetical and mytho-historical intersections that critically question personal, poetical, historical, and national identity.[98] In the end it becomes a kind of "therapeutic anamnesis," a healing way of recalling the past.[99]

In "The Grauballe Man" from his 1975 collection *North,* Heaney has an eye for both the anachronistic and anamnestic natures of bog bodies. But to him they are not absurd. Rather, they are made to represent key elements in the production of poetry, and they are asked to point to pivotal concerns in what it is to be a human being in historical (and personal) time and place. Heaney is often less interested in the bog bodies' past than in their *presence* in the present.[100] In "The Grauballe Man"—unlike in his poem on Tollund Man—he is less intent on finding symbolic answers to the violence in the present, the Irish conflict and killings in Ulster. Instead he allows the bog body to pose a question of commonality, of shared identity.[101] Notice how face and photography once more provide the trajectory:

> I first saw his twisted face
>
> in a photograph,
> a head and shoulder
> out of the peat,
> bruised like a forceps baby,
>
> but now he lies
> perfected in my memory,
> [. . .]

The photographic stillness gives way to movement, in language, so that the birthing also takes place by way of the poet's pen. In the end, and with its own strange finality, the body's parts and the long row of simile it presents are marked by such violence that the birth of the body is a mnemonic still image. The fetus-like forefather, newly born yet long since dead with "his rusted hair / a mat unlikely / as a foetus's," comes about by way of fusing photograph and memory—in words. Heaney calls upon the affecting and pathetic in the emergence of the striking mummified figure. And the odd correlation between modernity and antiquity is articulated as a liquefying of solid matter:

> As if he had been poured
> in tar, he lies
> on a pillow of turf
> and seems to weep
>
> the black river of himself.

Poetry to Heaney (as he discovered after spending a year in Berkeley) "was a force, almost a mode of power, certainly a mode of resistance."[102] And in "The Grauballe Man" this force and resistance is pictured as a pouring from the body of ink-tar, as if engulfing the body and creating a kind of landscape around it with the body's own properties. The ancient man's turf is a pillow on which he not only sleeps but also weeps ink. Ink and body—wrists, heel, instep, foot, hips, and spine—are fused into one, and the mud bath out of which the glistening head of Grauballe Man rises is both beautifying and deadly.[103]

Heaney has been criticized for using references to a brutal and mythical past as an apology for the violence of the present, a symbolic partaking in or acceptance of bloodshed through "silence." Yet he is acutely aware of the fact that his present, as British literary scholar Thomas Docherty has put it, is "a moment in flux, his spatial present as a moment bifurcated, divided, a moment when space has gone critical, differential, historical rather than antiquarian."[104] The transhistorical and mythical aspect that so often is linked with Heaney's poems (his place within the neo-romantic and modernist traditions) must then be questioned. According to Docherty, Heaney is not out to "discover an archaeological remnant of the past in its antiquarianism, but rather to write in the interstices of history itself, to be historical and to be aware of the flow and the movement of history, history as 'becoming' even as he writes-or because he writes—the poem."[105] Irish literary scholar Edna Longley in turn points to Heaney's use of "history as an experience rather than a chain of events." Heaney, she writes, "dramatizes his own imaginative experience of history" and focuses poetically on myth, memory, history, self, and nation. In "Bogland", for example, "metre, sound, and rhythms enact a descent through layers."[106] Others have argued that Heaney's poetry reveals his awareness of the danger involved in glossing over atrocities through the "artistry of myth"—that he, in the words of another Irish literary scholar, Elmer Andrews, "is uneasy about the power of language to identify him with, or to deify, the ancient bog victims."[107]

But "The Grauballe Man" is, as Docherty suggests, also "strangely androgy-
nous" because of the "linguistic slippage or ambivalence" in which metaphors
such as ball and egg are merged and produce "the theme of pregnancy"; it is a
kind of *anamorphosis* "as the male character mutates into something female."
Thus the man "gives birth to himself from the female bog" just as "the poetry
is in a sense also giving birth to itself, originating itself or authorising itself in
this peculiar act."[108] Reality, then, takes the shape of reversals: "what seemed
to be a tomb is a womb; what seemed a man gives a kind of birth while also
being the baby itself."[109]

Clearly, the inherently political aspect in Heaney's bog poems is not just
about national politics but also about sexual politics, about eroticizing the bog
realm and its "spawn," and about positioning writing (with the pen as a "spade"
but also as a sensualized extension of the hand) to uncover and reveal how "that
which ought to have been hidden but has come into light" again and again
brings us copious amounts of creative stories about illicit transgressions.

CHAPTER 4

Erotic Digging

On October 20, 1835, the mummified corpse of a woman was found in Haraldskjær Bog in the eastern part of Denmark. The body was swollen and blackened by the peat water, the hair dyed red-brown from the ferruginous properties in the bog. In a forensic report by the district physician from Vejle, J.F. Christens, published a year later, we hear that the body was that of a rather corpulent middle-aged woman with good but well-worn teeth, large breasts, clearly visible nipples, and long, thick hair. The hands and feet were fine and small, "hardly belonging to someone of the working class."[1] Her manner of death seemed to have been violent. She was, Christens writes, "probably nailed into the mud while still alive" since "her facial expression almost clearly could be seen as despair."[2]

Although Christens's description reveals some uncertainty as to how to read the body, his observations and particularly his use of the word despair [*fortvivlelse*] would echo in most writings on the Haraldskjær bog body from this point on. Otherwise abstemious in tone, his report prudently lays out the gradual process of dehydration and subsequent shriveling of the body, the coloring and preservation by the bog water, the body's dimensions and internal organs—or lack thereof. Christens concludes that "the deliberate and meticulous way in which the corpse was wrapped seems to play witness to something extraordinary," and he determines that the deed was hardly possible for one or two people to complete in any short period of time.[3] The mysterious circumstances surrounding the woman's death, the anguished face, the naked body—still voluptuous when found—soon made her a celebrity, even in centuries to come.

The bog body was soon, and mistakenly, interpreted as the Norwegian Viking Queen Gunhild, who according to tradition (saga literature) had been lured to Denmark by the Danish King Harald Bluetooth to become his wife, only to be drowned in the bog around the year 970. Known to be cunning, power-hungry, and untrustworthy, Gunhild led a dramatic life. In *The Bog People,* Glob explains: "Historical sources describe Gunhild as a beauty, refer to her love of pomp, and characterize her as shrewd, witty, clever, merry and eloquent, friendly and open-handed to everyone who would do what she wanted, but cruel, false, malevolent and cunning if anyone crossed her. She seems also to have been dissolute and domineering to a high degree."[4] This interpretation took hold in the public imagination not least due to the thrilling prospect that here was a body which had returned from the dead to speak visually about past passions and crimes—amongst the royal. A host of literary scholars, antiquarians, poets, and dramatists quickly saw her remains as a chance for eroticizing the past.

But before we look closer at the erotic potential in Queen Gunhild, let me briefly summarize an important controversy within academia in the years after the discovery of the body. The controversy centered on the provenance of the body and had to do with conflicting views on authenticity. It pitted traditional literary scholars against those who worked hard to pry material culture out of the stronghold of literary interpretations. While literary scholars, represented by the professor of Nordic philology N.M. Petersen, turned to selected textual authenticity (excluding Tacitus and *Prose Edda*) in an effort to ascertain which text might most accurately locate and authenticate the body, archaeological scholars, represented by one of the fathers of the archaeological discipline, J.J.A.Worsaae, rejected not only the truthfulness of the literary sources but also objected to what Worsaae saw as arbitrary connections between literary texts and material artifacts. Worsaae spoke in defense of the object as authentic witness in its own right, and saw himself as maintaining a healthy perceptual distance from Petersen's imaginative storytelling, which in his mind produced a clash between seeing things as one would have it and seeing them as they really were. To Worsaae, Petersen violated the material evidence and constructed a "beautiful poetical hypothesis," albeit an unhistorical one built on unreliable written sources.

The dispute between these two men illustrates how precariously the past was constructed. Looking in different directions—one at the *physical body*, another at the *textual body*—they seem not to have been sharing in the same past. If the body was identifiable, if it had "name" and mooring in the national history, then it would help legitimate not only the status of written material but also the accuracy of a knowable history tied intimately to the present. If, on the other hand, the body were not easily identifiable, it would open the door to a past that could potentially challenge national identity and the status quo. The interpretation of the bog body was not, then, just about material objects versus writing, but about a new way of understanding the past as different from and perhaps even challenging the legitimacy of the present. As a professional academic discipline, archaeology had grown out of disputes such as the one

Fig. 4.1. Laid out in her oak casket, the Haraldskjær Bog Body—also known as Queen Gunhild is both a "royal" figure whose name earned her a prominent place in the church, and a museum object. Photograph © Vejle Museum, Denmark.

between Petersen and Worsaae over the Haraldskjær bog body, and the study of these ancient remains was tied to—or at least not devoid of—connections to the development of a national consciousness. Petersen, morally enflamed, gave voice to what he saw as the "nemesis of history." Worsaae, in turn, argued that Petersen was using unhistorical sources as part of a historical argument.[5]

Although Worsaae would in time be proven correct in his assumption that the bog body could *not* have been Queen Gunhild, the irony is that without Petersen's literary interpretation, the material remains of the body would not have survived. Impressed with the archaeological find and the body's proposed royal origin, the Danish King Frederik VI commissioned a lavish oak casket at St. Nicolai Church in Vejle, where the body still rests. Without the literary misnaming, it would no doubt have turned into a paper body, preserved in writing only. The physical remains would have disappeared in an unmarked grave, as did many other such finds.

Queen Gunhild's royal standing did not shield her from continued public curiosity. In the church at Vejle the casket was open and hundreds, perhaps thousands, of visitors came to see the now emaciated body and grimacing face of the formerly illustrious and infamous queen. No longer voluptuous and full-breasted, the naked corpse brought proximity for ordinary men, women, and children to a past and a royal body whose physical attraction and legendary power were lost, but whose allure continued. Resting both in the oak casket and in the story of what she might have been, the queen's legend flourished for decades and would linger well into the twentieth and even twenty-first centuries.

A Bog Queen

In 1841 the Danish poet Steen Steensen Blicher published a poem in the journal *Brage og Idun* called "Queen Gunnild," in which he appeared to subscribe to the queen theory wholeheartedly. The poem is structured between a past and a present, a *before* and an *after.*

> Before
> Then you were clothed in sable and marten,
> And decked with precious jewels,
> Gems and pearls in your golden hair,
> Wicked thoughts on your mind

> Now
> Now you lie naked, arid, and foul,
> With your bald head

Blacker far than the oak stake
With which you were wed to the bog.

Before
You shone like a first-rate star
And thousands obeyed your command
Lovesick sighs followed your flight
When lovely you danced on the deck

Now
Now you lie still on your death linen
With hard and withered hands;
With this rigid and icy grin
You light no hearts afire.[6]

A *femme fatale* in the past, a temptress decked in luscious fur and jewelry, full of depraved desires, the queen, the poem suggests, has gotten what she deserves. Stripped of her seductive powers, as well as of her clothes, she has become abject and disgusting, a cold and rigid monstrosity joined in marriage to a realm that corresponds to her character: foul darkness. She is to be feared as a woman with seductive powers in the past, but also as a corpse in the present; she is at the mercy of the poet's pen, unable to defend herself from his verbal immuring.

Before
Then you gladly wielded the bloody axe
King Erik you thus unburdened,
You divided mountains, oceans, and land
And your time between flirtation and murder

Now
Now you feel quite dry and cool,
But grossly your mouth is grimacing
The death-cry, when you were drowned in the bog
Bogged down by the executors' stakes[7]

Doubly incarcerated in bog and words, the powerful saga-queen has become a disempowered bog-queen. In a like manner, the past/present, then/now structure works to underline the erotic play. The contemporary present "I" addresses the historically past "you" not only as a way of taking back voice and control from the "outrageous" woman, but also as a way to engage in a ritualized verbal spanking of the transgressor. Her ending in the bog is her own doing; it is her own inappropriate mixing of sexuality and politics that has backfired, and the poem's "I" revels in her misfortune.

> How could you believe Harald Bluetooth thus?
> You did your own undoing;
> Now you lie bluetoothed on straw
> While he takes a nap in the grave.
>
> Many years have passed
> While you have been in the muck
> I wonder how things are down there
> What news, Queen Gunnild, from the mire?
>
> Were you burned in hell?
> [. . .]
>
> In thousands of years—I now recall
> Your grace has been pinned down
> You will rise again to dawn
> You have long held death hostage.
>
> You appear quite like an antiquity
> From olden heathen days
> Good night! Sleep well, your majesty
> The black lid over you glides[8]

Throughout the poem the "I" is allowed to take verbal revenge on behalf of his sex toward a historical figure who has resurfaced defiantly from the bog to testify to her own depravity and corruption. In the present, she is finally "whipped" into place again—strapped down, not by the rods in the bog but by punishing and "justified" words which are ironically couched as being warranted because they benefit from the historical perspective of knowing about the queen's misdeeds, or so it seems. Historical distance serves here as a safeguard against the potential danger of facing the queen. The poem seems to want to close the lid of the coffin on her, to silence her grossly grimacing mouth with its death cry. The gloating tone with which the bog body is approached, or *re*proached, amounts to an indictment of the powers of someone who, according to legend, not only usurped the realm of national politics but used her erotic powers in doing so.

At the time when Queen Gunhild was found, she emblematized the exact opposite of the classical beautiful body defined by most aesthetic theories in the eighteenth and early nineteenth centuries. She did not possess, as the ideal dictated, an "uninterrupted skin-line of the softly-blown body"; she did not own a hidden unseen interior.[9] Quite the opposite: her body was literally opened for inspection, her anguish seen in her open mouth, her depravity and unseemly seductiveness located in her full breasts and in her erect and clearly visible nipples. Her skin folds and wrinkles were that of an old woman; she was bloated and soft. In the early nineteenth century these were all aesthetic taboos, potentially threatening not only to aesthetic pleasure but also to ethical order. Yet, fascinatingly, by the time her body surfaced, the aesthetic paradigms of classical beauty had already been flayed by the emergence of romanticism's "dark" aesthetics of shock and its taste for intense stimuli.

Consequently the new pleasures found in beauty's antonym—ugly and revolting bodies—gave Queen Gunhild's physical remains a paradoxical kind of *abject charisma*. This *abject charisma* counts in part for the erotic fascination with her. Her presumed scream—visible to the eye but unheard by the ear—added to her role as someone astounding, piquant, shocking, "romantic." It is interesting to note, therefore, that at the time of her excavation in 1835 the German writer Ludwig Tieck equated the term "romantic" with "disgusting" and "putrefaction."[10] In his attempt to distinguish between healthy (German) romanticism and unhealthy (French) romanticism, Tieck writes of the latter:

> Even in men who are blessed, a transport of rapture flashes within the joy
> of cognition itself; how it happens that the soul so often then voluntarily
> *plunges* from enthusiasm to passion, is an eternal riddle and eternal mystery.
> Now the intellect speeds forward, as if in spite of itself, along the fiery path,
> scorns the light as ineffectual, and *sinks* and *deepens* into what is most offen-
> sive to its nature, now believing it has found its most characteristic quality
> in the wild, rough, and injudicious. Now it lives in lies and untruth, sinning
> against beauty and holiness as if these were the lies. From excessive trans-
> port of freedom, the intellect must now become a slave of ugliness, and the
> tighter the chains that bind it, the more it boasts, scornfully laughing, about
> its boundlessness.[11]

When man is a slave of excessive boundlessness he "plunges," "sinks," and "deepens"—if not into bogs *per se,* then into their equivalent: the wild and rough, passionate and mysterious. While the German school of romanticism held onto ideals of classical beauty, Tieck deplored the French romanticists like Victor Hugo for writing "of vice, putrefaction, monstrosity, and the works of darkness" and of reveling "in the decay of vice, and [being] drunk on the disgusting."[12] It seems obvious that the kind of erotica offered by bog bodies has traveled far beyond the index of the aesthetics celebrated in the archaeological erotica of the eighteenth and nineteenth centuries, such as that associated with Pompeii. In fact, as was the case with Queen Gunhild, bog bodies in many ways stand opposite the beautiful body's construction; they threaten the pleasures that contours and firmness of form offered in classical statue-erotics. With this in mind, it would seem that under the Danish writer's pen the Norwegian saga-queen comes suspiciously close to looking like a depraved seductress in French romanticist makeup and disguise. Freud, had he known, would no doubt have found Queen Gunhild (archaeologically) uncanny.

More than a century and-a-half later, a very different literary incarnation of Queen Gunhild comes to light in the Danish author Camilla Christensen's 2002 novel *Jorden under Høje Gladsaxe*[13] (The Earth under Høje Gladsaxe). Here Queen Gunhild, "the mother of the huldres, strangled, drowned, staked down, a ghost, a restless corpse," takes the shape of an old dying woman, Gunhild, in a nursing home located in one of the high-rise apartment complexes at the outskirts of Copenhagen. Built on archaeological soil, on the remains of Iron Age settlements, on bog land, these buildings represent the precariousness and fragility of the present.[14] They are under constant pressure of the past. As was the case in the Danish Dogma 95 auteur Lars von Trier's film *The Kingdom (Riget)*— in which the past, harbored symbolically in the swamps, *Blegdammen,* would undermine and rock the very foundation of the modern (also in the shape of high-rise buildings)—Christensen's novel suggests that Høje Gladsaxe, in spite of its newness, is a place with a past that "whispers," leaves traces, and is potentially dangerous. Høje Gladsaxe is already almost a ruin itself and will eventually, the novel predicts, disappear just as did the other worlds that are now archaeologically layered under the high-rise structures (not only Iron Age settlements but also, in closer temporal proximity, the nineteenth-century garden in which a young Søren Kierkegaard supposedly fell from his uncle's pear tree and injured his back).

The archaeologists who listen and look for traces of the past in Christensen's novel are hopelessly myopic and unable to lift their noses from the ground (a characterization for which the author apologizes in her epilogue). Instead of digging, instead of taking the primarily archaeological trope from excavation images (although these traditional archaeological images are present in the novel as well), the author turns to an aerial photograph as visual marker. This cartographic perspective is used to rectify archaeological shortcomings—by offering a view from above—and by providing a different map for "digging"— i.e., reading. The Høje Gladsaxe world is seen both from a distance and up close, both from the present and with the presence of various pasts. In this place of time (past, present, and future), things are ephemeral; they are never permanent but are instead—like time itself—changeable, always on the move.

In the nursing home where Gunhild lives, she browses the computer and falls into the peculiar "time travel" provided by the Internet. As she homes in on the face of the Grauballe Man, time—his, hers—is seen as mere micro-, nano-, split seconds. In this warped and elastic temporality, bodies are left behind and should, as Gunhild indignantly demands, be left alone. "Let the dead have some peace." Yet she, like an archaeologist, like us, is caught in compassion coupled with fascination:

> She felt sorry for Lenin, who for three-quarters of a century had lain on *lit
> de parade* in the mausoleum in Red Square, while his brain was placed in a
> drawer at the medical institute three streets away. She felt sorry for the pha-
> raohs and their wives, who were pulled from the pyramids and coffins and
> looked at in Egyptian collections of all kinds of museums, and for the Egt-
> ved Girl and for her own namesake, Queen Gunhild, the bog body in Vejle;
> but most of all she feels sorry for the Tollund Man, as he lies in his montre
> in Silkeborg Museum and is looked at from all angles, he is beautiful, he is
> moving, he is peaceful (in spite of the exposure), a sleeping beauty, who is
> brutally exhibited to curious gazes, and now the far less appealing Grauballe
> Man (and thank God it is him and not the Tollund Man) is to be exposed to
> the Danish health system, he is no longer just to be viewed from the outside,
> but also from the inside.[15]

The bog bodies are probed and penetrated, and in spite of Gunhild's anger and abhorrence over the disrespect of the dead; regardless of her own fears of being probed in the nursing home, she cannot help but be mesmerized. "She lets one link follow the next and *seeks and finds* [higer og søger][16] more information, first about the Grauballe Man, but soon about the other European bog bodies from the Iron Age, and through the night she moves from the local Jubii out into the global Google and learns just about anything there is to be learned about preserved bodies, not only bog bodies."[17]

Another resident in the high-rise, Arthur, can see through things—literally, not as insightfulness but quite manifestly and materially. Gunhild, on the other hand, is nearly blind but has insight. She sees things through time. When the two come face-to-face, his X-ray gaze and her ability to penetrate reality mix: time becomes transparent to him and will, as the novel ends, finally come to an end for her.

They set out on an erotically charged time-travel in which the bog and cyberspace are meshed. As "his gaze penetrated her body" and he grasps her multilayered and multi-temporal essence, "she saw the confusion, the alarm, the bewilderment, the uncertainty in his face, and she *felt* it, felt his gaze as a somersault in her chest, her abdomen, and, even if she will not admit it (not even to herself), she felt an otherwise almost forgotten suction between her hip bones."[18] Gunhild, once a young woman—and in the logic of the novel *many times over* a young woman—falls in love. Her first love in her first incarnation was illicit, and in a Tacitus-like bog drowning she was punished with the limbo-existence she has since suffered. Information on the Internet about death-ice holes gives Gunhild in her latest incarnation the key to understanding how to put an end to her continual cycles of resurrection. Again lovemaking is a solution to the problem. Arthur is asked to slip through the portal of the bog some 2,500 years back in time, penetrate its temporal void to access the time and place where Gunhild died her first death (the "disgraceful, pitiful bog-death), find her there in her first incarnation, love her (again), and through this lovemaking finally release her. As they approach each other in the final scene of the novel, Arthur is able to see Gunhild as the blonde-haired beauty she once was many centuries and incarnations ago. But he is also able to perform an imaginary reconstruction:

> Arthur looks straight into her and constructs an image of her skeleton, as it once was; he rebuilds her, reconstructs her from the many layers of bones, until she is not just [. . .] harmoniously built, but flawless, impeccable; and on top he places ligaments, veins, muscles, flesh and fat, and from the decay a woman emerges at the prime of her life instead of the ancient form [*oldtids-fund*] with hip fractures from countless lived lives.[19]

Here Arthur serves as a kind of software program through which Gunhild is reconstructed, layer by layer, in her first incarnation while she repeats the passion and transgression that made her a bog body in the first place. Gunhild is dead at last, *and* erotically released. In cyberspace, invisible energies have replaced her corporeal matter. Gunhild and the bog body known as Queen Gunhild are blissfully de-materialized.[20]

Christensen's use of the Internet as a realm that corresponds with the bog (a place to search and "dig") does not display nostalgia for the *past;* on the contrary, there is a certain kind of nostalgia for the *future* as if it can remedy the traumas of the past. Yet to search the net for bog bodies is also to tap into a realm which, like the bog, struggles with its own organic "vernacular" of worms and viruses; it is a postmodern digging into a realm in which things can be found, combined—and lost again.

Archaeo-erotics

By now, we have encountered several examples of how bogs and bog bodies are eroticized. Before we continue it is important to note, therefore, that most of the sexual or erotic fantasies we can locate in fiction about bog bodies are tied to at least three interrelated repositioning or realignment strategies that seem to be necessary to making bog bodies objects of desire. One is a *temporal* repositioning strategy, which suggests that the physical or amorous appeal of the bog bodies is tied less to the remains as they *are*—after all, they are hardly the most attractive objects of physical desire—and more to visions of what they *were*. The abjectness of dead bodies is tacitly overlooked and the erotic potential comes instead from the tryst between what the bodies are *now* (mummified and vulnerably exhibited) and the apparently titillating vision of what happened to them *before* the transformation, *before* they became bog bodies. Or, said differently, their erotic appeal is not always nested in their visual presence, but lies in the verbal contexualizations that place the bodies in the past, in stories and histories, and—by way of words—return them to their essential humanness.

The second sexual-erotic strategy applies both to *textual* and *material* excavation. To dig into and search through layers of textual matter, to tease out objects (often in the form of metaphors), to reconstruct, to find surprising and hidden meanings, to make visible that which is not readily available to the untrained eye, is traditionally seen as the task of literary criticism but could equally well describe the archaeological imagination. Some archaeologists have seen the experience of excavating and unveiling layers of soil while "penetrating" time as a kind of sexual act, a striptease.[21] The British archaeologist Colin Renfrew, for example, talks about "the aesthetics of the excavation process" as a sensual "pleasure of digging." The physical and material reality of the work lies not only in making order out of chaos, he says, but: "the physical satisfaction, as you hold the trowel, of feeling the layer that you are trying to separate and peel away neatly and easily from that which lies beneath, so that one senses that one is not inventing a sequence of arbitrary levels, but really examining the record of the past, layer by layer, as it was laid out."[22] This kind of vocabulary and the sexualizing of the language of archaeology have been criticized by feminist archaeologists as harking back to and operating within conventional phallic tropes of conquest—and, as we shall soon see, conventional sexing of the bog (womb being the most used term) and the bodies in them has inspired counter-voices in some feminist fictions and poems.[23]

The third repositioning strategy is tied to the bog bodies' *spatial* and *visual* materiality, their potential for being erotic art. The visceral (and in some cases psycho-phenomenological) production of visual art from bog bodies has already been suggested in the comparison between marble skin and "leather" skin in our encounter with the archaeological uncanny in chapter 2. Yet there are, as Gaston Bachelard proposes in his investigation of *terrestrial imagination*, "secret pleasures" associated with the depths of bogs, swamps, and other wetland places that can be transported to production of visual art. Bachelard's importance for bog art will be looked at in more detail in the next chapter, but, as we shall see when we come to the end of the present chapter, there are poets who articulate positions on gender while they help "move" bog bodies into museums of modern art.

Adulterous Archaeologists

Now let us turn to adultery. Tacitus's description of the horrifying consequences for unfaithful women has been radically rethought in three fictional texts, published within the span of five years in the late twentieth century: a horror novel from 1986, a short story from 1991, and a poem from 1989. Far from being about "guilty" women, these fictions, each in its own unique way, have turned the tables on Tacitus and put adultery into play in ways that mirror late twentieth-century sensibilities (if not realities).[24]

In the horror novel *The Bog* by the American author Michael Talbot, archaeological digging turns out to be overtly masturbatory.[25] The plot unfolds sometime in the 1980s when David Macauley, a visiting American professor of archaeology at Oxford and a world authority on bog bodies (he has been trained in bog body archaeology in Denmark) is called to the spectacular discovery of a young female bog body:

> She lay on her back, her head twisted to one side and her left arm outstretched. Her right arm was bent up against her chest, as if defensively, and her legs were lightly drawn up, the left one over the right. She looked much like any young girl might have who had only recently settled down from a nap, except for the fact that her skin, once white, was now a shiny and resinous black. It was a jarring contrast, the perfect preservation of her features against the almost metallic and petroleum-like color of her skin. It was as if a talented sculptor had carved her out of coal and then polished the surface of his work to a high gloss. David was spellbound. Her every feature had been preserved, every pore in her skin, her nails, the whorls and lines of her fingerprints, and the gentle creasings of the skin at the bend of her wrists.[26]

Speculating that her eyes might actually have seen Caesar—that if "she was indeed as old as they believed, she had lived and breathed and smelled flowers nearly a thousand years from the Norman invasion"—David marvels at the fact that he is facing the "hair, the hands and the flesh of the person" who has actually *been* there—in the past.[27] To stand face-to-face with her grants him the authority to open a door to knowledge about the mystery of the bog people, and although her face also speaks an ill-fated language with "a look of unspeakable terror," David's scientific curiosity, his fascination with these "silent and awesome emissaries from the past," and not least his (ultimately fateful) new archaeological methodology—"the philosophy that to truly understand an ancient people, to see the world through their eyes, one had to [. . .] try to put oneself in their shoes, as it were"—prompt him to embed himself with his wife and two children close to the excavation site.[28] Although his thirst for knowledge carries its own intoxication, the archaeologist is soon thrown into lust of a more carnal nature. A fatal and destructive erotic force (we are hardly surprised to find) emanates from the bog, and to satisfy the archaeological craving for learning, a desire "to unravel and decipher all things unknown," as well as his sexual appetite, the archaeologist almost succumbs to the dangers of "digging" into the paradox offered by the bog realm:

> Everywhere he looked there were such juxtapositions, the beautiful contrasted with the deadly, the mist mingling with the thorns. In short, he realized that like all great things, like the ocean, the night, and even life itself, *the bog was a paradox.* The greatest bulk of its substance was dead vegetation, and yet it behaved as if it were curiously alive. It expanded and contracted. It reached out with sinewy tentacles and took and entangled and digested. And it even stirred occasionally in its slumber, groaning and emitting the most mournful and unearthly sounds, presumably from the peat settling, but to many who had heard them [. . .] it seemed more like the ruminations of some great beast, the restless rumblings of the living bog.[29]

Talbot uses the fertility-sacrifice hypothesis to the extreme. Alive with anthropomorphic, ominous abilities and mournful desires, the bog's capacity to take life and give life—to kill bodies and preserve them—is depicted as the "restless rumbling" of a massive earth-womb caught in labor-like contractions and spasms. And the scientific preparation of one of the female bodies found in the bog soon takes on not only a particularly masturbatory undertone but also a decidedly necrophiliac tenor during a scene that can only be seen as a symbolic ejaculation on (and thus violation of) the corpse. Using a hose to wash away "the last of the peat that still obscured the finer details of her anatomy," David allows his graduate assistant Brad the privilege of doing the final cleaning. Explicitly a gesture of passing academic acknowledgement from mentor to student, the cleaning of the body is also couched as an erotic favor. "There was nothing more he [David] wanted than to be the one to wash away the final patina of mud from the body of the young girl," but the young assistant too clearly takes pleasure in the act: "Overwhelmed, Brad took the rubber tubing and positioned it over the woman's abdomen as David went behind the table and once again released the spigot. The clear liquid quickly snaked through the translucent tubing and gushed out the other end." As the liquid streams out, "Brad moved the hose back and forth in slow and rhythmic sweeps, and both men watched closely as the last remaining peat broke away from the woman's flesh and collected at the bottom of the polyurethane tub."[30]

David is both excited about and disturbed by his "stupendous archaeological discovery." The body, it turns out, is covered with bizarre bite marks as if some strange animal has been feeding on her. It is as if it "had enjoyed her anguish and had lingered and caressed her with its bites in a way that seems less animallike and more . . . well, almost passionate.'"[31] The feeding on the body is a sexual act, and soon desire for knowledge is juxtaposed with sexual tension; the archaeologist must struggle between two kinds of temptation, carnal and cerebral. This calls not only for continued digging in the bog, but also for digging into actual bog bodies. Yet "for all his drive and yearning for knowledge, he always felt mixed when it came to cutting into one of the bog bodies [. . . .] Part of him viewed it as an extraordinary sacrilege to slice into them as if they were no more than just another specimen for dissection."[32]

David, unbeknownst, is already enmeshed in the paradoxical powers of the bog: when he digs into the anthropomorphized wetlands, he digs into the territory of the living corpse of a mysterious bogeyman. At a turning point in the novel, the bogeyman—guised as a wealthy marquis(!)—offers the archaeologist, his wife, and the graduate students the powerful aphrodisiac of bog myrtle wine, ostensibly made according to a recipe in Tacitus's *Germania*. The archaeologist's "burning passion to understand every unknown facet of the world" allows him to keep his (academic) head clear and cool.[33] But his wife succumbs to the aphrodisiac and is soon powerless "to do anything but acquiesce to the incredible-desire coursing through her." Her animal-like coupling with what she believes to be her husband's assistant soon turns out to be with that of a supernatural manifestation who at the "moment that he had climaxed [. . .] vanished, instantly and without a trace."[34] The completed intercourse is a moment of impregnation, and in a series of short interspersed flashbacks to 53 BC we discover that the archaeologist's wife is but one of many women who, over the centuries, have been seduced and impregnated by the bogeyman in disguise.[35] The archaeologist, whom the bogeyman has bribed with the power of infinite knowledge about the past, must of course save his wife from carrying the monster baby to term, but this means having to forsake and sacrifice his craving to know everything about bog people—a hard choice indeed for this archaeologist. When two more bodies surface in the bog, David slips into the excavation and "tingles" with an excitement that is both academic and sensual:

> This was the moment that drove all archaeologists on, that brief starburst of exhilaration, perhaps akin to the feeling a painter experiences when he puts the master stroke on a great work of art, or a photographer who, after years of work, captures the one ineffable moment on a roll of film. He savored the electricity that now coursed through his body as if it were the finest wine, for he knew that in years to come he would thirst for its memory.[36]

In the end, the choice is clear. David must either satisfy his archaeological craving or plunge a knife into his wife's abdomen to kill the monstrous fetus and put a stop to the evil of the bog realm. "He thought of all the puzzles of the past that he would be able to solve—how the pyramids had been built, how Joan of Arc really died, and what had caused the Mayan civilization to vanish [. . . .] and for a moment he almost gave in."[37] But the archaeologist sacrifices infinite knowledge and rescues his family so that the world will not "become stagnant, like a bucket of water that you just let sit and sit"—a dark and (sexually) destructive immortality.[38]

Adultery vis-à-vis bog bodies takes a different and more ironic turn in Canadian author Margaret Atwood's short story from *Wilderness Tips*, "The Bog Man" (originally printed in *Playboy*).[39] Here the bog bodies' propensity for sexual bursting does not mean that they are open for invasion; nor are they lacking in contour and firmness of form. In fact *they* are not ambiguous or erotically dangerous; but (male) archaeologists *are*. Atwood's short story, archaeological at the very core of its narrative structure, is fashioned as a retrospective "revision" by Julie, who as a young college student had an affair with her archaeology professor, Connor, a married man with children. The revision that Julie is engaged in (the story *we* read) is a ongoing repetition and retelling of the love affair that itself finally becomes an archaeological thing "which shines at this distance with a soft and mellowing light. The story is like an artifact from a vanished civilization, the custom of which has become obscure." The lovers, we learn, must leave the Orkney Islands in the midst of their investigation of the famous stone circles, due to news of a discovery of a bog man that must be examined before he is "ruined." At first Julie is reluctant to leave because she does not "connect" with the bog bodies as she does with the standing stones—since the bog people "aren't much to look at, judging by the pictures of them," dissolved, squashed flat "so that they resemble extremely sick items of leather gear"—but her archaeologist lover sees them as titillating testimony from "a sexual orgy of some kind" involving "voracious" nature goddesses.[40]

At the bog excavation site she feels dissolved, overlooked by her lover, and squashed flat (like the bog photographs) when accosted by a self-important, married (adulterous) Norwegian archaeologist who "looks like something out of a Viking movie."[41] In contrast to the two archaeologists the bog man, seen no longer through the flatness of photography but as heroically molded in and by time, has a profound and visceral affect on her and makes her wonder whether "this digging-up, this unearthing of him" is a desecration: "Surely there should be boundaries set upon the wish to know, on knowledge merely for its own sake. This man is being invaded." Julie suspects that she herself is being invaded by her archaeologist lover, like "other things that get moulded. Steamed Christmas puddings, poured-concrete lawn dwarfs, gelatin desserts, wobbly and bright pink and dotted with baby marshmallows."[42] When the Viking flirts by way of the bog man, she feels especially violated: "Some have said the dead cannot talk," he says to Julie with a twinkle. "But these bog men have many wonderful secrets to tell us. However, they are shy, like other men. They don't know how to convey their message. They must have a little help. Some encouragement. Don't you agree?"[43] Julie's lack of response pushes the Viking to ask whether "things of the flesh" disgust her. Julia, indeed, *is* disgusted—less by "things of the flesh" than by the archaeologist's adulterous behavior. Against the backdrop of such sordid manners, the bog man offers an "authentic" alternative, so that "of all of them at this moment, she would rather be with him. He is of more interest."[44]

Disgust, as it were, is projected away from the (rotting) corpse and the soft (and malleable) woman (who finally stands up for herself), and onto the squalid and sordid behaviors of adulterous men. While the archaeologist falls off his pedestal, the bog man is given shape, substance and dignity:

> His hands have deft, slender fingers, each fingerprint intact. His face is a little sunken-in but perfectly preserved; you can see every pore. His skin is dark brown, the bristles of his beard and the wisps of hair that escape from under his leather helmet are an alarming bright red. The colours are the effect of the tannic acid in the bog, Julie knows that. But still it is hard to picture him as any other colour. His eyes are closed. He does not look dead

or even asleep, however. Instead he seems to be meditating, concentrating: his lips are slightly pursed, a furrow of deep thought runs between his eyes.[45]

His "furrow of deep thought" befits a man whose feet were accidentally severed by a peat-digger but are now "placed neatly beside him, like bedroom slippers waiting to be put on."[46] This man is going nowhere, but can be trusted to stay in his "own bed."

Julie's disenchantment with her lover peaks as he impatiently lectures her on the difference between swamp and bog: "*Swamp* is when the water goes in one end and out the other, *bog* is when it goes in and stays in":

> But Julie prefers the sound of *swamp*. It is mistier, more haunted. *Bog* is a slang word for toilet, and when you hear *bog* you know the toilet will be a battered and smelly one, and there will be no toilet paper.[47]

The bog is a place of disgust, "smelly" and stagnant, just like the decaying love affair with the archaeologist. So she breaks off the affair "*in the middle of a swamp*" where the water can run out, and not in a bog where it stays in. In the end, because of Julie's obsessive retelling, the adulterous archaeologist-lover loses substance and becomes "flatter and more leathery" while Julia is "fleshed out" and gains solidity and contour.[48] The irony and crux of Atwood's short story is that while bog bodies offer a counterimage to the decadence of "living" contemporary people, the adulterous love story has been repeated so often that it is finally "told to death."

But if the adulterous story is told to death in Atwood's short story, it is given a different spin in the British poet Sylvia Kantaris's poem from 1989 entitled "Couple, Probably Adulterous (Assen, Holland, Circa Roman Times)." The poem's object is the postmortem embrace of the so-called Weerdinge Couple found in Assen, Holland, in 1904—pressed flat and shadow-like, looking almost like photographic prints made in nature's own darkroom. The bodies were originally interpreted as being those of a man and a woman, nicknamed by popular imagination Joan and Darby van der Peat.[49] Glob writes (romantically) about them in *The Bog People:*

> Two bodies together, a man's and a woman's, were recovered at the end of June 1904 [in Holland]. They lay, naked and on their backs, rather more than eighteen inches down at the junction between the grey and the red peat. The woman rested on the man's outstretched right arm. Only his skin was preserved. He was five feet ten inches tall and in the region of the heart

there was something that looked like a wound. The woman's hair was long and very fine and a shiny brown in colour, as was her skin.[50]

Forensic examination has since determined that the couple is in fact the remains of two men, and it is easy to imagine that Glob's tender description of a tall, strong, and protective man with a wounded heart and a shiny-haired woman resting in his arms would have been tweaked differently, had he known. Without their faces and without the visual appearance of their sex, the flat-pressed lovers were seen as being "coupled" in conventional heterosexuality and, ironically, not grouped with Tacitus's *corpores infames*. In Kantaris's poem, too, they are seen as a pair of universal heterosexual lovers:

> Just another couple of old lovers dragged up
> from a bog and propped behind glass, cured,
> their faces slipping off, their ribs skew-whiff.

Pulled out of proportion and virtually faceless, they are given identity as adulterers with an illicit story to tell—a story of pining lovers, like Tristan and Isolde, coupled with crude leather sex:

> Note her split crotch and the scroll of skin teased
> stiff between his legs. A joke? ('You know the one
> about this bloke called Tristan and some other
> joker's missus?') It's a laugh a minute
> getting it together in black leather after death,
> even for monogamists.

The reference to Tristan and Isolde is quite brilliant because the story of these tragic lovers is known as the "greatest European myth of adultery" and, according to Denis de Rougemont's *Love in the Western World*, provides a "kind of archetype of our most complex feelings of unrest." De Rougemont's point is that the desire that drives the lovers is a desire for death: "In the innermost recesses of their hearts they have been obeying the fatal dictates of a wish for death; they have been in the throes of the *active passion of Darkness*."[51] In Kantaris's poem this darkness is made manifest, and the bog bodies' leathery presence as cadavers, caught in the museal prison, reflects the predicament of the poem's narrator. The scene of reflection takes place in the museum space. The narrator's companion (probably her lover) does not share in the experience: "'They leave me cold,' my friend says." But the narrator remains "fascinated, like a necrophiliac." Reflections of another pair of lovers, young and

Fig. 4.2. The embrace of the Weerdinge Couple, found in Holland in 1904, was originally seen as that of a heterosexual couple. It has since been determined that the bodies are the remains of two men. Photograph © Drents Museum, Assen, Holland.

uninterested in the morbid representations, meet briefly in the reflection of the showcase. The ancient, dead couple and the contemporary living couples who view them make "coupling" in the poem an intricately temporal affair. Couples—legitimate or adulterous, ancient and dead or contemporary and living—connect through time in an institution setting.

Unlike her companion, and unlike the young couple whose mirror images meet in the showcase, the narrator is spellbound and reflects on how the bog bodies "ought to be released," as she ought to be released from her husband's (or is it her lover's) smothering love. She moves in close to view "and concentrate on coupled carcasses / preserved beyond the grave like sacred relics / run to puffball dust." Like the carcasses, she needs to be released from her entrapment, which is articulated within parentheses—"(As if I'd passed over my grave)"—to show her identification with the trapped couple. In another parenthetical statement that functions like a secret, silent monologue we read: "(I won't come back. You smothered love with guilt. / Now picture us light-heartedly united / in the afterlife as in this sad museum / of the sporty risen. It's a sick joke.)" The physical realities of the dead bodies mix with the inner reality of our narrator, and the encounter with the ancient illicit lovers is so intimate that it cannot be shared with anyone; she is trapped in silence. When her "friend" (lover), who is otherwise bored because "'everything's been said about bog people,'" shows interest in the bog bodies' lifelike fingernails, our narrator hides her own "quick half-moons" and concentrates "religiously" on the "dead black imitations" in the showcase:

> I've nothing new to say; we know how words
> embalm us in old habits. 'Still, I'd like
> to buy a postcard for an old acquaintance.'
> We sift through all the pictures—swords and moths.
> Late season; adulterers are out of stock.

Though postcard images of the ancient couple, as the poem posits, can be purchased in the museum shop, the commodification of the bog body adulterers (perhaps acquired by other museum-going adulterers) is so successful that they "are out of stock."[52] But the point of the poem is that these postcard images and the words on them can indeed "embalm us in old habits." And they can, as in Kantaris's poem, symbolize a kind of mummification of the relationship between the sexes. The poem gives us a narrator whose illicit love story is silenced, just as the untold stories of ancient secret lovers were silenced in the bog. On picture postcards, words can be written and untold stories can be told and held in reserve for future readings—except of course for the fact that postcards with "adulterers are out of stock." Instead the poem itself becomes the unsilenced voice of adulterous lovers over time, linking the couple "in leather" inside the display case with a contemporary couple (or many such couples) who visit bog body museums and look at their prehistoric partners in crime. The "sacred relic[s]" dressed in leather, with faces slipping off, come face to face with other anonymous lovers, acting both as a mirror that leaves observers "cold" and as a necrophiliac "sick joke."[53]

"Strange Crop"

If we return to Seamus Heaney's bog poems, we find not only articulations of adultery and sexualization of bog bodies, but also, and more importantly, a deliberate sensual overlay between the hand that digs into the earth and the hand that writes the poems—an overlay that is also connected to sacrifice. Heaney points explicitly to the linkage between the physicality of touching and the process of transforming matter to words. The results of this, the poems, have temporal "elements of continuity," he says, "with the aura and authenticity of archaeological finds, where the buried shard has an importance." He calls it aptly: "poetry as a dig."[54] Digging is a sensual experience, one that corresponds with the process of writing poetry. This is most famously described in the poem "Digging," in which the finger and the thumb perform a kind of "intercorporeity" (or, as French philosopher Maurice Merleau-Ponty would have it, a performance of "touching the touch") in which a circle is formed which engulfs the physicality of both the toucher and the touched.[55]

> Between my finger and my thumb
> The squat pen rests.
> I'll dig with it.[56]

The noun "feeling" in the title of his poetics, *Feeling into Words,* also suggests how physicality of touch can mold the archaeological past into something personal and viscerally present. The hand does not touch the archaeological object—in this case the bog bodies—but the feeling that acts as proxy is both a practical and sensuous mediator. "Feeling" means both to sense and experience, to think and consider. Consequently it allows the poet the possibility of wedding archaeological hermeneutics with personal experience. Heaney's poetic voice is contingent on the materiality of place so that the preserving properties of the bog become emblems of poetic preservation; poetry as unforgettable, memorable, earthbound, grounded in the material world, and peopled with ancient bodies and artifacts as foils for people in the present.

The erotic birthing of "Grauballe Man" has already been discussed, but also "The Tollund Man" from 1972 (with its obvious adoption of Glob's thesis that bog bodies were sacrifices to a fertility goddess) shows quite creepily how the bog morphs into a devouring and "processing" womb:

> She tightened her torc on him
> and opened her fen,
> Those dark juices working
> Him to a saint's kept body,

Touch and the eroticism of touching are even more explicitly at center in Heaney's poem "Bog Queen" from *North* (1975), in which the female bog body is made to see herself as "braille / for the creeping influences." This image of the body-as-Braille, which has to submit to a hand that feels its way viscerally into reading—a hand that "sees" through "touching"—brings up associations with another famous hand-read body, namely the one in Johan Wolfgang von Goethe's fifth Roman elegy. Goethe wrote his *Roman Elegies* erotically beguiled while kneading marble bodies into writing: "Oft I have made poetry in her arms / my fingers softly tapping out hexameters / along her back" (*Oftmal hab ich auch schon in ihren Armen gedichtet / Und des Hexameters Maß leise mit fingernder Hand / Ihr auf den Rücken gezählt*).[57] As the lover-poet embraces the sculptural marble woman's body, the hexameters are used to tap rhythm and rhyme into the pulsation of lovemaking as poetry-making. Like Goethe's marble woman, Heaney's bog woman is seemingly a passive recipient of poetic touching. But in "Bog Queen," the point of view is shifted to the buried bog woman herself. Suspended between death and life, her voice is self-elegiac:

> I lay waiting
> between turf-face and demesne wall,
> between heathery levels
> and glass-toothed stone.[58]

The touch she experiences turns out to be not a human lover, but "hands" of nature. The sun has "groped" her and the "seeps of winter / digested" her. As she is Braille-read, her body is disfigured and partly dissolved by the very "hands" that have read her. The unseemly invasion by nature and time is made all the more invasive when we are told that the "hands" are "illiterate roots" which "pondered and died / in the cavings / of stomach and socket." Thus "misread," the bog queen lies waiting with her darkened brain "a jar of spawn / fermenting underground // dreams of Baltic amber." Clothed in the bog as in a "black glacier" and "knowing" (as in the biblical "knowing" between lovers) the cold winter "like the nuzzle of fjords / at my thighs—," she is finally released from her dormancy, only to be mutilated once more by the peat-cutter's spade, "barbered and stripped." She is robbed of her hair as the turf-cutter is bribed to give it to "a peer's wife": "The plait of my hair, / a slimy birth-cord / of bog, had been cut." So when at last she rises from the dark, she is reduced to "hacked bone" and "skull-ware."

"Step by step," as the American literary critic Helen Vendler puts it, "the buried woman is undone until she becomes a geologic rather than a human phenomenon."[59] Although Vendler's reading differs on some points from mine (she sees a debt to Sylvia Plath's "Lady Lazarus," while I suggest a tacit response by Heaney to the archaeological nomenclatures of marble erotics in the works of Goethe and others), her sharp observation of the relation between body and writings is worth quoting:

> As the bog queen describes her slow changes, she has the equanimity of the dead, and she reaches almost the unintelligibility of a script in a lost language: as the two-thousand-year-long disintegration is narrated, her equal and far more surprising underground resistance to disintegration is not mentioned. After all—despite the "creeping influences," the "darkening" and "fermenting" and "reducing" and "wrinkling" and "soak[ing]" and "fray[ing]"—the bog queen, once exhumed, is still unitary, recognizable, present. Heaney's even-handed attentions to brain and nails, pelvis and breasts, thighs and skull, hair and feet, "realize" the body entire, with the blazon fuller than the convention normally allows.[60]

An Irish body found in 1780 or 1781 supposedly inspired the bog queen poem, and in Vendler's reading this contextualization underscores the lack of sacrificial imagery. Her royal status and lost diadem are "the witness [of] that civilization of torcs and gemstones that Heaney had once rejected in favour of elk and butter as bog-treasure."[61] Clearly the source is ultimately an archetype, but let me offer an alternative context: is it not equally possible that the story of Queen Gunhild looms tacitly behind the royal association and offers an extra perspective and layer of context, perhaps even an explanation as to the royal title? Queen Gunhild must have been known by Heaney from Glob's description, as must Glob's mention of the misinterpretations by Danish literary scholars and authors in the 1840s. Just as they did, Heaney sees his bog queen as a casualty of politics and hostilities.[62]

Three other poems from *North* are explicitly erotic: "Strange Fruit," "Bone Dreams" and "Punishment." In "Strange Fruit" we are presented with the head of a bog girl as a visual display, as if on a fruit plate, as an offering:

> Here is the girl's head like an exhumed gourd.
> Oval-faced, prune-skinned, prune-stones for teeth.
> They unswaddled the wet fern of her hair
> And made an exhibition of its coil,[63]

"They" made a museum exhibition, but the poem, too, "exhibits" the head—indeed, offers it up for erotic consumption.[64] The eyes can feast on the "fruit," on her "leathery beauty," and hold an amorous affection for its soft fatty substance, a "pash of tallow" for the "perishable treasure." Fruit is to be eaten; and here the strange fruit of the bog girl's head is to be eaten by the eyes in museum display. The beheaded bog girl's own "eyeholes," however, are "blank as pools." She no longer has what could have been the petrifying (and seductive) Medusa gaze suggested by the coils of her hair.[65]

But this Medusa reference aside, Heaney's poem also shares its title, "Strange Fruit," with a song made famous by Billie Holiday in the early 1940s, a song that mourns the fate of the lynching (sacrifice) of black men in the southern United States. Here southern trees bear the strange fruits of terror with blood on their leaves and blood seeping deep into their roots.[66] In Holiday's mournful rendition and heated intonation, the song became a powerful testimony to the force of remembrance. It has been celebrated for its "genesis, impact, and continuing relevance," not only on behalf of the African-American victims of lynching, but also as a complex layering of other ethnic injustices.[67] The lyric ends: "Here is a strange and bitter crop."

Heaney's own strange and bitter crop may or may not have taken its cue from Abel Meerpol's lyric, but it is unlikely that the poet would *not* have been cognizant of the song and its genesis when he added Catholic victimhood to the roster of "strange fruits." The gendered "fruit" on his display platter, which ends with these lines: "Murdered, forgotten, nameless, terrible / Beheaded girl, outstaring axe / And beatification, outstaring / What had begun to feel like reverence"— problematizes such a Catholic "beatification." In the words of Vendler: "The bodies do not want to be beatified (religious language is inadequate to them), nor did they exist to be murdered (the language of violence is inadequate to them). What they claim now, and claimed in life, is what all human beings want: existence on the same terms as their fellows."[68] In the end, "Strange Fruit"—ripe with multilayered contexts and implied iconography and myths—presents itself as a closed form within the contour of a sonnet, as if it, like "an exhumed gourd," could be placed (and read/eaten) in the museum of writing.

"Bone Dreams" is not a bog body poem proper but an archaeological-erotic encounter in which lovemaking and ancient history are combined, and once again we observe a couched reference to Goethe's eroticizing of the marble woman. In Heaney's erudite sleight of hand (he would later translate *Beowulf*) the beloved is now a bone-woman as he calls upon "philology and kennings" to "re-enter memory / where the bone's lair / is a love-nest / in the grass," and where the narrator, like a latter-day Hamlet, can hold the skull of his "lady's head / like a crystal." Indeed, as the narrator touches and makes love to her he essentially "ossifies" himself:

> Soon my hands, on the sunken
> fosse of her spine,
> move towards the passes.

> V
> And we end up
> cradling each other
> between the lips
> of an earthwork.[69]

While the eroticism of "Bone Dreams" is innocuous (these are indeed "lovely bones"), "Punishment," also from *North,* is far more problematic. The object is the so-called Windeby Girl, now known as Windeby Boy. As was the case with Weerdinge Couple, recent DNA results have suggested that Windeby Girl is in fact male, a gender change that has cast erotic projections in a new light. Nevertheless, since the poems inspired by this bog body refer to Windeby Girl, I will stay here with the faulty gendering.

The approximately 14-year-old "girl" was found in a bog in northern Germany in 1952, naked and blindfolded with a band around her eyes, a leather collar around her neck, half of her hair shaved, and the rest of her hair cropped short.[70] The cranial remains show clear signs of the force involved in the scalping, which has been interpreted as punishment, perhaps for premarital transgressions.

Heaney positions the bog body in a transhistorical setting, evoking both the ritual execution of ancient women for adultery and the contemporary punishment of Catholic Irish women for dating English soldiers.[71] Here the bog girl is a "barked sapling / that is dug up / oak-bone, brain-firkin." The poem's narrator first aligns himself with the girl:

Fig. 4.3. Windeby "Girl" is the most eroticized of the bog bodies. Recent forensic evidence shows that "she" was a boy. Photograph © Stiftung Schleswig-Holsteinische Landesmuseen Schloß Gottorf, Germany.

> I can feel the tug
> of the halter at the nape
> of her neck, the wind
> on her naked front.[72]

Then he turns from empathy and the tactile to the visual, and makes us (with him) observers or voyeurs:

> It blows her nipples
> to amber beads,
> it shakes the frail rigging
> of her ribs.

The narrator can "see her drowned / body in the bog," observe her "shaved head," and note that "her blindfold" is "a soiled bandage." As his compassion for her brutal fate builds, he speaks directly to the sacrificed girl. "My poor scapegoat," he calls her in a tone of voice that mixes paternal endearments with erotic titillation. The noose with which she has been found is seen as a ring that stores "the memories of love."[73] But in line with Tacitus, these love memories are adulterous ("little adulterer"). With the shifted point of view, the narrator makes himself both complicit in and implicitly responsible for her misfortune, both as someone who might have "known" the girl

> before they punished you
>
> you were flaxen-haired,
> undernourished, and your
> tar-black face was beautiful,
> My poor scapegoat,

and as someone who observes and condones, even enjoys, the spectacle of her sacrifice as the scene turns to necrophilia:

> I almost love you
> but would have cast, I know,
> the stones of silence.
> I am the artful voyeur.

The sacrifice is a necessary one (French critic and philosopher René Girard's analysis of sacrifice as restoring order to society springs to mind), just as the punishment of her "betraying sisters, / cauled in tar" (a reference to contemporary Irish women dating soldiers from the oppression) is made understandable as "the exact / and tribal, intimate revenge."

Bad Bog Babes

Sacrifice is *not* an option in the American performance artist and poet Lori Anderson Moseman's 2003 poetry collection entitled *Persona*. In twelve pieces that range from concrete poetry and traditional prose poetry to hyperpoetry, with titles such as "Bad Bones," "Bog Girl On Belay," "Badland Babes," and "Bog Girl Goes Bowling," Anderson Moseman stages Heaney's "little adulterer" Windeby Girl as an unpretentious yet self-confident Bad Bog Babe who responds to the projections to which she and other bog bodies have been submitted. "Shinbones showed how starved she was," we read in "Bog Girl," and with a sense of audacious futurism the poet suggests that the malnourished bog body from the past be sent to "Biosphere II for food."

Anderson Moseman is concerned both with the materiality of the bog body and with its potential for posing questions about the subject. She is not glossing over the surreal, bizarre, or anachronistic in the remains. Yet while she holds onto artificially projected personas and a perceived sense of unreality, she also sees aspects of our humanity in these things. In the opening poem she fuses a contemporary girl called Hog Girl with Bog Girl, and casts her as "a new container for the infamous resurrected peat body."

> *Hog Girl's* favorite bag-o-bones in P.V. Glob's *The Bog People* is Windeby
> Girl [. . . .] This girl, *Hog Girl* thought, had to be more than the adulterer
> Seamus Heaney posits in "Punishment." Child labor could be the subject.
> *Or.* Pre-Christian fertility. *Or.* The way monopoly is played in the Iron Age:
> no plastic hotels for Park Place—just boulders and girls to bury.[74]

The references to Glob and Heaney continue in a number of her other bog poems, and her unsentimental approach can be directly compared to Heaney's pathos. Whereas Heaney saw moments of continuity (past-present) in the bog bodies, Anderson Moseman emphasizes discontinuity, the incongruous, and the absurd. While Heaney in "Punishment" tied the band around Windeby Girl's eyes to sexual fantasies, Anderson Moseman sees the same band ironically: "she started weaving a sprang band to bury as a cure for her husband's glaucoma" ("Mrs. Anderson Moseman Rereads *Mosefolket*"). Like Heaney, Anderson Moseman relates explicitly to Glob's *Mosefolket* (she deliberately uses the Danish title to play off her own last name), but unlike Heaney she cuts directly and irreverently into the archaeological account.

She does so in a number of ways, none more graphic than in the poem "Empirical Collar." Here she uses lines from both Glob's text and the Canadian poet Erin Mourés's *O Cidadan* (2003) and creates a typographically layered poem in which her own writing is supplemented by italicized cuts from Glob's text and bolded cuts from Mourés's.

> *Inside next to the skin* her *fur* scalped then
>   sewn to ox-hide (Nerthus needs extension in space)
>
> **(("this vertex of skin occupied the climate of order"))**
>
> *Inside next to the skin* no evidence of strangulation[75]

Glob's "*inside next to the skin*" is cut from his depiction in *The Bog People* of how the leather collar touches Windeby Girl's skin, and is then repeated and stitched together with Anderson Moseman's own observations about the bog body. Anderson Moseman mimics the archaeological world as image and structure, and allows the three different poetical voices to create a graphically recognizable layering which the reader is asked to dig through to find what at times turns out to be empty parentheses, as seen here after Glob's italicized words:

> *We must suppose. We must suppose*
>   ( )( )( ) (( )) (( )) (())
> *We must suppose* what we learn through thorough
>   examination becomes a double collar for us
>   binds us to the bog acids we interrupt[76]

The empty parentheses can be seen as an interpretative vacuum, or as an arabesque-like pattern mimicking the leather strap around the girl's neck. Interpretation, then, if indeed it is possible, is tied to the abstract visual image, where the parenthetic lines paradoxically and satirically expose their potential emptiness as they point to that which cannot be said or cited about the bog girl.

In the poem entitled "Eleven Lines on Windeby Girl's Shinbone," Anderson Moseman places the physical remains under a microscope.

> fall: oat harvest a hope against rye's ergot
> winter: rerun sun hides her daily vitamin
> again: winter sun hides her daily vitamin
> again: winter sun hides her daily vitamin
> again: winter sun hides her daily vitamin
> again: winter sun hides her daily vitamin
> again: winter sun hides her daily vitamin
> again winter: sun hides her daily vitamin
> spring: first time she's ever bled there
> summer: late light sprang weave easing
> fall: blame cock's spur for St. Anthony's fire[77]

The temporal and spatial arrangement of the poem, the suggestion of seasons, the rhythmic and tautological use of the adverb "again," and the marked horizontal and vertical lines mimic the so-called Harris-lines, named after the Welsh anatomist Henry Albert Harris (1886–1968), who discovered how opaque lines on radiographs of bones could pinpoint lack of vitamins and stunted growth in human remains. The words are compressed into a poetical slide, the artistic mimicry of an anatomic slice into the young girl's remains. In its visual-verbal configuration, the poem deliberately resembles a quadrangular X-ray of a body fragment in time and space whose story can be read in and as a scientific and poetical combination. Here, the ancient body is identified in minute detail and placed in a time travel scenario which finally lands it under the modern poet's microscope-pen.

In yet another poem with the title "Mrs. Moseman Rereads *Mosefolket,*" the poet's own last name, acquired through marriage (*moseman* in Danish means bog man) is employed with self-irony: "Having practiced both promiscuity and homosexuality without fatality, middle-aged newlywed Mrs. Anderson Moseman resumed her study of Bog Girl." And further: "The Mose Mrs. vowed to gather sufficient data to chart in ways her husband, the behavioral researcher, could respect." In front of the mirror she arranges her hair in a so-called Swabian knot, a hairstyle found on some male bog body remains, and which Tacitus comments on in his *Germania* as being characteristic of the freeborn from the north who used the coiffure not out of vanity but to look taller and fearless in the eyes of the enemy.[78] Mrs. Moseman, in her poetic impersonation of bog person or persona, ponders whether she should become "A bog beautician?" or a "forensic cosmetologist?" But in "Bad Bones" any kind of direct narcissistic mirroring in bog bodies is challenged:

> We crumble when names are etched upon us.
> We undermine pedestals, dissolve
> All plaster. We are fragile communicants:[79]

The plaster which crumbles is an implied reference to the archaeological face reconstructions which have taken place in the past few years and, by analogy, to other sorts of plastic surgery in contemporary body culture (Windeby Girl was the first bog body whose face has been reconstructed). But in Anderson Moseman's universe, and in the poem "Bad Bones," reconstruction is not simply attacked—"We are bad bones, and we hate our replicas"—but is also seen as part of poetical imagination and the verbal attempt to humanize the bodies.

> Literary fancy is
> simple. Liberation is multiple. Reconstructed
> from the grave: wax and wigs are guesses—
> attempts to humanize—as are words.[80]

Moseman sees the bog body as a "semiotic simian" but also as a mnemonic space, "Cell memory / making us mirrors" ("Badland Babes") which resist, almost like counter-memory, any fixed meaning or place. The use of such an open model of interpretation can be seen even more clearly in the cyber-poem "Door Where Carol Merrill Is Standing: Goat? Goat? Car?," an interactive poem in three rounds, each with three doors that can be unlocked to reveal textual or visual constructions and readings of the bog girl built around an American TV game show, *Let's Make a Deal,* which was broadcast during the 1960s, 70s, and 80s.[81] Here the host, Monty Hall, and the hostess, Carol Merrill (a former beauty queen), entertained viewers by allowing studio guests to pick from three doors in three rounds in order to win valuable goods, such as cars, or risk losing the prize by selecting a different door.

If you click on Anderson Moseman's cyber-poem door, you can read about the obedient Carol Merrill; known by a generation of Americans as the quintessentially smiling but mute TV game hostess whose gesture toward the doors (in Moseman's optic) becomes a spectacle of sacrifice, not of the bog girl but of the participating "victim" in the greedy play of consumerism. If you choose the wrong door and win a goat instead of a car you are, so to speak, sacrificed to the laughter of the audience and viewers. That's the name of the game. In that game, the bog girl challenges the slick host, Monty Hall, as the poem insinuates a carnality of sacrifice different than the one to which she was submitted two thousand years ago.

Poetic Autopsy

Sexuality, sacrifice, disgust, adultery, and homosexuality have become powerful tropes in the literary and poetic contexts that bog bodies dwell in. Add to that the inherent and complicated voyeurism at play in displaying dead people in the museum space, and we must, as seen in most of the examples above, make a direct confrontation with modernity. In the British poet Geoffrey Grigson's "Tollund Man," from 1969, this confrontation does not take form as the kind of "therapeutic anamnesis" Thomas Docherty located in Heaney's poem on Grauballe Man; there is no healing in Grigson's recalled past.[82] Rather, he articulates how modernity has subsumed the ancient body and claimed it as its own.

Grigson first casts his gaze toward the north from "Upper Norwood of neatness, where Pissarro / took refuge and painted the Crystal Palace." The bog man's features are not those of the "bearded old French-Jewish painter," but "simply a neighbour?" The evocation of the painter, whom he does *not* resemble, vaguely implies (at least in part) the fate they share in the museum world. The bog man is neither a painter nor a painting, but he is placed in a museum where the "heath has retreated, / bogs have been cut, suburbs extended." From the earth's "Goddess of Growing" he is "exhumed, then examined, / kept out of wet, black, in a case, / in a Jutland museum, focused—." Ripped from the earth's "belly" into modernity's suburban reality, he is submitted to the examination of pedagogues and tourists, made equivalent to other attractions, ogled and consumed:

> they tramp all the time up the suburban
> creak of the stairs—through huge-hooded
> lenses by teachers of physical culture (wives
>
> looking on, not at him) from violent
> Dallas, eaters of Natural Foods on a
> *Fairy Tale Tour of H-C-Andersen Land:*

American tourists feeding on "natural food" insensitively or distractedly invade the space of the bog man. But the (sensitive) poet in turn offers his hand (and pen) as support against this touristy carne-voyeuristic modernity, addressing the bog man in familiar terms and with a sense of shared defeatism:

> No consolation after millennia, my friend,
> sacrificed for a future, to stir in this way
> by your shiny, silky-black shape; your
>
> fluidity hardened, more natural than
> Lenin; your stomach post-mortem'd; head cut from
> your body; like ours, your finger-prints taken.

The comparison between Tollund Man's "shiny, silky-black shape" and Lenin's embalmed corpse serves to highlight the artificiality of the modern in contrast to the naturally hardened fluidity of the ancient body. Consumed by "natural" food eaters (those distasteful tourists!), Tollund Man's own naturalness has ironically placed him on par with the inauthenticity of the modern. He has been dismembered (the head removed from the body in display, as we have already seen) and his stomach has been "post-mortem'd."

Grigson's poetic autopsy dovetails in this respect with the work of many other poets and prose writers who all, in one way or another, use the bog body's stomach content as the innermost material proof for the relevance of a trope for time which has the interior corporeal at its center—literally inside the center of the body, in the guts and digestive channels (and behind all is Glob's description of the last meal). In Grigson's poem the bog man is

[...] fed by rite
on dry seeds, flung in sodden-weed winter
to grow in black earth, a gift

for the Goddess of Growing grasping
her belly, for her being in spring,
gaily-served—or they say so—great exacting

Mama of Increase;

Compare this with William Carlos Williams's "A Smiling Dane," from 1955, in which "the cast of [Tollund Man's] features / shows him / to be / a man of intelligence."[83] The poem makes use of information from press releases and the rudimentary interpretation of the bog man's death shortly after his discovery, but Williams also points to the anachronism in the presence of such a lifelike figure from a pre-Christian era:

The Danish native
        before the Christian era
            whose body
features intact
        with a rope
            also intact
round the neck
        found recently
            in a peat bog
is dead.
        Are you surprised?
            You should be.

To be surprised over a pre-Christian-era man's death is certainly somewhat surprising. But the point Williams's poem makes is that the intactness of Tollund Man's well-preserved face makes the confrontation between the ancient man and his various latter-day beholders an exercise in *anamnesis,* "presence of the past." In his two-thousand-year-old face, time has been challenged— brought to a standstill. But more telling, as a temporal image, is when Williams matter-of-factly points to how food, indeed the very concept of the *last* meal, has not only material resonance but temporal importance: "His stomach . . . its contents examined / shows him / before he died / to have had . . . a meal / consisting of local grains / swallowed whole / which he probably enjoyed / though he did not / much as we do / chew them." Food changes over time in the stomach and is the key forensic "clock" in most autopsies; a tell-tale of the time of death from *within* the body. The clock stops and "freezes time" when the digestive juices are brought to a standstill. The rather unpoetical matter of half-digested stomach content becomes the ultimate trope for the passing and the stopping of human time.[84]

Williams cunningly moves our attention from the diggers who found Tollund Man ("Frightened / they quit the place / thinking / his ghost might walk") and onto his executioners, whose terrified features are recorded on his face as if by photo-magic:

> And what if
> the image of his frightened executioners
> is not recorded?
> Do we not know
> their features
> as if
> it had occurred
> today?
> We can still see in his smile
> their grimaces.

The burden of his death is finally moved onto us, the readers of the poem and beholders of the bog man: "Are you surprised? / You should be" at the unexpected and sudden return from the dead. The juxtaposition of *his* smile with *their* grimaces offers a reversal of fortune even if, as the opening line proposes, the cast of his intelligent features "did him no good." "The Danish Native" has been positioned vis-à-vis frightened and surprised beholders (who "expected more"), and brings with him the most rudimentary of human knowledge. "What his eyes saw," Williams speculates, "cannot be more / than the male / and female / of it—/ if as much."

The male and the female of it, instead of being a reference to the bog body fertility sacrifice theory (a more common use of this reference), has been taken out of its mythical context and made laconically brief. Human relations, life and death, victim and executioner, are condensed to a common denominator of male and female—"if as much." Williams's poem has a slightly elegiac tone, not one that mourns the death of *this* man, but rather one that grieves for the poverty of human relations. In the end, the smile from the bog man becomes an overbearing smile from a man who knew (and knows) that when it comes down to it, the confrontation between his own corpse and his various latter-day confronters is to be found in the simplicity of such essential dichotomies.

In conclusion, Heaney's "artful voyeur[s]," Kantaris's "sad museum ... adulterers," Moseman Andersen's "fragile communicants [on] pedestals," Grigson's "shiny, silky-black" display objects, and Williams's surprised onlookers viewing "the cast of [Tollund man's] features" pull us from the realm of the poetic into the representational challenges of visual media. It is time to change our focus from the world of bog body writing to that of bog body art.

CHAPTER 5

Bog Body Art

In 1971 the German artist Joseph Beuys performed a ritual that included running into a bog, wrapping himself in its mud, and all but disappearing into its darkness. At one point, with his signature hat floating on the surface of the bog lake, only his hands were visible to the camera that recorded the performance. This so-called bog action (Eine Aktion im Moor) would later be described by eco-theorists as an *ecovention* (ecology plus invention).[1] Beuys's aim was to call attention to the precariousness of bogs and the dangers of drying them out; they were, he claimed, "the liveliest elements in the European landscape, not just from the point of view of flora, fauna, birds and animals, but as *storing places of life, mystery and chemical change, preservers of ancient history.*"[2]

More remarkable for the present study, however, is the fact that two decades earlier—in the very year of its discovery, 1952—Beuys had already been deeply engaged with Grauballe Man and had thus become one of the first, if not *the* first, to see the potential for visual artistic expression in bog bodies.[3] Although bog mummies were known and used in literature and poetry, as has been shown in the previous chapters, they had not found their way into works of visual art (with one exception, a now lost life-size oil painting from 1835 of Queen Gunhild by the amateur archaeologist A.F.A. Lassen—most likely the product of his archaeological curiosity and not artistic aspiration) until international publicity, generated by photographs of Tollund Man and Grauballe Man, sent currents of interest across artistic communities.[4] Only days after Grauballe Man's discovery, while he was still being watered every few hours to avoid dehydration, the Associated Press distributed accounts of his first preliminary public exhibit in the Prehistoric Museum in Aarhus. From Boston to Kyoto, people could read about the astonishing discovery and reflect upon photographs of the body in situ (see fig. 1.1). The artistic potential was clear. Although far from a classical sculpture in a traditional sense, Grauballe Man *did* have "sculptural" qualities in his own right. Resembling a prehistoric Michelangelo sculpture trying to escape its marble prison, his upper torso struggled to emerge from the tight embrace of the peat grave. More importantly, he and his "cousin," Tollund Man, arrived in the public eye at a time when art had long since reclaimed, in the words of Peruvian writer Mario Vargas Llosa, "everything in human experience that artistic representation had previously rejected"—indeed, at a time, post-World War II, when a heightened interest amongst artists in finding ways to deal with wounds and trauma had become ever more pronounced.[5]

Beuys' *Grauballe Man* only vaguely resembles the human figure from which it takes its title. The title, Beuys claimed, was the only "direct reference to the event" of the bog body discovery. Rather, the installation is to be seen as a general comment on objects that survive through time: "an eternal thing called Field Character, the magnetic energy field which mobilizes human powers in a physical and transcendental sense."[6] It consists of a wooden casket with a series of coiled copper rings poised between abstract and concrete material expressions, and it seems to fuse and compress the perceived corporeality of the artifact together with the physicality of the bog body. Here, Beuys informs us, the "power field is interpreted as the upper, middle, and lower regions of the human body expressing thinking, feeling, and will powers, all the things that later appear in my theory and actions."[7] "Copper," he reflects: "is used to suggest the quick conducting potential of the human body as an antenna or transmitter."[8]

Although it is confined in its casket-like entrapment, Beuys' bog body is meant to articulate a sense of dynamic proximity with earth. That the remains are "earthed in asphalt" is not to be read as an image of stasis but as an expression of human evolution, of humans in process: "It is important that we do not have too abstract an idea of human evolution or an understanding that is restricted to positivistic and materialist science," Beuys argues. "Evolution is a dynamic anthropological and morphological biography-biology that needs

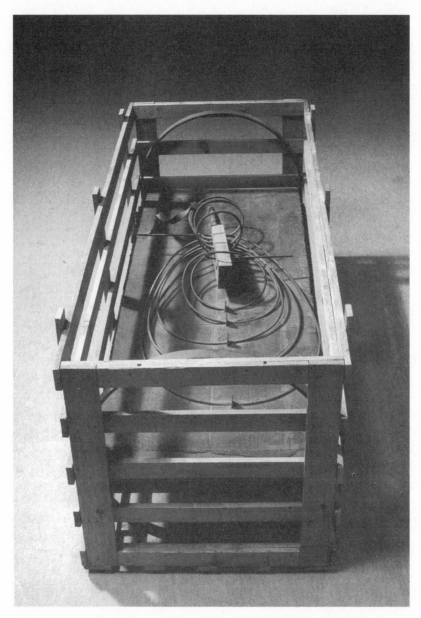

Fig. 5.1. In 1952 Joseph Beuys built an installation called *Grauballe Man,* in which the bog body is represented by copper circles and is "earthed to the asphalt surface" by the metal associated with the age in which the bog man lived: iron. Photograph © 2008 Artists Rights Society (ARS), New York / VG Bild-Kunst, Bonn.

dynamic images to express it."[9] His view on materiality and the animal world is further expressed in this reflection on death:

> Yes, and death—death is quite a complicated thing. The purpose of West-
> ern thinking and the science that grew from it was to reach material, but
> one only does that through death. If you take the brain as being the material
> basis of thought, as hard and glossy as a mirror, then it becomes clear that
> thinking can only be fulfilled through death, and that a higher level exists for
> it through the liberation of death: a new life for thinking.[10]

If death is material and material is the conduit for thought (an example of Beuys's penchant for poetic reversals), then the dead bog man can certainly work within the Beuysian logic as an agent for liberation of thought! In the forms of copper circles (copper's properties speak volumes about the material and abstract meanings of being a "medium"), the bog man has taken on a po-litical and philosophical dimension within Beuys's dynamic image of evolu-tion, and also as part of what he called an "anthropological and morphological biography-biology."

The "wound" is a central theme in Beuys's work, and his early drawings show how he uses circular forms as symbols of healing. In his *Grauballe Man,* the centric circles of copper articulate this therapeutic power directly from *within* the wounded body itself. Since circular forms embody a higher, un-chaotic realm for Beuys, the trauma of the past represented in the wounded *and* coiled body is an expression of the curative process of art. These healing processes are not distinguishable from the importance that Beuys places on physical (and corporeal) *matter.* In fact, the bog's particular property of pres-ervation and Beuys' interest in mythic time (where past and present are not consecutive layers-in-time but resemble what art editor Christopher Lyon has called "a kind of permanent present tense") can be seen in an installation called *Irish Energies,* from 1974, in which peat briquettes, Irish butter, and Welsh coal are layered to form sandwiches. In another piece, *Untitled,* from 1975, fragments of peat encapsulate a fossilized leaf and toenails—seemingly a "found object"—while yet another untitled piece echoes the same motif via use of felt, fat, and toenails.

The energies in the various materials are understood dynamically as "the cy-clical movement between life and death, between 'warm' intuition and 'cold' rationality, between the inorganic, crystalline world and the 'warm' organic world of living things."[11] The ability to turn warm fluids to cold solids in ma-terials like fat, honey, and blood has, of course, metaphysical connotations for Beuys. Likewise, the use of animal images as analogies to humans (such as his use of the stag, elk, swan, and hare—all "figures which pass freely from one level of existence to another, which represent the incarnation of the soul or the earthly form of spiritual beings with access to other regions"[12]) and the chromatic scales, gray and corrosive browns, are part of what he sees as anti-images and a way of implying the spiritual. From this also originates his concept of social sculptures that represent "how we mould and shape the world in which we live," where sculpture is "an evolutionary process" and where "everyone [is] an artist."[13]

In short, human evolution in Beuys's Rudolf-Steiner-inspired anthroposophist optic evolves from the age of myth (past) to the age of analysis (present) in which materialism and rationalism rule and dialectically convene in a utopian sun state (future) in which the warmth of creativity and humanity (without national borders) allows everyone freedom to exercise individual creativity (such as Beuys's social sculpture). The myth of prehistoric man—which Beuys emblematizes in *Grauballe Man*—and the channeling of (his) energies by way of the conduits of copper, iron, and concrete partakes in this evolutionary process. The utopian possibility is materially *and* abstractly coiled into the form of a body already restored, or in process of healing, from the "wounds" of trauma.

As his biographer Caroline Tisdall has observed, Beuys evoked a shamanic aspect in his work; his 1957 drawing *Girl Astronaut,* for example, indicates "a special kind of future [moonwalking]," in which "the figure with her sharply reduced limbs strains upwards from a stretcher-like base similar to the one that supports the *Grauballe Man.*"[14] The bog body, then, seems to perfectly fit Beuys's use of artifacts as having both physical and metaphorical resonances; they exist "in a metaphoric field—on a continuum of fluid connections and associations from which metaphors emerge and radiate."[15] When he conceptualizes the "passage of things" from something raw and unstructured to something processed and molded (as in the bog body interpretation) he works within a semantic attentiveness where creative forces and social practices are intertwined. In this extended sense of creative thinking and practice, the materials used in Beuys's various sculptures, multiples, and installations are carefully selected for their potential as personal, universal, and therapeutic conduits which are meant to share and disseminate ideas concerning inter-humanity.

The material properties of the conduits are used in turn as concrete metaphors meant to articulate visual transmissions between temporal and spatial realms. Within such logic it would seem rather obvious that a bog body such as Grauballe Man should interest Beuys, since an abundance of elements for conduits is to be found here. The bog body's skin and hair, infiltrated with other bog matter, is not far from Beuys's favored materials: earth, felt, fat, blood, and dead animal bodies. In fact, his customary use of found objects resonates in this case with the found-ness of the bog man and vice versa. Likewise, the echo of human suffering or sacrifice in early times like the Iron Age similarly corresponds with Beuys's engagement with and articulation of social and historical responsibility after his own traumatic experiences in World War II. Art is not intended to *reflect* the world, but to *remake* it.[16] Thus, the recovery and remembrance of a traumatic past in the novels of Michel Tournier and Anne Michaels, to name but two literary examples used in this study, were already forecast visually in Beuys's work on Grauballe Man.

Kneading Hands

While Beuys (and by association the Fluxus movement) used Grauballe Man to press the onlookers to "face the reality of their own historical experiences,"[17] Tollund Man found a similarly important place in the works of a Cobra-influenced artist, the Flemish photographer, painter, and ceramist Serge Vandercam. Cobra (an acronym made up of Copenhagen, Brussels, and Amsterdam), founded in 1948, was formed by a group of abstract expressionists as a reaction to World War II atrocities. The Cobra artists subscribed to spontaneity and communal experiments and were inspired by primitive and ancient art forms along with an interest in folklore; they saw themselves as being in opposition to the dominant idiom of geometric abstraction. One of the most prolific members, the Danish painter Asger Jorn, was particularly inspired by archaeology and from 1949 well into the 1950s he worked closely with P.V. Glob (who had a fervent interest in visual art) on a large-scale twenty-eight-volume project on ancient Danish art from the Stone Age and Bronze Age. Glob and Jorn shared a passion for ancient art symbols found in the deep Danish past, and their grand interdisciplinary project was conceived to present the general public with ancient Danish art forms.[18] After a series of conflicts, the painter and the archaeologist's relationship eventually soured; their planned collaborative project never materialized, but fell prey to what looks to have been a rather conventional clash between conflicting perceptions of archaeology as science (Glob) or art (Jorn).[19]

Jorn's interest in Glob's Tollund Man would, however, lead to a number of visual art works, not by Jorn himself but instigated by him. In 1962 through 1964, and later again in the early 1970s, Jorn encouraged Serge Vandercam to use Tollund Man as a motif; Vandercam would in time produce paintings, sketches, and ceramics named after Tollund Man and articulated in the vernacular of the Cobraists.[20] Yet even before Jorn introduced him to Tollund Man,

Vandercam had been interested in the bog as a liminal and spiritual space. Together with the Belgian poet and painter Christian Dotremont, with whom Jorn had also collaborated, Vandercam had made a number of works inspired by the bog, called *Boues* (dirt), in which pieces of clay were ripped into rough-edged fragments and inscribed with words.[21]

The Cobra group's interest in earth, depth, and the historical past was steeped in the influence of the French philosopher of science Gaston Bachelard's theory of the elements; in fact, Jorn painted a portrait of the philosopher in 1960. Bachelard's thoughts on materiality, alongside Marx's theory on materialism, resonated with Cobra's aims. His *Earth and the Reveries of Will*, in particular, served as the theoretical backdrop for the group's artists, as did—albeit to a lesser degree—his earlier *Water and Dreams: An Essay on the Imagination of Matter*. Bachelard's idea was that the elements (air, fire, water, and earth) pressed upon the human imagination not by mapping themselves onto powerless material, as if to make their mark on a blank page, a tabula rasa, but rather by allowing that material to facilitate the making of images and to permit the imagination to reshape it. The human hand therefore became an important agent in the dialectic between soft and hard governing images of terrestrial matter. In fact, the hand engaged the resistance of the material: "Material imagination tends to see mud as the original substance, *la prima materies*."[22] Bachelard celebrated the way earth gives substance to slowness and joy to hands "that knead [and] gently [work] the sluggish elasticity of matter until the moment of discovering the extraordinary sensation of gumminess, the secret pleasure of matter's tiny, connecting threads."[23] Terrestrial imagination (as a variation of material imagination), Bachelard proposed, experiences the *passing* of time materially: "It is able to follow the slow, notorious intimacy of the passage from liquid matter to thickening matter to matter which, solidified, bears the whole of its past within."[24] This resistance from the earth and the gratification of slowness it brings seem perfectly fitted to describe the bog bodies' unhurried journey through two thousand years of time. When René Girard later protested in his 1972 *Violence and the Sacred* that Bachelard's concept of material metaphor was but "a poetic recreation of little real consequence," he seemed to be wide of the mark when it came to the impact Bachelard's terrestrial imagination had on the Cobra movement and also on Vandercam's bog body art.[25]

In Vandercam's oil painting of Tollund Man from 1964, two large dark eyes dominate and stare sideways out of the frame, as if caught in a moment of petrifying angst. Like other Cobra artists (including Jorn, Constant and Karel Appel), Vandercam was drawn to the subconscious—not as a cool analytical exercise, as was the case with much surrealist work, but with broad and loose

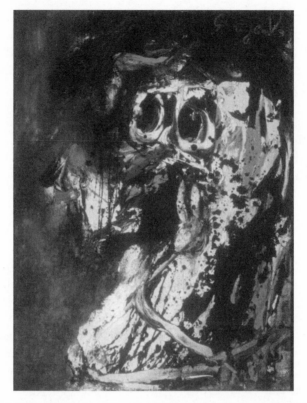

Fig. 5.2. Serge Vandercam depicts *Tollund Man* (1964) as the embodiment of an anxious scream. The palette is thick brown and black with red, yellow, and cream added to create a sharp contrast with the earth colors. Photograph © 2008 Artists Rights Society (ARS), New York/ SABAM, Brussels.

expressionistic strokes. His Tollund Man painting is informed by this subconscious expressionism and it showcases the trademark Cobra image: a fantastic being with round owl-like eyes. Yet compared to most other Cobra figures, which are childlike and naïve, Vandercam's *Tollund Man* appears spectre-like and ominous. There is no clear mimetic resemblance to the bog body in the painting, but only a transposition from an identifiable human shape back into an archaic and decomposed form.

This gloomy aspect is also present in a group of serigraphs in which Tollund Man's sacrifice is placed in a realm where he seems suspended between Iron Age mythology and Christian iconography. The cover of Vandercam's serigraphic portfolio (with a photograph of Tollund Man's head on the back

Figs. 5.3 and 5.4. In Serge Vandercam's *Tollund Man* serigraphs (1963), black and gray strokes are brushed over with red. The red suggests blood on the body dangling from the noose and on the soaked earth, and it contrasts with blue "earth arms" that extend from below. Photographs © 2008 Artists Rights Society (ARS), New York/ SABAM, Brussels.

cover), shows a peat-brown base on which random red strokes lead to a small wide-eyed creature in the middle, and blue strokes spell his name in uneven and broken orthography. In one of the seven serigraphs, the bog man is seen reaching from the earth into the air as a pair of intimidating eyes (a heavenly father!) look down. In another, he is lifted by two giant earth-arms (resembling waves) that seem to execute the man by hanging him while at the same time forming a protective parenthesis around him.

In yet another serigraph, the bog man sits solemnly on what looks like a stick-bench next to a giant cross. In two others, crosses are grafted directly onto his face, the sign of crucifixion (formed by eyes or eyebrows and nose) suggesting a sacrificial death that conflates him with Christ. And in yet another image, the wide-eyed Cobra-signature owl looks from above as an ill-omened representation of a powerful master, with a red "umbilical" cord connected to the still earth-encased bog body.

Vandercam's *Tollund Man* series inspired the Flemish multi-artist and poet Hugo Claus to write a long poem, "The Man from Tollund" (1963), which was included as part of Vandercam's exhibit catalogue.[26] Claus starts:

Figs. 5.5–5.7. Serge
Vandercam's *Tollund Man*
serigraphs link Tollund
Man with Christ by placing
crosses around him or
directly on his countenance.
Photographs © 2008 Artists
Rights Society (ARS), New
York/ SABAM, Brussels.

> As a relative we seldom see in the family
> and sometimes sits in the corner of the room,
> a bitter king full of schizophrenic silence,
> he does not sleep
> but rests in silence.

The "bitter king"—i.e., Tollund Man/Christ—is here not only a victim who has been "strangled with a leather cord / and dropped in his property: the earth, and / in a time of ice and iron"; he is also the victim of present-day voyeurism: "The parasites are we, / with our desiring eye"—a parasitic gaze that feeds on the body in the museum where encasement is evidently less protective against the corrosive powers of the environment (the curious gazes) than was the case in the original bog setting.

The perspective and voice shift between those of the bog man and those of the onlooker/poet, and are interspersed with the pronouncements of a godlike creator who is also a rapist, as we learn from "his" (sic!) comments, which are placed in parentheses and italicized:

> (*'When I sat on you, powerful,*
> *I found the world fabulous*
> *Until I glowed from sorrow*
> *for the drowning of things.*
> *You screamed as a dog, a doe and a goat,*
> *when I made you a son.*
> *A dagger of birch wood was I,*
> *in your skin, that endless marsh.'*)

The bog once again is a womb, now "raped" into a space of horror (with rot and blood and gas and black thistles) from which the bog man (himself perhaps a rapist) must ejaculate his way out:

> Black as soot and a hole in it,
> I cannot breathe, not move, I won't nor can,
> unless I ejaculate on the ground and my bones,
> tiresome put together,
> released escape from the deep dark that swells
> as I gasp for light gas, ammonia and manure.

While Hugo Claus captures the complexity, and sometimes absurdity, of the voice from the bog (transposed back to primary matter, "blood," "snot," and "seed"), it is "*a man from clay / that speaks*"—and clay, as we know from Bachelard, offers precisely the kind of combination and imagination that is caught up in ambivalences such as those in the "combination" of Claus's poem and Vandercam's visual art; in fact Bachelard stresses this point thus: "Material imagination needs the idea of *combination*."[27] Water and earth combined provide what he calls *la pâte,* the basic component of materiality. *La pâte,* paste, or "dough" can be kneaded and modeled by the hand, and is "the point of departure for any description of the real and experienced relationship between formal and material causes."[28]

The poetics of paste, with its fluidity and pliability, becomes an important facet in the understanding of the "creative unconscious."[29] Bachelard is not arguing for a metaphoric understanding of the elements and their importance for the imagination, but is asserting what he calls a "direct poetic reality." His examples are drawn mostly from literary and poetic texts but it is easy to see how visual artists, like those of the Cobra movement, would find his ideas fruitful—particularly with respect to the tactile kneading of clay into ceramics. The tangible sensation of digging into matter, the pleasure of engaging with "intrinsic *solidity*" and the joy of the "malleability of whatever matter is to be subdued" is tied by Bachelard to a desire and need to allow the hand to assist the eye in melting the solid earth into fluidity, and into dreams and poetry. If, as Bachelard says, "poetry is to reanimate the powers of creation in the soul," then we "must come to understand that the hand as well as the eye has its reveries and poetry. We must discover the poetry of the touch, the poetry in kneading hands."[30] Vandercam heeds this call concretely when, as seen below, he forms "poetry" in the shape of a vase.

In Vandercam's ceramic vision, Tollund Man is made of petrified mud (*la pâte*) and emerges through a ripped opening in the cylindrical vase's smooth, earth-gold exterior. The play with interior and exterior—the raw inside and culturally inscribed outside—is further stressed by the incisions and lines of a chisel on the surface of the vase, which at times take the form of loosely scribbled writing. The three-dimensionality gives volume and suggests the possibility that the vase is a funerary urn holding the bog man inside. The hand of the artist has engaged the malleability of matter and kneaded earth and water, *la pâte,* in a vessel that holds the contorted visage of Tollund Man and forms his vanished body in clay. The raw surface allows us to sense the artist's active hands; these are not passive hands, which according to Bachelard imagine nothing. We see how the material or terrestrial imagination bears a resemblance to what Bachelard calls the act of taking "sticky substances and *impos*[*ing*] a hard future upon them according to a calculated timetable of progress. In fact hands think only as they squeeze, only while kneading, only while in action."[31]

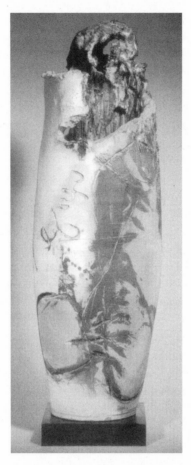

Fig. 5.8. In Serge Vandercam's massive ceramic vase *Tollund Man* (1962) is pressed out of the vessel as a raw and jagged remnant from the deep archaic past, surrounded by a smooth gold surface inscribed with images and signs. Photograph © 2008 Artists Rights Society (ARS), New York/ SABAM, Brussels.

The Touch of You

As I have shown so far, a direct correlation between physical matter and the human (artist's) hand plays an important part in the making of bog body art. The art historian Aloïs Riegl has a similar observation: "The human hand fashions works from lifeless matter according to the same formal principles as nature does. All human art production [Kunstschaffen] is therefore at heart nothing other than a contest [Wettshaffen] with nature."[32] Riegl's point is well taken, although the relationships between art and nature in most of the bog body art examples I have located appear to be less of an adversarial contest and more of a utopian combination. This can be seen very clearly in my next example: the Canadian artist Kathleen Vaughan's works called *Bog Series* and *Bog Fragments* (1995–96).[33] Here we find the privileging of the artist's hand, the thematic of the healing of wounds (in the vein of Beuys's enterprise) and an attention to the palpable physicality of bog bodies and body parts. Vaughan's expressed objective is to highlight a sense of shared human physicality with the past. She writes: "We are all united by the sentience of our common human flesh."[34] Her use of organic textures, linen, and wax and her repeated evocation of human skin are meant to imply that touch and tactility are privileged over vision and aesthetic distance. In fact the onlooker is asked to abandon vision in favor of touch, to engage in a "full-contact kinesthetic relationship with the piece."[35]

Again Glob's bog/book works as source. Vaughan tells us that "*The Bog Series* came into being as the result of a happy accident. One night in the library, my glance strayed to a book called *The Bog People* by Danish archeologist P.V. Glob. Who couldn't be intrigued?" She finds the photos "remarkable, beautiful, extraordinarily moving." She "was hooked, and began a substantial research project to find out who these people were [. . .] and how they came to be in bogs."[36] Her working method, built on what she calls "physiological aesthetics," has several archaeological features. In a mimicry of archaeological layering she mixes wax, paint, and cloth to form "organic flesh analogous to our own," and to introduce several types of reality she makes use of life-sized photo images of the bog figures, as well as full-scale drawings from models. Vaughan writes, "I hope these works will lead the viewer to muse upon the frailty of human life, the pleasures and terrors of being inside the bog, and the mysteries of living and dying within human communities so different from our own."[37]

A "poetics of touch" is clearly articulated in Vaughan's three-dimensional sculptured book called *The Touch of You.* The piece folds by way of hinges into four red leaves that allow the spectator to see both inside and outside. It features an X-ray of a bog body hand made visible on both sides, while a photographic image of Grauballe Man's clenched hand (seen from within) appears inside the book; on the opposite side of the same leaf, the artist's own hand (seen from the outside and spread open) is molded into the surface—"redhanded."[38]

The Touch of You is a way to make prehistory visible and to inscribe the artist's own personal history into the provenance of the bog body—as hands meeting hands across time. Vaughan intentionally stresses the double meaning of the word redhanded: as metaphor (to catch someone redhanded—in the act of touching?) and as the color flooding the book-sculpture. The thick

Fig. 5.9. In *The Touch of You* (1995–96), Kathleen Vaughan uses a variety of materials—oil, acrylic on encaustic paint, photographic emulsion, acrylic casting, and xerography on acetate,—on canvas to create a thick relationship between the bog body and its articulation. Photograph © Kathleen Vaughan.

red paint is equally symbolic of blood and of healing, and Vaughan deliberately seeks out a chromatic scale of muted colors, ("earth/copper/iron-based colours to evoke the feeling of murky bog water, [and] of human flesh") as well as a bright blood-red hue to evoke "wounding and blood-letting" and "redemption through sacrifice." The fusing of hands and the suggested touch across time is palimpsestic in nature. There is permeation at play, almost an osmosis between historical and personal ages that makes it impossible to pry apart long-established dichotomies between subject and object. Melted into each other as two sides of the same coin, the hands—meant to be read *and* seen *and* touched as one—not only are merged by material and imagination but also, as implied in the title, create a relation between subject and object that is emotional and erotic. The mirror effect, the doubling of the hands, makes touching both the touch of the other and the touch of the self. The book that opens also (potentially) closes upon itself and tells a story of erotic and autoerotic encounters.[39]

The centerpiece of Vaughan's bog body works is a large double canvas entitled *Bog Series 3*. Frozen and shriveled in his fetal position, Tollund Man is placed in the upper part of the right canvas, embedded in a field of thickly layered red paint. Some of the paint spills in over his body, but the rope around his neck is untouched and remains monochromatically grayish. On the left-hand side of the canvas a life-size drawing of a man in the same chromatic scale flaunts a classically beautiful sculpture-like body: toned, youthful, and vibrant. He looks as if he is floating in the field of red, sleeping sensually behind closed and dreamy eyes.

While Tollund Man's body is contorted and closes itself off from the onlooker's gaze, the other body is posed in a position meant to "echo works by Michelangelo and Dürer." His back is arched and his arms fall on either side of his torso in a relaxed gesture. The two parallel bodies deliberately reflect a play between past and present. As the viewer looks back and forth, a shift of past and present tense (he "is," he "was") springs into view; the present is seen on the left in the form of Tollund Man as he looks now, while the sculpture-youth on the right emblematizes his possible past.. Before and after, now and then, the two bodies are located within two archaeological/artistic traditions and trajectories: one made from photography (and linked to Nordic prehistory)

Fig. 5.10. On the left of Kathleen Vaughan's double canvas *Bog Series 3* (1995–96) we recognize Tollund Man in a life-size reproduction of the photography from Glob's book; it is paired with a sketch of a conventional classical body on the right. Photograph © Kathleen Vaughan.

and one drawn from Renaissance painting (and linked to the Greco-Roman tradition of the sculptural body). Again, they are connected within one field of thick red paint; thus two kinds of bodies from two different fields of archaeological-iconographical traditions are tentatively "fused" into one.

But violence and human sacrifice are also articulated in a three-dimensional rope that mimics the one in the photograph of Tollund Man. The rope, saturated in red, eerily fixes the feet as it cuts across the ankles as if to sever them from the rest of the body. In fact, another piece in the series (*Bog Fragment 2*) shows a foot, which allows us to imagine that the perfectly contoured youth beside Tollund Man is in fact fragile and in danger of fragmentation—ready to be sacrificed. As was the case with Beuys, Vaughan's work shows how trauma and wounds can be healed through the visceral experience of art. Visual art, she says, is breathed in through the body, so that the close relationship we imagine we have with the photographed bog bodies sealed in their "moist verisimilitude" can serve to heal the rupture imposed on us by technical manipulation and the bombardment of the senses in modernity and postmodernity.[40] Healing by way of "breathing life" into art is, in Vaughan's view, an overcoming of the Cartesian split: "Thus, artistic practice would be sensitizing to aesthetic perception, the reverse of the necessary self-anaesthesis proposed by Benjamin. Indeed, over time artistic practice would restore 'perceptibility' and promote aesthetic knowing—certainly a healing."[41] Finally, Vaughan's own reflections place focus on neuro-philosophical memory theory in which, as she cites the words of artist and writer Robert Emmet Mueller, the experience of art is "an event of consciousness that we do not want to forget; the intention of great art is to be memorable and remembered forever, so that past life will be mnemonical also."[42]

Once again, photography as source material is used as a deliberate reference to the fact that Tollund Man's physical presence has been preserved photographically while the body itself has partly disintegrated. "I find it highly appropriate," Vaughan writes, "that the mid-1950s photo of one Danish man, abducted from his 2000-year-old gravesite to be photographed and then disintegrate, should be part of a mid-1990s mixed media work about ritual of death and the frailty of the body."[43] She is cognizant of photography as a kind of "thanatography," a death-drawing connected to "fetish, memory, and past presences."[44] But at the same time the life-giving aspect of the photographic process is stressed. With calculated care, Vaughan uses the bog's paradoxical power of preservation/destruction and its similarity with photography: "I've re-created these photographic images on canvas using Liquid Light (black and white photographic emulsion suspended in gelatin). With poor surface adherence and unreliable light sensitivity, this medium is as quixotic and unpredictable as bog preservation, in itself entirely dependent upon the pH balance of the water. The partial readability of Liquid Light images corresponds to the fragmentary knowledge we can re-create of these individuals and their culture."[45] The presumed truth value of photography is undercut by the deliberate manipulation of its color and texture so that different realities and different time periods (Nordic prehistory and Italian Renaissance) are allowed to mesh. Vaughan's interest in theories of science also becomes evident in her use of magnetic resonance imaging (MRI) as inspiration. The visual penetration offered by modern medicine demonstrates the artistic intent of wringing visual resonance out of the physical remains of bog bodies and the "complicated images of internal soft tissues, identifying how organs have shifted as the body condenses under the weight of the peat"; the "electron spin resonance" that is "applied to the stomach's remains to ascertain the content and its status (raw and cooked)" gives the opaque bog body a physical transparency which corresponds to a metaphysical inside story of inter-humanity. Onlookers can, as Vaughan muses, find their own "inner bog man."

Paleo-Modernity

Trauma and wounds, remembrance and restoration take a very different shape for the Dutch sculptor Désirée Tonnaer. In her large 1993 monument entitled *Vastelaovesmonument,* modeled on the Weerdinge Couple, she poses a distinctive set of questions about the relationship between the archaic and the modern.

The sheer size of the sculpture and its combination of prehistory with modern life is linked to the sculptor's interest in archaeological material—also manifested in some of her other art pieces, which combine fossils of dinosaurs with religious relics. The use of masks to both personify and hide the identity of the faces in the Weerdinge Couple monument is particularly intriguing in that the original archaeological remains lack faces (see fig. 4.2). The faces are reinscribed and reestablished in Tonnaer's work by reference both to the bodies' rather morbid predicament and to the artist's imaginary and imaginative rehumanization of the couple.

The sculpture is directly derived from the remains of the two deceased humans (as a kind of *prima materia*) who have survived only as shadow-like images.[46] In the artist's vision, they have been filled out as larger-than-life townspeople among other still living townspeople. Although they have been made mimetically recognizable through the use of a flesh-like material, the sculpture, in its three-dimensionality, comes about as a contour of abstraction inlaid with smooth bronze. The bronze functions as a polished seam of

Fig. 5.11. Désirée Tonnaer's *Vastelaovesmonument* (1993) transforms the diminutive Weerdinge Couple into larger-than-life figures carrying baskets filled with maize and bread to symbolize fertility. The sculpture is made from cloth dipped in wax and cast in bronze. Photograph © Désirée Tonnaer.

the modern in contrast to the mimetic, historic, and archaeological aspect of the sculpture. The abstraction of the modern highlights, I think, an intended sense of anachronism. It is both a representation of the Weerdinge bog bodies and a *concept* of representation. Placed as dressed-up townspeople but elevated on a rounded pedestal, the couple is performing a pantomime through time, as if amusing themselves on their way from the bog through the bricklayered anonymity of modernity, hovering over the other people in the town of Sittard in Holland.

Tonnaer's Weerdinge Couple sculpture is interesting for a number of other reasons. On the one hand it is linked to the practices and preoccupations of some late twentieth-century artists who staged confrontations with the realities of the body in its radical physicality. On the other hand, it subscribes to a modernist aesthetic, using the boundaries between abstraction and figuration to articulate a body *removed* from its own fleshy reality. As such it poses (literarily) as a curiously synchronized affirmation and negation of the presence of the past: *anemnesis*. I have mentioned a number of times how when an archaeological artifact is (or rather *has been*) a human being, the tension between presentation and representation is both muffled and intensified. When the figures embody humans like ourselves and become representations of someone like us, time is compressed so that their past and our own future past are imagined to be alike. It is precisely this combined identification and alienation that gives bog bodies—and other mummified human remains, for that matter—a different resonance than other archaeological objects. And it is precisely this resonance and tension that Tonnaer's sculpture uses as an affecting and ironic attribute. I would suggest that the common understanding of authenticity which sees the presence of the original as a requirement is rearticulated, literally refigured, here in Tonnaer's sculpture as a negation. It is a negation in that Tonnaer's figures are extra-human in their superimposed largeness and artistically layered flatness, as art objects made of bronze and clothed in garments, and as resembling but not being human remains.

For Tonnaer the bog bodies are not merely symbols of the archaic, but resemble W.J.T. Mitchell's definition of paleoart in which the archaic is an image of "the modern seen as dialectically equivalent to and turning into the archaic—and vice versa."[47] The couple is modern and ancient, walking through time, clothed in bodies from the Iron Age with masks from commedia dell'arte. While one could argue that natural objects such as bog bodies are less vulnerable to the mechanics of reproduction (and also less prone to the potential for losing aura), the use of human remains, in the examples discussed here, causes us to restate our concern with what is natural and authentic and what is not. In the end, Tonnaer's giant shoppers are relatively unbounded by

Fig. 5.12. *Vastelaovesmonument* in situ at a square in Sittard, Holland. The feasting carnivalists are seen on a pedestal of freestone, a fossilized organic material. The setting, in a modern urban environment, emphasizes the blending of archaic and modern already present in the sculpture itself. Photograph © Désirée Tonnaer.

the facts and particulars of the archaeological world and the historical claims made by the archaeological remains. On the other hand, these unique remains and their extraordinary presence in her sculpture suggest the possibility of re-questioning boundaries between past and present: they offer new ways to question our humanity.

Bogland: Earth Art

Unlike in Tonnaer's contoured bog couple sculpture, where modernity and irony work as antidotes to nostalgia, nostalgia is in other instances used deliberately—particularly in the ecologically and ethnographically oriented bog art that turns to marshy landscapes and concepts of origin. The American art historian Hal Foster has called attention to what he sees as an ethnographic turn in contemporary art, one that involves in sequence "the material constituents of the art medium," "its special conditions of perception," and "the corporeal bases of this perception."[48] In the last part of the twentieth century he observes an inclination across media and disciplines to "redefine experiences, both individual and historical."[49] This return of the real converges, he says, with the return of the referential and with a rebirth of nostalgia and humanism—often couched in "the register of trauma."[50] It is a project which he (with reference to American anthropologist James Clifford) calls "ethnographic self-fashioning": a fashioning which becomes, he maintains, "the practice of a narcissistic self-refurbishing." American cultural critic Lucy Lippard is more positive about what she calls "prehistoric art," an art which evokes a premodern mystical era.[51] She discusses the echoes through time from various pasts, and proposes the term "overlay" for visual metaphors that emerge as a "sensuous dialectic between nature and culture." That which we have "learned from mythology, archaeology, and other disciplines is the overlay's invisible bottom layer." This layer is essentially the "gap" to the past. Bog bodies too can be placed in such a sensuous dialectic; a continuous production of overlay as spatial metaphors for temporal distance. When Lippard argues that there is a "mysterious, romantic element to wondering about the past, however critically one goes about it," her comment resonates closely with much of the "bog body art" I have located. But she also shows how romanticism attached to ancient sites and images can encompass rebellious potentialities as "outlets for the imagination that can't be regulated, owned, or manipulated like so much contemporary art because so little is or will be known about them."[52] Such a subversive "romanticism" is also to be found in the work of some artists who use bog bodies to articulate a type of mythological archaeo-art sutured to the landscape and its ancient inhabitants.

Cases in point are the sculptures that resulted from the *Bogland Symposium* in 1990 arranged by the Sculpture Society in Ireland under the patronage of Seamus Heaney. The symposium took its key from Glob's *The Bog People* and its influence on both Heaney and on the Irish painter Barrie Cooke (who had illustrated some of Heaney's bog poems in 1975) and it counted artists who were "interested in the extraordinary peatland environment and the 'black butter' of peat itself, with which Ireland is so generously endowed," as Irish art critic Aidan Dunne reflects in his opening essay to the symposium catalogue.[53]

The event produced a number of works in which environmental interests were joined together with aesthetic production.[54] The artists were essentially working to help protect and sustain an archaic landscape under pressure from modernity. Tim Pearce, the conservation officer of Ulster Wildlife Trust, writes: "We must prepare ourselves to campaign and lobby for the retention of our peatland heritage; its historical significance, the evidence of ages past buried in its depths and the peculiar sense of space, the wilderness that is our bogland."[55] Eighteen art pieces are featured in the *Bogland Symposium* catalogue, each aligned with the tradition of what is known as earthworks, earth art, or land art; some, in fact, quote more or less directly from the works of Robert Smithson by echoing his spiral forms or circular shapes carved out of or into the earth, bringing to mind his description of earthen objets d'art as being "charged with the rush of time, even though it is static."[56] All the pieces are concerned with nature as a primary reality, a physical presence of elements and matter and—as in the work of Smithson, the American "father" of earth art—an "anti-museum argosy of 'earthworks.'"[57]

Smithson's own archaeological imagination did not extend explicitly to bog bodies, but there are implicit links. It is a curious fact, for example, that the poet William Carlos Williams, whose own interest in earth sciences resulted in his poem on Tollund Man ("The Smiling Dane") was Smithson's pediatrician. One might even venture to say that Williams's poetry, which Smithson no doubt was familiar with, places bog bodies into propinquity with earthworks. Smithson is worth mentioning not only because he might have known Williams's "Smiling Dane" (discussed previously) but also because his earthworks, as seen amongst the Bogland artists, have had direct influence on some of the contemporary artists who have worked with bog bodies.[58]

Two of the Bogland artists, the Dutch sculptor Remco De Fouw and the Irish artist Ann Henderson, form human faces in the bog. De Fouw's piece entitled *Let Sleeping Bogs Lie* suggests, as do many other bog art pieces, that the bog is dormant, sleeping, ready to wake up and release its prisoners.

We see the upper part of a face, the closed eyelids and the nose, shaped in smooth turf and uncovered as if in still uninterrupted sleep, surrounded by a blanket of bog plants. In the accompanying text De Fouw describes his walk into "the wilderness of the bog" as a "metaphor for a journey of mind, inward or maybe outward to an unspoilt primal ground. The open landscape of spongy blanket bog consists of a bed of layers of years upon years of its own sleeping memories, consciousness turned to matter maybe."[59] The slumbering face is resting in a space of quietness, in an "unseen depth of soft black substance" but precariously close to the noise and "hustle and bustle of Dublin city center." The sleeping face is part of a "primitive and eerie" realm, which—if we allow ourselves to see the title as having a double meaning—advises us to leave it alone, let it "lie," and to sanction its "underground" deception, let it "lie."

While De Fouw's piece does not point directly to a known bog body, its exposing of a face in the bog clearly suggests such an association. Henderson, on the other hand, makes the bog body link explicit. She, too, constructs an image of a half revealed head in the peat, but hers is shaped more closely like that of a dead body. In her work, called *The Whisper,* bog plants surround the head and the face appears to have a halo made of protruding roots and fern

Fig. 5.13. Kathy Herbert's *Container* (1990), an example of earth art–inspired bog art. In the catalogue, Herbert adds this text: "The aged bog / time worn and sculpted / makes a laugh of little art. / Scale marks the imposition / of imported form. / Strength of elements / Unity of perfection / Simplicity of being / strip learning bare / Falling on resourcefulness / answers creativity." Photograph by Christine Bond © Kathy Herbert.

Fig. 5.14. In Remco De Fouw's *Let Sleeping Bogs Lie* (1990), the face of a man melts into the bog as an image of how consciousness turns to matter. Photograph by Christine Bond © Remco De Fouw.

leaves. A white (silvery looking) material flows from the half-open mouth as a visualized whisper. Her image is accompanied by a poem that starts:

> Stripped to my animal bone
> wind, rain and elements
> Yet raised to the spirit
> High

And if we follow the poem's directive, it is meant as a representation of Lindow Man: "Chosen Lindow head of princely sacrifice to / voice the outcry / from within your depths-silver message to the / ear." This "princely sacrifice," which seems to draw from archaeologists Anne Ross and Don Robins's interpretations of Lindow man as the "Druid Prince," is seen as triumphant composite of peat and skin, "Precious joy of you passing, place and people."[60]

In another Bogland contribution, called *Wicklow Nerthus,* the Irish artist Catherine Harper performs a ritual which invokes the earth mother Nerthus, the Teutonic goddess of fertility to whom Glob assigned a privileged place in bog body mythology. The remains of a sheep are used to perform a ritual offering, and through "the centre of a triangular [. . .] construction," Harper writes, "I birthed the broken sheep up out of peat" and "through the performance, affirmed my own vitality and mortality."[61] She finds a particular spiritual connection and affinity with the landscape and the bodies in it, and in a statement about her art, entitled *A Beginning,* she explains how her "interest in the natural history, the archaeology, ecology and mythology" of the bog

Fig. 5.15. In Catherine Harper's *Nerthus* (1990), the dense brown earth is separated from the triangle of entangled branches, roots, and skeletal parts by a sharp white field that cuts into and lifts the knotted material up into the air. Photograph © Catherine Harper.

landscape she grew up with became an interest in "the physical and spiritual aspects of bogland."[62]

Harper's work with bog bodies goes well beyond the Bogland Symposium and is inspired directly by Glob and Heaney. The sensuality of the images of bog people allows Harper to experience an emotive connection. The fact that these were people, she writes, makes her reflect on her own place as a woman; they literally make her see her own face in the bog face. For her the materiality, or here more correctly the corporeality, of the past is at the fore. The use of muted colors, of earth materials, creates a sense of there-ness, proximity, and shared fate. Harper's pieces are manifested symbolically and employ an iconography that weaves the materials of the earth together with a pictographic depiction of the female reproductive organs. In a mixed media piece called *Nerthus,* for example, Harper describes bogland "with its folds, faults, intrusions, rifts and crevices of sod and clod, hidden places, wombs and layers" as "certainly female, producing life, growth and eventual decay in the continuous cycle of life and earth."[63]

The bog as a vaginal canal, a trope familiar to us by now from literary and poetic works discussed elsewhere, provides Harper and other women artists with material for a kind of archaeo-feminist imagination. What happens in Harper's piece resembles what Hal Foster has termed the artist "primitivized, indeed anthropologized"—that is, she embodies the *other* in the bog and essentially puts her "self" on display (she makes a self-portrait of bog material). Yet as she does so, she reflects carefully about the nature of auto-archaeological memory work, and also about "realignments" between the bog people and contemporary people and the parallels this presents in regard to the human condition.

Nostalgic Overlay

Self-display—or self-archaeologication—is even more pronounced in a photographic bog series by the Dutch artist Trudi van der Elsen. In *Fall into the Bog,* a cycle of black-and-white still-shots which resemble freeze frames of an interrupted motion film, we find Tollund Man in the shape of a woman and witness how he (as a she) walks into the bog, submerges, and finally rests in the prenatal position of Tollund Man. Here the bog man-as-woman illustrates the intentional and/or unintentional gender bending that happens in much bog body reception (particularly in the representations of Windeby Girl-Boy and the Weerdinge hetero/homosexual couple). Van der Elsen deliberately draws on the intimate analogy between archaeology and art photography and articulates a moment of time shared between the contemporary subject and the prehistoric object. Her photographs take the structure of a visual story plucked as "quotations" from Glob's book, and they echo its photographic illustrations. In this iconographical mimicry van der Elsen shows how a woman walks into the bog lake as if mesmerized, and finally assumes the position in which Tollund Man was found.

If human sacrifice is the key to understanding the bodies in the bogs, the artist here gives us a photo-performance in which the sacrifice is restaged as nostalgic self-sacrifice. In one photograph the woman's face is actually depicted as if grafted directly onto that of Tollund Man.

By mimicking the ancient face, which in the words of Glob is the best preserved "to have survived from antiquity in any part of the world," and one on which "majesty and gentleness still stamp his features as they did when he was alive," van der Elsen changes the archaeological narrative into a subjective vision, evoking a sense of "mythography" to borrow a phrase from Mieke Bal.[64] That is, van der Elsen's archaeological-artistic mimicry is wrapped up in a coding of identity vis-à-vis alterity, as a performance of actuality, a strangely romanticized challenge for the modern subject. On the one hand there is an urgent sense of identification; on the other the play with quotations is both ironic and melodramatically self-conscious. Interestingly, van der Elsen not only quotes from a photograph of Tollund Man but also from Carravagio's painting *Narcissus:* her gender-altered version is a woman on her knees bending forward over a bog lake as a mirror, like the Narcissus we see in Caravaggio's painting. In *Quoting Caravaggio,* Mieke Bal offers a comment that could be applied to van der Elsen's bog-mirror (although her book does not mention van der Elsen's work) when she says that "unlike the bodily skin, the mirroring surface *touches* visually, not physically. This does not make the act of seeing any less 'touchy,' sensitive, or formative to the subject. Only a semiotics focused on the production of meaning in coevality can theorize the structural similarity between touching and seeing that is important here. For seeing is a semiotically informed act of indexicality, of reaching into space."[65]

As a result, Van der Elsen's overlay of Glob's archaeological photo-text transforms the two-thousand-year-old bog body from a materialization of a national culture into a personal memory. In her photo series, the bog-man-as-woman morphs into nostalgic artistic manifestations and reminds us that nostalgia, as Svetlana Boym has argued, is about "repetition of the unrepeatable," and as such can become a way to overcome trauma.[66] By walking back into the bog, back to a place of origin, and by repeating the traumatic event, trauma *can* be released and repaired.

Figs. 5.16–5.19. In Trudi van der Elsen's photo series *The Fall into the Bog* (1991), she performs a gender shift in which the powers of the bog implicitly transform the woman who walks into the bog to the (Tollund) man who resurfaces. Photographs © Trudi van der Elsen.

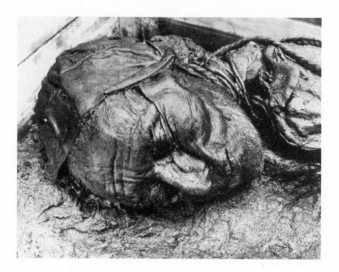

Figs. 5.20 and 5.21. Notice the carefully choreographed resemblance between the semi-profile of Tollund Man and Trudi van der Elsen's bog woman. *Untitled* (1991) © Trudi van der Elsen. Photograph of *Tollund Man* © Silkeborg Museum, Denmark.

Figs. 5.22 and 5.23. The two Narcissuses—separated by time (1645/1991), media (painting/ photography), and gender (male/female)—are united by the motif of self-reflection. Caravaggio's *Narcissus* (1645). *Narcissus* (1991) © Trudi van der Elsen.

Disgusting Bodies

In much of the visual art that makes use of bog bodies, (and I would like to stress that the body of visual work on bog bodies is far more voluminous than what I have been able to include here[67]) the artistic focus is on corporeality and a great deal of attention is paid to skin (wrinkles, lines, protrusions, gaps, wounds, and softness)—skin as surface but also as depth, skin as a kind of amalgamated exterior-interior. This is also seen in a work by the German artist Etta Unland called *Austrocknung* (Withering).

Austrocknung, which was exhibited in Oldenburg Museum along with other art pieces as part of the 1992 exhibit More Moor, homes in on the fragile and sensuous quality of the organic material from the bog. Here the fabric's transparency reveals imbedded peat segments that resemble two shadowlike forms that look not unlike X-ray photographs of amputated legs and feet projected into the modern world and onto the museum walls. The linkage of culture and nature fused in the material and in an image that resonates with technology (capturing the past as snapshot) is abstractly anatomical in form. But what is even more telling in this piece is that neither title nor content tells us that the art piece represents bog bodies.

This brings me to an important point, namely that all the visual and verbal artistic representations discussed here are bereft of the actual corporeal bodies they describe. No matter how intimately the poet's hand "Braille-reads" (Heaney) or the visual artist's hand "kneads" (Vandercam), the bodies are never *there*.[68] (We need to wait patiently for our museum visits in the next chapter to find them face-to-face.) While this point may seem obvious, since bog bodies are protected by institutional boundaries, the changed idiom of visual art in recent decades has allowed for more radical inclusions of body material. Because a bog body so easily occupies a liminal space between that of being a natural-archaeological object and that of being a fabricated art object (as I have shown in the chapter on photography), it also provides an opportunity to look at the oftentimes fuzzy borders between museums of art and museums of science or archaeology.

The Spanish philosopher José Ortega y Gasset has noted that in the museums we find "the lacquered corpse of evolution" where the flux of centuries of visual art has been "congealed as in a refrigerator."[69] The corpse is easily revived, he says, by placing the artwork in a certain order, a chronology of time. But in museums this "evolutionary corpse" is often broken down into fragments or "crystallized facts, frozen to position." While Ortega uses this observation to meditate on what he calls the tactile quality of proximate vision, in the context of bog bodies the breakdown of the evolutionary corpse (with its

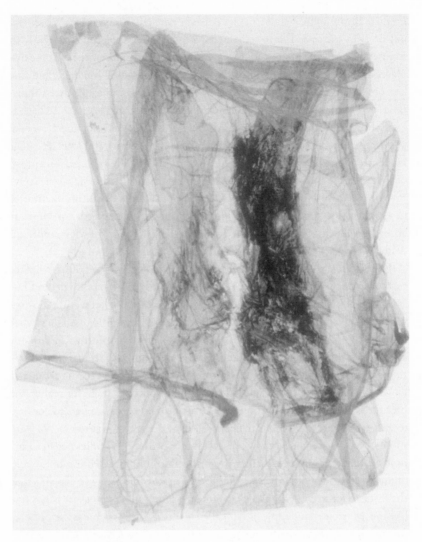

Fig. 5.24. Etta Unland's *Austrocknung* (Withering; 1992) uses transparency as a primary trope and makes archaeology accessible as a see-through filter that reveals a plethora of textures and objects. Photograph by Norbert Gerdes © 2008 Artists Rights Society (ARS), New York / VG Bild-Kunst, Bonn.

implied chronological trajectory) has a very particular resonance. Whether a bog body is "processed" into artistic representations or is placed in the "raw" in a museum setting, it has been submitted to an act of displacement.[70] These displacements into institutional frames of natural history on one hand and fine arts on the other, different as they may seem, are oftentimes rendered indistinct: bog bodies can look like art, and art can look like bog bodies.[71] In fact, as I will show in the next chapter, a number of bog body exhibits deliberately take part in the kind of blurring of "institutional codings of art and artefacts" that Hal Foster has discussed in his *The Return of the Real*. While Foster is sceptical of what he sees as a "quasi-anthropological paradigm" because it often "strays toward self-fashioning, from a decentering of the artist as cultural authority to a remaking of the other in neo-primitivist guise," the curators of some archaeological exhibits embrace the concept from the opposite viewpoint and deliberately place the viewer in a position of puzzlement.[72]

But before we enter the museums and discuss display strategies, let me return to Ortega's meditation on the elimination of the human figure and human metaphors in abstract painting. When he says that "wherever we look we see the same thing: flight from the human person," he calls this (somewhat misleadingly) a "dehumanization" and "disgust for human forms" in modern art.[73] This dehumanization is stirred by an aversion to traditional interpretation of realities, and he asks: "What is behind this disgust at seeing art mixed up with life? Could it be disgust for the human sphere as such, for reality, for life? Or is it rather the opposite: respect for life and unwillingness to confuse it with art, so inferior a thing as art?"[74] Ortega's questions, written in 1925, are specific to a discussion that took place decades before bog bodies began to appear as human figures in visual art. Even so, his call for an attention to the role of the human person in art is helpful for my discussion here. He explains that the distaste for mixing human elements into modern (abstract) art is not unlike the aversion that people of (so-called) culture harbor against the popular wax models at Madame Tussaud's. Here human reality is tested and generates an "awkward perplexity" in the viewer because "we do not know whether to 'live' the things or to observe them." The discomfort is tied to the wax figures as liminal objects:

> The origin of this uneasiness lies in the provoking ambiguity with which wax figures defeat any attempt at adopting a clear and consistent attitude toward them. Treat them as living beings, and they will sniggeringly reveal their waxen secret. Take them for dolls, and they seem to breathe in irritated protest. They will not be reduced to mere objects. Looking at them we suddenly feel a misgiving: should it not be they who are looking at us? Till in the end

we are sick and tired of those hired corpses. Wax figures are melodrama at
its purest.[75]

Ortega uses the wax figures to illustrate and stress the point that "lived realities" can be experienced as "too overpowering." To avoid what he sharply calls
the melodrama of "hired corpses" (we may with some right see bog bodies as
participants in such a melodrama), modern art (he essentially means abstract
art, but also refers to music and poetry) in the first few decades of the twentieth century opted to purge the canvas of human figures, and consequently
take out elements of lived reality. Yet, as I have shown, to see bog bodies as
objects in modern art production does not mean that we look away from their
particular humanness. Neither does it mean that we abandon the linkage between bog bodies and their metaphorical value of being "frozen in time." But it
does mean that their specific humanness is under pressure when represented in
modern art. In order to understand this pressure, it is important to remember
how contemporary aesthetics, as Mario Vargas Llosa points out, "has established the beauty of ugliness, reclaiming for art everything in human experience that artistic representation had previously rejected."[76] I would suggest
that the consideration of bog bodies as material for modern art seems to be
positioned within the tension of these two points: the erasure of the body and
the return and reinstatement of a body not beautiful.

Ortega's wax figure examples are perhaps odd in that no one would argue
that wax figures are fine art objects in the traditional sense of the term, yet they
are also telling in that he uses them to show what happens when the viewer
experiences a lack of distance in seeing. If this distance is missing, there is no
"derealization"—and since art needs this "derealization" to transfigure its object, the human figure and its implied lived reality are wiped away. Or so it
seems. But is there another line, a connection, between the lived reality and
modern art besides the one outlined by Ortega? To make a distinction, as Ortega does, between "living" the thing and "observing" the thing, the human
form was for a while at risk in modern art. It would soon reemerge, albeit not
in the shape of traditional interpretations of reality. With this reemergence,
human forms such as bog bodies became useful and poignant material.

This very aspect of realness and mimesis lies at the core of the work of sculptors such as the Americans George Segal and Duane Hanson, whose life casts
and resin and wax works embalm bodies to stress the tension between original
and copy. Later we shall see how facial reconstructions (the bog body's equivalent of being a wax figure) are similarly removed from the real. Yet ironically
such reconstructions produce reality effects by fleshing out and giving the bog
bodies "real" faces, even when the new real faces are recognized as artificial, as

simulacrum. While furthest from the aura of the real in terms of the raw and authentic, these reconstructions bring a different sense of realness into play.

The questions remains: to what degree do bog bodies test the boundaries of the use of dead bodies in modern art? The following description by Dutch art historian Ernst van Alphen of a number of contemporary paintings could well be seen as describing bog bodies: "The distortion of bodies and faces, the dissolution of boundaries, the pain, the mouths opened as if screaming, the agonies, paralysis, cries. The figures appear as nervous systems laid bare and struck."[77] The bodies in question, however, are found not in archaeological museums but in British artist Francis Bacon's well-known paintings. He, like many other visual artists in the twentieth century, takes issue with the conventions of Western art and the cliché of the beautiful body.[78]

Discomfort with beauty in modern art—if we subscribe to this cliché—does not mean that unsightly bodies or cadavers are uncontested material for public viewing. In April 1997, for example, British-American critic Alan Riding could report to the *New York Times* about the arrest of the British sculptor Anthony-Noel Kelly on suspicion of illegal possession of corpses. Kelly made moulds of heads, torsos, and limbs given to him "under the cover of darkness," and the police found "plaster casts of dead men and women as well as some 30 body parts when they raided" the artist's studio. "Mr. Kelly's work first came to public attention," we read, "when several gold-and silver-coated body casts were presented at the London Contemporary Art Fair in Islington. One cast of an old man's head and shoulders had a price tag of $7,000; it found no buyer." Was Kelly's artwork a dignified rendition of death, or a work macabre? Clearly he was not the first to use cadavers as models. From the tradition of death masks through the annals of art history, dead bodies have been used in the art of Leonardo da Vinci, Michelangelo, Holbein, Mantegna, Caravaggio, and Géricault, to name the most famous. As mentioned, trends in contemporary art from the mid-twentieth century showed increased interest in the corporeal, uncanny, and weird, the untimely and bizarre, the crude and mystical. Above all the body as *material* became interesting for a number of artists; the realness in fragmented and distorted bodies is by now almost a cliché. Different as they may be, Bacon's distorted bodies and grimacing faces, American artist Andres Serrano's morgue photographs, fellow Americans Gwen Akin and Allan Ludwig's sliced faces, British artist Mark Quinn's self-portrait in frozen blood, American sculptor Robert Gober's hyperreal dismembered limbs, and British painter Chris Ofili's *Shithead* are but an infinitesimal segment of the ongoing renegotiation of the body in modern art.[79]

Interest in the transitoriness of the body is not without its detractors. In a 1992 exhibit at MIT's List Visual Arts Center called *Corporeal Politics,* to give

Fig. 5.25. The best known version of Kiki Smith's *The Ice Man* (1995–96) is made in translucent fiberglass and resin, but the rendition in bronze, seen here, presents dark contours of corporeal remains that look like photographs of bog bodies. Photograph by Katherine Wetzel © Virginia Museum of Fine Arts, Richmond, and the Sydney and Frances Lewis Endowment Fund.

just one example, a number of works by American artists Kiki Smith, Robert Gober, Louise Bourgeois and others demonstrated "how the fragmented human body is an appropriate vehicle to convey the feel and focus of the moment," as the director of the exhibit, Katy Kline, put it. The exhibit was denied funding by the National Endowment for the Arts, and in the accompanying catalogue American historian Thomas Laqueur wrote how *Corporeal Politics,* by "making manifest the body in all its vulnerable, disarticulated, morbid aspects, in its apertures, curves, protuberances where the boundaries between self and world are porous," also "demands the return to public discourse of pain, sickness, fluids, and the meaning of artifacts. Bodies and things do not speak for themselves in some magical, unmediated fashion" but are allowed "manifest presence in the art."[80]

Kiki Smith's fiberglass-and-translucent-resin sculpture *Ice Man,* from 1995, is particularly interesting in this respect because it is inspired by an older "cousin" of the bog people, the frozen Neolithic man from the Ötztal Mountains between Austria and Italy who was found in 1991 "in the position of someone who had stretched out to take a nap" with one arm spread across his chest, perhaps by the slow movement of the ice.[81] The man had died in the Alps and somehow was encapsulated in the ice for 5,300 years. Smith's oeuvre as a whole takes inspiration from mummies, relics, and the grotesque body—and in her *Ice Man* sculpture, which, when exhibited, is always hung from the wall above eye level and often as part of an environmental whole with effigies of dead crows scattered on the floor, the near transparency of the body and its transcendent rise up the wall is supposed to suggest a kind of psychic space. The *Ice Man's* body with its crucifix-like gesture seems intended to linger between matter and immateriality.

Crossing the Line

If bog bodies occupy a liminal place as bodies-turned-into-things, they can lead us back to the question posed by Ortega in regard to the elimination of the body in modern abstract art: "What is behind this disgust at seeing art mixed up with life?"[82] This question has become particularly pressing due to the artistic pretensions of new curiosity shows such as the traveling exhibits known as *Body Worlds* [Körperwelte] created in 1995 by the German doctor Gunther von Hagens. Dr. Hagens's plastinated real cadavers in *Body Worlds* give ethical questions surrounding the display of dead humans a new edge and make the exhibits of most bodies (as we shall see in the next chapter) seem rather innocuous. In order to preserve and display the cadavers, Hagens has developed a process of plastination in which fats and fluids are removed before the bodies are preserved in various poses and tableaux. A plastinated body becomes in a sense an eternity model of itself, but it also presents an uncomfortable mix of science and art. In fact, in March of 2002 a man "enraged after watching a father guide his 5-year old daughter around the exhibition, pummeled one of the specimens with a hammer" because the dignity of the dead had been violated by such public display.[83] American critic Verlyn Klinkenborg takes a different view and notes that there is a pleasant aspect in the exhibition. "That may sound like a strange thing to say about a display of some 200 flayed corpses, their muscles peeled back, brains exposed, nerves articulated. But what makes the exhibit palatable is the entire strangeness of the objects on display. We see these bodies, with one or two partial exceptions, after dissection. Identity—familiarity—has been shed with the flesh."[84] The artifice in the display is intentional, and many of the bodies mimic poses known from art history, such as the Flaying of Mayas.

While Hagens has been called a "Frankenstein," he would no doubt prefer to see himself as an updated Renaissance anatomist or, as American critic Mary Ore has suggested, compare himself to "Leonardo de Vinci, who dissected human bodies to draw the muscular, vascular and skeletal systems."[85] Or to Andreas Vesalius, who in 1537 began his famous dissections of cadavers and depicted them in classical attitudes, images of which—it should be no surprise—are used on promotional posters for *Body Worlds* to remind museum visitors of the historical precedents for such displays.[86] By connecting anatomy with popular culture, Hagen's explicit intent is to "re-democratize anatomy" and to break down the taboo in observing and learning from the dead. Yet the commodification aspect, as was the case with bog bodies, can not be overlooked: "The gift store," as Mary Ore reported from the exhibit in London in 2002, sold "watches with faces showing the skin-holding man and mouse pads decorated with the chess player plastinate leaning over a computer."[87]

But there is an important and vital difference between bog body displays and the plastinated bodies. Bog bodies, as I have argued so far, bring the *past* into the *present*. Hagens's anatomical bodies, in contrast, bring the *present* into the *past*. His bodies come from "freshly" deceased people who undergo plastination only days after their deaths. When displayed shortly after, they are placed under illustrations of historical anatomical theaters and banners with metaphysical quotes from the Bible and philosophers such as Kant, Goethe, etc. This allows Hagens to *drag* the bodies back in time, bestow historical depth on them, and connect them to a family tree of multiple other anatomical specimens. The metaphysical and historical contextualization, in other words, serves to gloss and to legitimate the display of bodies so recently dead.

The crux of the matter here is the problem with the *recently* dead. The repulsion against the use of corpses in art has not only a corporeal element but also a temporal aspect. In the use of bog bodies in art, the *sediments of time past* become the *sentiment of time present*. This explains to some degree why bog bodies in visual art are seemingly inoffensive. Because they always already are "canvases of conflict," and because this conflict (violent death, sacrifice) is part and parcel of their frozenness in time, temporal distance once again functions as a kind of buffer or gloss. The offensive cadaverousness is circumvented by time. The bodies' status as archaeological objects gives them safe access to the iconography of modern art because time and science (in this case archaeology) effectively protect us, the onlooker, against the risk of "contamination." The bodies are sanitized by way of the refrigerator of time, and as a result they manage to steer clear of the putrefaction and repulsiveness of recently deceased bodies. Just as with photographs, frozen time (the very nature and essence of archaeological specimens) provides a kind of varnish and protection against the raw and atrocious. Time itself has prepared and preserved the human body as if it was already a piece of art. In a sense, visual artists therefore have to "quote" not only from the museum of the earth, a kind of terrestrial museum that has always held title to the bodies, but also from the museum of archaeology, where the bodies are kept and catalogued. The result of this is twofold: by the time the bodies have traveled to museums of art they are twice removed from their original essence as dead bodies; this in effect safeguards bog body paintings and sculptures—at least those I have located and examined here—from overstepping ethical boundaries. The bodies are lifted from the already twice enshrined. But, we have to ask, what about the "real" bog body remains?

After being unearthed, it is the fate of most bog bodies that they need to find other homes than those in which they were found. If scientists and museums are able to employ appropriate conservation strategies, the physical remains do not melt into disappearance. As a matter of fact, if these strategies are successful, the museums and the archaeological records help guarantee that the bodies' frozenness in time becomes permanent. In this embedded permanence and from within the safe boundaries of institutional "eternity," bog bodies can gradually (or sometimes rapidly) unfreeze into the fundamentally unsafe territory of the imagination.

But how do museum curators negotiate the potential "danger" in displaying dead bodies? Let us turn from the museums of art to the museums of archaeology.

CHAPTER 6

Museum Thresholds
and the Ethics of Display

With increasing intensity in the last few decades, bog bodies have traveled from museums of archaeology to museums of fine art and onto the walls of art galleries. On this journey the line between archaeological specimen and fine art object is oftentimes obfuscated. To illustrate this, let us visit the exhibit "Weder See noch Land" (Neither Water nor Land) at *Landesmuseum Natur und Mensch,* Oldenburg in Northern Germany (henceforth referred to as the Oldenburg exhibit). When in August of 2004 I entered the bog exhibit I was met by four large rectangular "canvases" on the walls leading from the first to the second floor. At first glance they looked to me like muted variations on the French artist Jean Dubuffet's *art brut* (raw art) paintings, or like the German artist Anselm Kiefer's use of earth-like organic material and a grainy, almost monochrome palette which often reveals contours of human shapes embedded in the canvas. In fact I was so puzzled that I stopped a museum guard and asked whether or not the "canvases" were what I believed them to be: genuine imprints of bog bodies in the earth. He confirmed that they were indeed "echt und authentisch"—the real thing. I had read about these imprints and often thought about their equivalence to the Shroud of Turin, which had also been seen as a kind of photographic negative. And despite looking forward to seeing how the earth coloration revealed faint traces suggestive of photographic prints accidentally made of bodies by random lightning, I was confused all the same. My confusion was prompted not only by the framing of the slices of earth as if they were paintings, but even more by the fact that one of the "canvases" had the thin tubular outline of a body attached to it (fig. 6.1).

The contrast and sense of pressure between the thin neon-like tube protruding from the surface and the nearly monochrome brownish earthen surface gave the distinct impression that we were looking at a piece of modern art. It spoke in tune, I thought, with the bilingual visual language in which the fabric of the archaic and the smoothness of the new intermingle. I was impressed with the boldness of the curators. Their conscious play with the limits and parameters of museology, and their deliberate sandwiching of the bodies in the ambiguous territory between ethno-archaeology and the fine arts, made

Fig. 6.1. In the Oldenburg exhibit a plastic tube highlights the contour of a bog body. Notice how the barely ascertainable outline of the remains and the small curator caption, at left, mix art with archaeology. Photograph by Karin Sanders (from Landes Museum für Natur und Mensch, Oldenburg, Germany).

me pause and reflect on what I saw. The kind of found-ness that has taught audiences for decades to see natural objects as pieces of art also resonated here and pressed the philosophical problem of the difference between art and nature to the fore.

A bog body is not a sculpture, of course, nor is the trace of a bog body in a slice of earth a painting. Yet when we experience a fuzzy separation between art *proper* (which is intended and culturally fabricated) and archaeological *foundness* (which is unintended and accidental), such as the one I have described at the Oldenburg exhibit, the very core of institutionalized and historical déjà vu takes on extra meaning. The act of "looking back" and reexperiencing can be seen as a cognitive pendulum between frozen and unfrozen pasts. Sometimes, as I have suggested in the previous chapters, the blurring of borders between fine art and archaeology becomes a kind of deliberate re-archaeologization, making the present (and the present self) archaic *and* modern in one fell swoop. This re-archaeologization can take place within individual artists' oeuvres *and* at institutional levels.

Self-Archaeologization

Present-day museum practices demonstrate an increasing interest in such *re*-archaeologization in the form of *self*-archaeologization. Remodeled museums like, most famously, the Tate New Modern in London and the Museum of Modern Art in New York now display the archaeological strata from their own construction sites, in both cases under the direction of the American artist Mark Dion. The *Tate Thames Dig* from 1999 takes form as a large double-sided mahogany chest with drawers and shelves, placed in the middle of a room. Viewers are invited to walk around the drawers, touch them (something rarely allowed in a museum context), open them, and look at the artifacts inside, which are assembled taxonomically in groups (broken glass, bolts, knobs, cups, bottles, bricks)—all remnants from the layers of ground under the museum. On a wall in the room appears this instruction: "By refusing to follow a more orthodox, historical approach to categorization, Dion encourages their own imaginative and poetic association."[1] The grouping of artifacts according to function, shape, and matter, and not according to specific time periods, allows viewers to imagine the layers under the museum as being simultaneous and synchronized with the present.

In 2000 the Museum of Modern Art in Manhattan invited Dion to perform archaeological excavations under the museum's garden, to dig into the remains of former townhouses of the Rockefeller family and the Dorset Hotel, all of which had been torn down to provide space for the museum. In this case the museums' self-archaeological display, called Rescue Archaeology, has been placed "in depth"—in the basement—giving it a sense of banishment from the rest of the museum, but also a sense of actuality and *place*. The lost architecture of previous buildings is suggested by the remnants of walls, pillars, and ornaments, and made intimate to spectators by showing the shifting fashions of wallpaper, ending with some quite close to the present—thus allowing an experience of personalized history within museum history. Once again a treasure cabinet is placed in the room holding ephemera, or "relics," as the museum texts call them. Finally, in one of the latest examples of this trend, the reconstruction of the Neues Museum in Berlin incorporates the scarring and neglect that building has endured since the Second World War. Damaged parts are kept as a reminder (quite apropos for a museum of archaeology) that the architecture of the building itself is an organic part of its historical process.[2]

This kind of self-archaeologization dovetails in some respect with what the German literary theorist Hans Ulrich Gumbrecht has called "production of presence." Productions of presence in his optic emphasize space, "for it is only in their spatial display that we are able to have the illusion of touching objects that associate with the past."[3] Gumbrecht sees the growing popularity of museums and archaeology as being tied to this desire for "presentification" (Gumbrecht's term). But it can also be seen as a part of the quasi-anthropological gestures of self-fashioning of which Hal Foster is sceptical. Foster proposes that artist envy amongst anthropologists has turned to ethnographer envy amongst artists, and has resulted in works that "aspire to fieldwork." He sums up the involvement with the *other* in twentieth-century art and points out that most "are primitivist, bound up in the politics of alterity: in surrealism, where the other is figured expressly in terms of the unconscious; in the *art brut* of Jean Dubuffet, where the other represents a redemptive anti-civilizational resource; in abstract expressionism, where the other stands for the primal exemplar of all artists; and variously in art in the 1960s and 1970s (the allusion to prehistoric art in earthworks, the art world as anthropological site in some connectional and institution-critical art, the invention of archaeological sites . . .)."[4]

My experience with the bog canvases at Oldenburg would seem to bear out the point made by both of these studies; it has allowed me to see the "canvases" through a bifocal lens, as pieces of art *and* as archaeological specimens. The resemblance to some works by Kiefer in particular was uncanny, as I have already mentioned, perhaps not least because his work resonates so clearly with the effects of the Second World War—a staple in bog body representation. In his early works (influenced by Nordic mythology) the archaic landscapes are often dark, almost monochromatic, and as American art historian Mark Rosenthal has pointed out, "we experience the earth as if our faces were pushed close to the soil and, at the same time, as if we were flying above the ground, but close to it."[5] Kiefer, Rosenthal goes on, often "employed the transformation of land as a metaphor for human suffering"; his series entitled *Marsh Sand,* consists of photographic volumes in which landscapes are gradually covered in sand. "His integration of tangible substances with photographed or painted images at once unites means, subject, and content into an intensely physical presence."[6]

This intense physical presence in Kiefer's work seems to me to resonate in the "canvases" on the wall in Oldenburg. But the history of which it spoke was not one embedded intentionally, since no artist's hand had been at play. It was not an artist's deliberate amalgamation of history and personal story, such as in Kiefer's work, and my moment of puzzlement was soon replaced by the recognition of an underlying display strategy, of the curator's deliberate exercising of the museumgoer's capacity to recognize that the museal space places the artifact, as the Australian cultural critic Tony Bennett phrases it, in "thickly lacquered . . . layers of interpretation." You cannot, Bennett maintains, distinguish between the "rhetoric of the relic and the reality of the artifact," since

> No matter how strong the illusion to the contrary, the museum visitor is never in a relation of direct, unmediated contact with the 'reality of the artefact' and, hence, with the 'real stuff' of the past. Indeed, this illusion, this fetishism of the past, is itself an effect of discourse. For the seeming concreteness of the museum artefact derives from its verisimilitude; that is, from the familiarity which results from its being placed in an interpretative context in which it is conformed to a tradition and thus made to resonate with representations of the past which enjoy a broader social circulation.[7]

This implies that any authenticity the object might have had is instantly circumscribed and its meaning (in the present) derives from its placement vis-à-vis other artifacts. Museums, Bennett goes on to say, are "repositories of the *already known*" now placed in a fictional landscape of telling. "This has the obvious consequence that the same signifiers may give rise to different meanings depending on the modes of their combination and the contexts of their use."[8]

The "canvases" at the Oldenburg exhibit, with their dim outlines of ancient bodies helped into view by a curator's daring gesture, offer, I think, just such a combination. We could with some justification call them exemplars of unintentional art—or, as British archaeologist Colin Renfrew would call it, "involuntary art." Their artistic potential, on the other hand, is undoubtedly contingent upon conventions of the representation of human forms in visual art as they have developed in the twentieth century. In other words, the lens through which I (and, I feel certain, many other museumgoers) saw the slices of peat with inlaid body contours and mistook them for paintings is beholden to modern sensibilities. These sensibilities would, it seems, grow in part through the elimination of human forms from modern painting. Furthermore, Kiefer's evocation of memory in the form of layered soil that holds the bodies of victims of past German atrocities can ultimately parallel the use of bog bodies as images for a repressed past. While Kiefer did not make *direct* use of bog bodies, other artists similarly interested in the wounds of history, and in the layering and excavation of memory objects, have done exactly that. In their work, as I have explained in the previous chapter, we feel (at least if we follow Bachelard) that we are allowed to *touch* the past.

Do Not Touch!

Yet, in reality we are rarely or never allowed to touch objects in museums. American poet and critic Susan Stewart astutely observes: "In museums today, when we turn quickly from the untouchable artwork to the written account or explanation placed besides, we pursue a connection no longer available to us: the opportunity to press against the work of art or valued object."[9] Although museums are about sense experience, the museum visitor is caught in what Stewart calls a "paradox of materiality" because culturally and/or aesthetically valued objects are guarded so vigorously against the erosion of time by separating them from and protecting them against the spectator's touch.[10] Stewart's point is well taken but it is important to understand that if museum artifacts consist of human remains such as bog bodies, the risk of contamination also runs counter. That is to say, while the museum body is guarded against unwelcome touches from the visitors and protected against wear and tear, the museum visitors in turn are protected against the (uncouth or uncanny) feel of the corpse.

The possibility of touch and contamination, and the risk of bringing humanness into peril through the display of dead bodies has been described with emotion and disdainful fascination in Mary Wollstonecraft's *Letters Written during a Short Residence in Sweden, Norway, and Denmark,* from 1796. Although she does not address the subject of bog bodies, her remarks about mummified remains in a grave chamber in a Norwegian church, St. Mary's in Tonsberg, are worth citing because her reflections on the display of human remains resonate with the reactions of modern museumgoers to bog body exhibits. After a visit to the vault at St. Mary's, Wollstonecraft writes:

> A desire of preserving the body seems to have prevailed in most countries of the world, futile as it is to term it a preservation, when the noblest parts are immediately sacrificed merely to save the muscles, skin and bone from rot-

tenness. When I was shewn these human petrifactions, I shrunk back with disgust and horror. "Ashes to ashes!" thought I—"Dust to dust!"—If this be not dissolution, it is something worse than natural decay. It is treason against humanity, thus to lift up the awful veil which would fain hide its weakness.[11]

Without the soul, "the noblest parts," the vacated and raw body disgusts Wollstonecraft. She minces no words: "for nothing is so ugly as the human form when deprived of life, and thus dried into stone, merely to preserve the most disgusting image of death." Unlike the contemplation of "noble ruins" which speak to the fate of empires and to the necessary changes of time, so "our very soul expands, and we forget our littleness"; the "vain attempts to snatch from decay what is destined so soon to perish" bring only pain. "Life, what art thou?" she cries. "What will break the enchantment of animation?"[12] Her lamentations of treason against humanity are an objection to the deliberate intentionality behind unnatural preservation, which in her view shows a corrupted view of humanity. She grieves: "For worlds, I would not see a form I loved—embalmed in my heart—thus sacrilegiously handled!—Pugh! my stomach turns." The mummified remains, she goes on, have become "a monument of the instability of human greatness." In such a state the bodies are robbed of the nobility of form and are compromised as dignified vessels of and for the soul; hence "it will require some trouble to make them fit to appear in company with angels, without disgracing humanity."[13]

For Wollstonecraft, the impropriety of the mummified remains is twofold. She sees the bodies as liminal, out of order; and she sees their deliberate display as unnatural and disgusting. She implicitly recognizes what Bakhtin would later tell us about the grotesque body—namely that it "transcends it own limits" and stresses those of its own parts "that are open to the outside world": apertures and convexities.[14] The decaying corpse is, to quote another scholar of the grotesque, the German literary critic Winfried Menninghaus, "not only one among other foul smelling and disfigured objects of disgust. Rather, it is the emblem of the menace that, in the case of disgust, meets with such a decisive defense, as measured by its extremely potent register on the scale of unpleasurable effects."[15] Disgust, then, is part of a reaction of self-preservation; it is a "state of alarm and emergency, an acute crisis of self-preservation in the face of an inassimilable otherness." Such a strategy for self-preservation can be necessary when the interior turns exterior, when it becomes monstrous, disgusting, and painful to look at or touch.

While the bog, with its unlimited "wetness" far removed from the walls of institutional "dryness," invites a sense of "digging in,"—both *as such* and in a metaphorical sense, as we have seen ample examples of—this kind of close contact is an absolute negative in the logic of museum display: "Do not touch!!" Even though the most tangible way of approaching material culture is through the physicality of touch, such sensory nearness remains a vague promise (or threat!) in the museum space. Here, as Stewart says, we are "guided against all practice of touching, ranging from breathing on the artwork to stealing it." American cultural theorist Barbara Kirshenblatt-Gimblett has attributed this kind of compartmentalization of the senses to "sensory atrophy" in the Western tradition. Sensory atrophy, she maintains, is particularly pronounced in the museum space, where it is "coupled with close focus and sustained attention. All distractions must be eliminated—no talking, rustling of paper, eating, flashing of cameras. Absolute silence governs the etiquette of symphony halls and museums. Aural and ocular epiphanies in this mode require pristine environments in which the object of contemplation is set off for riveting attention."[16]

If the museumgoer is asked to remain at a distance from the displayed object and to follow an etiquette of silence—almost as if in a funerary home (the museum-mausoleum analogy is hard to escape)—this means that senses *other* than those of touch are enhanced. In fact, the "customary velvet ropes in museum display may be seen as compensatory in this sense," proposes Stewart, since "the museum organizes seeing into looking and so organizes a passive into an active relation—one capable of transforming the motion of the spectator into emotional response."[17] In this respect the museum "retains a vestigial relation to touch as the primary sense for the apprehension of powerful matter and material" as visitors have the option to turn from the untouchable artifact to a more clinical contact provided by written accounts (captions, information narratives, and so forth) and hence can pursue another kind of "touch" by proxy of words.[18]

Yet in spite of sensory atrophy and the suspension of touch, ordinary museum experiences are not just about the ocular sense. They are also about sharing a space with displayed objects. And when these objects are authentic human remains, the curatorial options must inevitably consider the instability that Wollstonecraft lamented, or the uncanniness that Freud pointed out. This is particularly urgent when the displayed bodies are victims of violence, as is the case with bog bodies, in that this violence doubles the "normal" experience of death as a violation of the living. If looking at human remains is a matter of "proper" respect, the parameters of *how* such respect is to be maintained is clearly complex. In fact, as I will show here, if dead bodies challenge the ethics of viewing, it might be because museum displays of humans inevitably "teetertotter on a kind of semiotic seesaw," as Kirshenblatt-Gimblett has pointed out: "equipoised between the animate and the inanimate, the living and the dead."[19] In ethnography these displays (she discusses both living peoples exposed as cultural specimens and dead remains of peoples exhibited as ethno-artifacts) have oftentimes blurred the lines between "morbid curiosity and scientific interest."[20] To understand how this has bearing on bog bodies, let us visit a number of different museum displays selected to show a range of curatorial choices, each of which illustrates in its way how the liminality of *humans* turned *things* is tied to an ethics of display.

The Real Thing

Our first *real* museum visit to bog bodies returns us to Grauballe Man. In his original display he was laid to rest on a bed of turf under glass, available for all to see at the Museum of Moesgaard in Aarhus, Denmark. Placards on the walls told the story of his find and contextualized his exhibited body with quotes from Tacitus along with photographs of his original location, excavation, and conservation. But after spending decades in this conventional display mode, his body was reexamined in 2003 and 2004 and then revamped in a freshly curated environment. Since 2004 (henceforth referred to as the 2004 exhibit) he has been in a cave-like room at the end of a long wooden walkway. In the curator's notes we read how his body is set out for careful viewing:

> Museum exhibits depend on authentic material, no matter how sparse and scattered it may be. Grauballe Man is this exhibit's absolute central object, which the public can view close-up in a quiet and condensed atmosphere with opportunity for contemplation. He is to be displayed with dignity and respect for death as nature's law, which the small chamber will convey to the visitor. In this room nothing will be explained or described. Here, one is alone with the human from the bog.[21]

The curator's notes in some sense subscribe to a convention that regards the museum as a kind of "tomb with a view"; it accepts that the body is reinterred and that the primary sense in the visitor's meeting with him is a combination of visceral proximity and the ocular sense. It also accepts that the body has been permanently displaced far from its original home. Grauballe Man, in other words, is in a suspended see-through capsule, seemingly floating in time and space.

As a hushed nucleus, the entombed body is freed from immediate "contamination" of context. The intent is to give both the bog man and his visitors pause ("room to breathe") and to promote a kind of etiquette of silence. The visitor is expected to be primarily a viewer, but one for whom the ethos of anthropological-archaeological display encourages not distance but familiarity. In place of contextual thickness, the visitor can experience a sense of wonder. After the deliberate separation of inauthentic materials from authentic, Grauballe Man's now-decontextualized body floats in a metonymic realm, a kind of temporal void, providing an abstract, even existential, commonality with the beholder standing in the museum.[22] Contextualization, in fact, is described in the curator's notes as something that breaks the circle of attention and disturbs both the peace of the dead and the meditation opportunity for the visitor. Without the need to read notes, labels, and so forth, a "condensed atmosphere" and sense of wonder is generated and wrapped around the viewer.

The result illustrates what American art historian Svetlana Alpers has called the museum effect, which "turns all objects into works of art" and heightens the object's visual interest, often by isolating it; it is "a way of seeing" which museumgoers are implicitly encouraged to tap into.[23] In point of fact, Grauballe Man's refurbished body now "resting in peace" *does* look increasingly like a sculpture. This is not just because of "the museum effect," but because the effect of seeing has been attended to in a number of other ways. The use of lighting in the chamber, for example, makes the surface of the bog-man's skin glitter like a silvery-black polish (something less obvious in the previous display, although it is apparent in the photos of the man on his old peat-bog bed, which are still sold as postcards). From various spectator viewpoints—both within the chamber and from above, as seen through a skylight—the body appears as if suspended in space and, implicitly, in time. If this sculptural quality makes him look less authentic as a corpse and more like an artifact, it is (in part) a paradoxical consequence of the decision to remove him from any immediate contact with inauthentic materials.[24]

The focus on authenticity, ironically, cannot help but be compromised by an eerie sense of reproducibility; while the body's natural state has been preserved, this preservation of naturalness is done artificially.[25] So while the body *is* still natural and original, it also is *not*. The consequence for our experience of authenticity is manifold, and the importance of this altered experience hinges on our experience of humanness. The raw and visceral experience of being in near contact with a corpse is emphasized *and* glossed over, not with the usual in-context labels and captions, but with meditative and contemplative metaphysics.

This paradox reflects the ongoing tension between aesthetics and ethics in the display culture of human remains. On the one hand, without the customary "fake" bog-peat environment in which most other bog bodies are displayed, the visitor to Grauballe Man can be "alone with the human from the bog" and get a heightened sense of closeness and respect for the dead. On the other hand, the decontextualized space in which he floats alone also allows the visitor to abstract from his corporeal humanness and imagine that the body is a produced objet d'art lingering in a zone between the natural object and the fabricated artifact; between art and archaeology. While contextualization and excessive use of labeling, in this instance, can hamper the effect of seeing, many bog body archaeologists fear that without contextual information, displayed dead bodies cannot help but be valued more for their aesthetic than for their scientific importance.[26]

The intent to display Grauballe Man "with dignity and respect for death as nature's law" is not only an attempt to separate him from inauthenticity in all its forms, but also to separate him from the "lack of dignity" and "disrespect" of his early reception.[27] This story, which filled the news media for years after his discovery, finally and ironically has itself become an artifact in the 2004 exhibit, a time capsule from the early 1950s in the form of selected newspaper clippings displayed behind glass as "context" through which the present-day museum visitor can look back to a time when archaeologists were challenged by a less informed public. When the visitor walks out of the bog man's isolated meditative nucleus and into the larger exhibit, she is once again free to encounter the conventions of museum informatics: the numerous didactically displayed points of information, newspaper clippings, and educational artifacts. An animated short film projected on a small screen shows a morose man being brought to a bog by a number of men and sacrificed there; a meditative soundscape evokes a mythical past and fills the air; and in a projection room, a music video combines artistically conceived images of landscape with bog body parts in a continual overlay. In all, Grauballe Man is both wrapped in his own reception story (allocated to the showcases) and protected from this very context in his secluded nucleus.

Indeed, as a last example of keeping authenticity away from contextualization, his "authentic" remains are kept at a safe distance from (protected against) the "artificiality" of his own facial reconstruction, which is located without fanfare in a corner of the 2004 exhibit. Visitors are invited to bend down and put their heads inside a box with mirrors on the sides. When they press a button, a video begins on the screen at the bottom of the box, showing how British forensic artist Caroline Wilkinson has made a facial reconstruction of the bog man. At the end of the video, a glass screen lights up and the actual facial reconstruction is seen, slowly rotating. It finally comes to a stop, bringing the viewer face-to-face with the man from the bog (fig. 6.2).

The facial reconstruction is "raw" in that no colors have been added to the clay, nor have any hair or inlaid prosthetic eyes been added to "give life." The last Promethean fire of animation (liveliness of color, hair, eyeballs, etc) has deliberately been left out. Instead the mirrors inside the box continually reflect

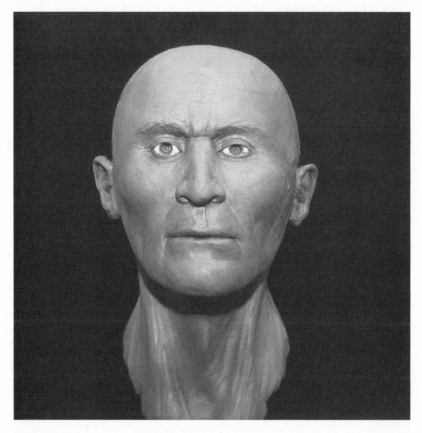

Fig. 6.2. The reconstruction of Grauballe Man brings the museum visitor face-to-face with a two-thousand-year-old human being, but it also functions as a mirror for self-reflection. Photograph by Jens Kirkeby © Moesgaard Museum, Denmark.

the face of the visitor, who is caught in a position where she cannot help but see herself in the act of observing. Thus, curious visitors are forced to contemplate their own physiognomy vis-à-vis that of the ancient man, and are implicitly encouraged to engage in a kind of self-archaeologization. Face-to-face with the countenance of the bog man, the visitor comes face-to-face with herself. Although she cannot touch the face with her hands, she is asked to "touch" the past through her (mind's) eye.

The viewing box is designed to offer the possibility of recognition and identification while still maintaining a sense of estrangement (*Verfremdung*). When the visitor places her head in the box, she is bifurcated not only into these two viewing positions (mirroring and alienating) but also into an epistemological schism involving the nature of humanness. The face she sees is recognizable as a fellow human being. But it is also an artifact, a *thing.* The visitor is not seen in return. Therefore at the very moment she identifies with (and implicitly reanimates) the countenance in the box, she must also acknowledge and accept the artificiality of this reanimation—accept that it is partly a fabrication of the imagination. Inside the box, the spectator finds material for re- and self-archaeologization, and also for the production of presence; this material illuminates the very liminality that fuels a modern archaeo-artistic concept and practice. This self-contemplation is different from that which takes place vis-à-vis the "authentic" body.

Thresholds

Our next museum visit returns us to the Oldenburg exhibit and allows us to continue our investigation of authenticity and inauthenticity. In fact, this exhibit offers an example of how the effect of a radical in-situ display model as a key stratagem turns into production of in-situ inaccuracy—a deliberate meditation on the artificiality of "making it look real." The Oldenburg exhibit has an innovative way of emphasizing the temporal difference between the bog bodies as archaeological specimens and the contemporary museumgoer.[28] In a large museum room, a replica of a bog has been built in the form of peat layers. Inserted into the peat behind glass "windows" are several bog graves. Museum visitors can peer through the windows into the graves and imagine (or pretend to imagine) that the bodies lie there untouched, undisturbed, and unaware of the conversation surrounding them. Far from the danger of real wet bogs, the dry-shoed visitor can contemplate this safe containment of the dead and marvel at the depth from which the peat graves have been excavated. Each neatly lodged in its own bog-bed, the bodies of several bog men, most notably that of so-called Rote Franz, cannot be touched nor can they contaminate the visitor. From their "authentic" bog graves they participate, ironically, in a postmodern setting where the strata-slice of peat bog resembles a kind of earth art piece, moved away from its original location and into the very setting—the museum—against which most earth artists rebel (fig. 6.3).

The curators of the Oldenburg exhibit are keenly aware of the potential of crossing thresholds, not only between living and dead bodies but also between archaeology and fine arts. They have opted, therefore, to mix in non-authentic elements: exaggerated props such as oversized black beetles that draw attention to small glass-incased exhibits in the museum floor, large shattered boulders

Fig. 6.3. Note the bog man barely visible in the slice of peat. Visitors can come close to the bog body—touch the glass wall to his museum grave—but also maintain a safe distance from which to contemplate the deliberate construction of the display. Photograph by Karin Sanders (from Landes Museum für Natur und Mensch, Oldenburg, Germany).

Fig. 6.4. On the back wall of the Oldenburg exhibit, slices of peat are displayed as abstract art. Each slice denotes a specific geological time period, thus allowing the viewer to reflect on the temporal aspect of the bog material. Photograph by Karin Sanders (from Landes Museum für Natur und Mensch, Oldenburg, Germany).

in adjacent rooms, segments of peat displayed like paintings, and Calder-like sculptures suspended from the ceiling (fig. 6.4).

The difference between the raw material and the purpose impressed on it when it is displayed in the museum gives spectators reason to pause. In his examination of the parallel visions of artists and archaeologists, *Figuring It Out,* Colin Renfrew has pointed out a similar kind of pause. He discusses a similarity between the cognitive processes of perception in archaeology and modern art, and claims that "there is an apparent analogy between the position of the observer, the gallery-goer who sees such works for the first time, and the archaeologist, who has excavated assemblages of artefacts from the past and has to make some kind of sense out of them. You the viewer and you the archaeologist are in much the same position. You are coming face-to-face with aspects of the material world, the man-made (or human-made) material world."[29] An assemblage of artifacts, Renfrew argues, produces a sense of puzzlement and mystification and places the viewer of archaeology and the viewer of modern art in the same position. Although the level of intentionality differs in archaeological objects and in modern art, the act of making sense of what is seen, he claims, is the same.[30]

While Renfrew focuses on the perception of the beholder, another study of the connection between art and archaeology, *Ancient Muses: Archaeology and the Arts,* by British archaeologists John H. Jameson Jr., John E. Ehrenhard, and Christine A. Finn suggests a similar alliance but shifts the focus from the beholder to the intentions of archaeological museum curators. These authors claim that the "cognitive connections between archaeology and art reflect an inductive approach in defining and explaining archaeologically derived information and making it meaningful to the public. An emphasis on using artistic expression in interpretations is consistent with a new direction in archaeological practice that challenges the positivist paradigm of processual archaeology and promotes the relevance and validity of inductive reasoning over deductive reasoning."[31] Whereas some archaeologies signal a practice that allows for a multifaceted and creative interpretative approach, some artists, both in text and image, are adopting archaeological *things* to either reimagine the past or comment on the present—or both.[32]

The Ethics of Display

The question of what happens when the displayed archaeological artifact is a human being has not yet been fully answered. Are there ways to solve the problem other than the ones already discussed? Is "our fascination with archaeological mummies ethically responsible or sensationally voyeuristic?" to borrow British literary scholar Jennifer Wallace's question.[33] Clearly, as seen above, a tension between presentation and representation is intensified. But if the parameters of authenticity and inauthenticity are so difficult to maintain, how does this have bearing on the ethics of displaying human remains?

In our next museum visit, to *Archäologisches Landesmuseum* at Gottorf Schloss (henceforth referred to as Gottorf Castle) in Schleswig-Holstein, northern Germany, the home of Windeby Girl and the Men from Rendswühren, Damendorf, and Österby, we find much awareness of the kind of ethical alarm that many visitors may experience when viewing cadavers. Here display ethics are addressed directly and didactically.[34] Before the visitor walks into the display room where the bodies rest in their showcases behind glass and on beds of turf-like material, she or he sees a bulletin board on which previous visitors have posted responses such as: "Bury the bog bodies," "I have the feeling that the bog bodies are unreal [unecht]," and one (in combined English and German) calls for: "More Leichen please!" The responses fall into two categories. One group sees the bodies as helpless and violated by archaeologists and curators. The other regards the bodies as vacated by life, as legitimate objects for pedagogical observations and learning (fig. 6.5).

The posted responses of museumgoers at Gottorf Castle are primarily concerned with soul and immortality. One visitor argues that if one believes in rebirth, then the soul has already migrated on to other bodies; hence the corpses, now devoid of soul and life, "should be exhibited."[35] Another visitor points out that the museum would never exhibit a corpse from 1950. Yet another writes that the exhibit is "unkind (*lieblos*)": "What would the bog bodies say, if they knew, that they lie, here and now, behind a glass-case like an animal in a Zoo, for all people to see?" Along the same lines, one museumgoer chastises fellow spectators and asks: "Is it really so much fun to look at the corpses of others? How would you feel if your last rest was disrupted and you woke up to find yourself 'encaged' and gazed at?" Another says; "I see nothing wrong in exhibiting 'real' people. It removes anxiety from death." One argues that it is a good thing that the bodies are made available for laity (*Laien*) and not just for the learned (*Gelehrter*). And another sees a similarity between viewing the corpses and watching the evening news on television. A letter from an English university student regarding Rendswühren Man asks: "Do you agree that there are no moral boundaries in archaeology? Have you ever experienced any problem or opposition from groups/people against exhibiting the Bog Body?" The placement of her letter amongst the many posted reactions obliquely asks the museum visitors to consider their role as spectators before, during, and after they view the bodies. Finally, a number of children's drawings also posted on

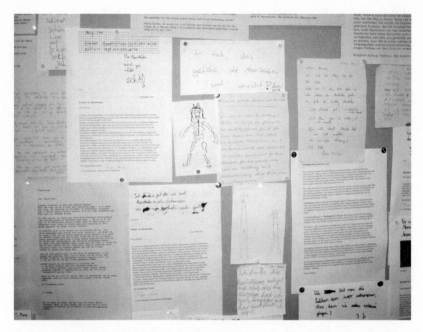

Fig. 6.5. On a bulletin board outside the exhibit room of Windeby Girl, visitors are encouraged to express their opinions and feelings regarding the exhibit of ancient human remains. Photograph by Karin Sanders (from Gottorf Castle, Schleswig, Germany).

the bulletin board at Gottorf Castle testify to the fact that bog body exhibits (both here and elsewhere) are frequently visited by groups of schoolchildren.

The main attraction in Gottorf Castle is the so-called Windeby Girl, now known to be a boy (as explained earlier, I will refer to the body as female to remain consistent with the responses that followed its discovery; see fig. 4.3).[36] One side of Windeby Girl's head was scalped before she was deposited in the bog. The cranial remains show clear signs of the force involved in the scalping, which has been interpreted as punishment, perhaps for premarital transgressions. She is young, around fourteen years old. One feature grabs immediate attention: the cord, a so-called *sprungband,* tied around her face. When she was found it was lodged under her nose, but it was later placed by archaeologists over her eyes. "We may feel sure that it had been used to close her eyes to the world," writes Glob.[37] According to Allan A. Lund, this is a possible misinterpretation. The string tied around her mouth would have prevented her from screaming, he writes, and "the garrote (or blindfold) might have served irrational motives: the intention would have been to prevent the doomed

Figs. 6.6 and 6.7. Two children's drawings transform Windeby Girl to an empowered "Skeleton Girl" on the left and a disempowered "Hilfe Girl" on the right. Both drawings show how young children negotiate the gruesome scenario of being suspended between life and death. Photographs by Karin Sanders (from Gottorf Castle, Schleswig, Germany).

from spewing her particularly effective curses. The dying girl's last words or gaze would have been specially dangerous and harmful."[38] Gaze or scream, the archaeological interpretations seem to suggest that the girl was silenced and blinded to avert retribution.

Outside, amongst the posted responses, a drawing by a seven-year-old (fig. 6.7) shows a terrified person sinking helplessly into the bog and shouting, "Help, help, help," while a shrieking black bird hovers above. Another child's drawing (fig. 6.6) shows a bog girl's face with a big smile and a bow in her hair on top of a body resembling a rather full-boned skeleton. Far from seeing bog mummies as frightening, this child finds in the ancient girl a content contemporary.

In *The International Council of Museums'* "Code of Ethics for Museums," which "sets minimum standards of professional practice and performance for museums," we read that human remains "should be housed securely and respectfully"; that they should "be available for legitimate study on request"; and that if they are exhibited, this "must be accomplished in a manner consistent with professional standards and the interest and beliefs of members of the community, ethnic or religious groups from which the objects originated. When sensitive material is used in interpretative exhibits, this should be done with great tact and with respect for the feelings of human dignity held by all peoples."[39] But how does one define "respect for the feelings of human dignity?"[40]

Virtually-Inside

The ethics of displaying human remains has developed over time from the old curiosity cabinets and anatomic theaters to modern ethnographic displays. In recent years the display of dead bodies has been diverted into two directions: one which leads to the virtual museum filled with virtual bodies and virtual dissections, geared toward the production of experience; and another which maintains the possibility of authenticity. Some bog body exhibits, although not the most conventional ones, partake in the reconfiguration (and implosion) of the museal space in line with what has been described by German cultural critic Wolfgang Ernst thus: "The museum is no longer the terminal for parcel post from history, art, and culture; instead the institution becomes a flow-through and transformer station. Its demand now is mobilizing, unfreezing the accumulation of objects and images in its repositories, making them accessible to the public eye by displaying the stacks or recycling them into the exhibition area [. . . .] The architecturally supported memory of museums is liquefying in an age that permanently transforms objects into images."[41]

In a large-scale traveling exhibit called *The Mysterious Bog People*, these questions are further complicated by the use of high technology. The visitor (in this case to the exhibit's tenure at Drent Museum in Assen, the Netherlands, 2005) is met by a voice from the floor: "Stop, don't come nearer." Looking down, you see a peat-digger looking up and addressing you from a video screen in the floor; he explains how bodies are sometimes found in bogs and how peat diggers such as himself have been scared by the prospect of spectres from the past. Further into the exhibit, a screen split into three vertically stacked sections shows high-tech images and CT scans of the bog body known as Rote Franz (here on tour from his original museum location) along with digitally fleshed-out images of his body. Finally the face of Franz is beamed onto the screen from top to bottom: while the emaciated and shrivelled bog face disappears at the top, the reconstructed face zooms into modernity below. This is repeated in a continuous flow interspersed with images of ancient bogs, making the meeting of bog and technology pregnant with time.

There is no affectation of authenticity at the opening of the *Mysterious Bog* exhibit, but instead a sharp awareness of the trope of time travel; the conventions of which highlight the fact that most museumgoers are familiar with them from film and TV shows. Further into the exhibit, through a darkened room, another voice resonates from behind a glass screen. In the darkness a face appears: first as a white fog-like presence; then projected as the image of a misshapen and warped bog body face which gradually morphs into various stages of face reconstruction. Finally the projected images yield to the "real" thing, the actual face reconstruction of a girl with long blond hair: the Yde Girl (fig. 6.8). The final trick of animation comes when the "head" itself is overlaid with digitalized images allowing the eyes to flash and the mouth to move to the sound of *her* voice. She *comes alive.* Then the light is dimmed and the show repeated. The narration is interesting in that Yde Girl tells the onlooker how she was found, probed, examined by a CT scanner, and interpreted. "They think they know why I was sacrificed. But do they know?" she asks in a ventriloquized self-examination by the curators. The voices of the peat digger in the video pit and the bog girl in her dark room overlap and create an interesting simultaneity, forcing visitors to ponder the temporal double projection of voices from different moments in time: a two-thousand-year old girl speaking in sync with a peat digger from around the end of the nineteenth century, both voices mixing with those of present-day onlookers.[42]

In the *Mysterious Bog People* exhibit, the animated speaking face of Yde Girl is deliberately separated from—but also works as a teaser for—the real, natural remains of the bog girl, which are not seen until one reaches the very end of the exhibit. Before museumgoers reach that far, however, they must wan-

Fig. 6.8. Through the use of digital projection, Yde Girl's face reconstruction was made to speak to visitors to the *Mysterious Bog* exhibit. The voice gave her an uncanny presentness that made the fleshed out face strangely familiar. Photograph © Drent Museum in Assen, Netherlands.

der through a two-sectioned exhibit. The first section is a kind of imaginary world. A large darkened room is filled with white gauze flowing from the ceiling to resemble the misty vapors from the bog (called "bog brew" in Danish, or "misty maidens" in Dutch) and a soundscape fills the room with tunes that underline a sense of mysticism; it is as if the natural space (bog) and the cultural space (museum) have coalesced. The room brims with bog treasures from different countries, including the bodies of the Weerdinge Couple and Rote Franz, each resting in darkness on peat-like material behind glass covers. On the walls throughout the exhibit are artist's depictions of scenes from people's everyday lives through various ages of the past. Many of the faces are modeled on those of the archaeologists, artists, curators, and technicians who worked to make the exhibit a reality. This moment of self-reflexivity can be seen as a clandestine wink of the eye (the public is not told that the faces on the walls belong to real and contemporary people) to indicate awareness of the fact that the archaeologists themselves are agents in the narratives and display stories they produce.

At the end of the exhibit, the *real* Yde Girl is finally found in the second section of the display, which is devoted less to the wrapping of the museumgoer in imaginative and suggestive atmosphere and more to the didactic. Yde Girl's natural remains lie simply on a gurney in front of a large white CT scanner. Harking back to the technological tour de force of visual display, this arrangement reminds us that the scanner's foremost function vis-à-vis bog bodies is to provide forensic information, not popular culture.

Even so, travel to the past is more than ever also a travel into the innermost details and fibers of the archaeological body, and oftentimes this travel is seen as an exercise in *event-making* more than *information-making*.[43] As was the case in the nineteenth century, archaeology has once again become fashionable and popular, both as experience and event. For example, in the British Museum's special 2004 show, *Mummy: The Inside Story,* the audience is asked to wear 3-D glasses for a virtual joyride "inside" the body of the mummy of Nesperennub, priest of Karnak in ancient Egypt. Sitting in a high-tech theater, the viewer is privy to a twenty-minute narration by the British actor Sir Ian McKellen while a three-dimensional mummy floats through time and space as a corporeal spacecraft (the time-travel and time-capsule connotations are palpable) and then gradually morphs from corpse into a CT-scanned image and ultimately into a face reconstruction which is finally superimposed upon a living actor who resembles the remade face of the ancient man. Alive and present on the screen, the virtual mummy and his contemporary stand-in *drag* the spectator back in time through maneuvers familiar to us from popular culture: 3-D hieroglyphs resembling video-game graphics float rapidly toward the onlooker, who may well feel the thrill of almost being hit by one. Egyptian mummies, arguably the most popular archaeological artifacts for museum-going children, have for a while become a theme ride akin to those offered to tourists at Universal Studios which feature special—and uncanny—effects like those known from horror movies and television shows like *CSI* ("crime scene investigation"). Like the *CSI* audience, the *Inside Story* audience is allowed to follow the archaeologist-detective-media designer as he penetrates the ancient mummy for clues and answers to unsolved mysteries.

Unlike the uncanniness of the now well-trodden path of familiarity/unfamiliarity proposed by Freud (an uncanniness which viewers of today's popular culture know to the core), the archaeological inside story offers educational and scientific thrills—as in the case of Nesperennub, when the MRI reveals a hitherto hidden secret in his skull: he appears to have had a brain tumor.[44] Abbreviations and acronyms like DNA, MRI, CT, and even 3-D and CSI function as a kind of modern-day hieroglyphic in which the operations of technology and the institutions of visual display culture (including those of television) feed the archaeological imagination with visual and verbal images of the "inside story" of human existence.

After feasting on the marvels of penetrating the mummy at the *Inside Story,* the British Museum visitor can wander to the old part of the building and experience the rather more quaint exhibit of the bog body known as Lindow Man (see figs. 7.10 and 7.11).[45] His remains are exhibited in a corner, under dim light in a glass showcase.[46] A hologram of his face reconstruction is mounted on the back of his glass case at a safe distance from the authentic body. Here the museumgoer can listen to the guides tell about a spectacular story of misidentification. It involves not Lindow Man but another bog body found close to his. A man named Peter Reyn-Bardt had been accused of killing his wife but had firmly denied the accusations of murder, and there was no evidence against him. Then in the middle of May 1983 a woman's skull—of the right apparent age, with tissue and hair still adhering—was brought in from Lindow Moss. It had come from a spot about three hundred meters from the cottage where Reyn-Bardt had lived. On being told of this discovery, Reyn-Bardt made a full confession, confirmed the information given, and said his wife had been killed in a fight over blackmail demands.[47] The police, in the meantime, had asked for an archaeological opinion of the skull, and a radiocarbon test dated it to be well over 1700 years old. Ironically—and herein lies the crux of this story—Reyn-Bardt was tried at court in December 1983 and convicted of his wife's murder. With the conviction, the untimely (improper) resurfacing of a bog body had once again both disrupted order and helped restore it.

While many visitors to the British Museum are entertained by this "old-fashioned" story of misidentification, such tales must now compete with new curatorial strategies that show an eagerness to break with convention and practiced tradition—or at least offer other opportunities for reaching an audience accustomed to fast-paced technologies. Although no bog body exhibit to date has been able to match the technical tour de force of the British Museum, the use of cyberspace and computer memory in exhibits offers a different model of stored time than the ones we have visited in traditional museum settings. "Digging" for bog bodies and looking at displays has, to a large degree, become virtual. In cyberspace we encounter a kind of ultimate presence unmoored from the temporal logic of historical time and disentangled from individual human beings' sense of irreversible time. Here, time offers possibilities for repetition, reversal, and high-tech experience.[48]

The virtual museum displays a spatio-temporal facet that transposes the corporeal bog body to repeatable moments. As Wolfgang Ernst has pointed out, virtual museology disrupts the chronological order on which museums were built. Instead of chrono-linear museum exhibition, we now experience a co-presence in which objects and testimonies are kept on the same digital "page." The museum is once again a curiosity cabinet. Ernst explains: "Beginning *in medias res,* the virtual museum visitor navigates on the monitor through the Internet where (s)he faces a kind of profusion of data that might deter traditional archivists, librarians, and museum directors. The digital wonderland signals the return of *temps perdu* in which thinking with one's eyes (the impulse of *curiositas*) was not yet despised in favour of cognitive operations. Curiosity cabinets in the media age, stuffed with texts, images, icons, programs, and miracles of the world, are waiting to be explored (but not necessarily explained)."[49]

Bog bodies' special kind of spatio-temporal being (and not least their humanness) certainly provides a kind of corporeal treasure cabinet. Not only does the analogy between art museum and archaeological objects and collections suggest that bog bodies grant the potential for intimate subjective investments with the past; they also project a sense of continuity and permanence because they provide a visualized historical linkage which makes it easy for the museum visitor to find herself transported in time and placed vis-à-vis ancient kith and kin. Yet when bog bodies are placed into virtual museum settings, the "normal" mnemonic museum frame implodes and the bodies are no longer governed by the logic of a discursive trajectory through which archaeological objects are otherwise usually historicized. Rather, the bodies are floating in a space that at least to some degree brings objects back to their essence as material. And as material without complex historical designs, the archaeological objects have been decoupled from the historical dimension.

Human or Thing

An often-quoted archaeological credo reads: "The archeological excavator is not digging up *things,* he is digging up *people.*"[50] The question is then: are bog bodies people or things? And can museum displays of archaeological human remains like bog bodies escape a sense of redundancy, given that these archaeological *things* already are (or were) *people*? Most museum display strategies, as we have seen, essentially aim to create firm contours of protection around bog bodies put on view. But they also aspire to overcome dichotomies between the dead material and its imagined past life, similar to what British anthropologists Elisabeth Hallam and Jenny Hockey have called the "object body (the dead body in the present) and the embodied person (the living body of the past)."[51] Although bog bodies too are bifurcated into "object body" and "embodied person" (the first is the material body that can be seen in the museum; the other is the imagined person who inhabited the body) curatorial strategies such as facial reconstructions can be said to suture "object body" and "embodied person" into one. In museum displays this re-suturing, and the rehumanizing that comes with it, is of vital importance.

To fully understand the ethical aspects of this, let us again call to mind George Bataille's thesis that the corpse always puts violence on view; a violence that represents the destruction of life. The viewer, Bataille argues, struck with a sense of "destiny" but also with a sense of "awe," protects himself or herself from the violation that the corpse represents. The distance we place between dead corpses and ourselves, in the form of various taboos or curatorial strategies, serves as a safety measure.

One way to safeguard against this experience of violence is to imagine that the human in the display case is not *quite* as human as the onlooker. This kind of encounter with corpses has been articulated by Ludwig Wittgenstein in his *Philosophical Investigations* as a query between the nature of live bodies and inanimate "things": "Look at a stone and imagine it having sensations—One says to oneself: How could one so much as get the idea of ascribing a *sensation* to a *thing*?" He goes on: "And so, too, a corpse seems to us quite inaccessible to pain.—Our attitude to what is alive and to what is dead, is not the same." Wittgenstein's aim is not only to show how we assign meaning through language to things, but how sensations such as pain are part of a shared (public) language that we can test by way of attributing it to something inanimate. "How am I filled with pity *for this man*? How does it come out what the object of my pity is? (Pity, one may say, is a form of conviction that someone else is in pain)."[52] The "someone else," in question here is "human behavior, which is the expression of sensation."[53]

It would seem apparent that whenever a human corpse is displayed in a museum, it is forced to occupy an intermediate space in which it is both a *person* and a *thing*. It may also be helpful to remind ourselves that something is experienced as a "thing" when, as American literary critic Bill Brown has pointed out, there is "a changed relationship to the human subject." That is to say, we "begin to confront the thingness of objects when they stop working for us." Things, to follow Brown's theory, own an "audacious ambiguity" in that they have both a sensuous force and a metaphysical presence, even if in a rather opaque way. In fact, he points out; the word "thing" tends to "index a certain liminality, to hover over the threshold between the namable and the unnamable, the figurable and the unfigurable, the identifiable and unidentifiable."[54]

But is that the case with bog bodies? Dead humans have in some ways lost "function" and can be said in this sense to have gained thingness. Yet to think of human remains as things suggests erased identity and facelessness—and because face and identity are key players in our perception and interest in bog bodies, it is naturally difficult to see the bodies as merely things. Exteriority (thingness) in a dead body is always imbued with humanness, and as such with implied interiority—lived life, a departed soul. At the same time, the museum visitor's gaze and visceral proximity helps re-ignite, re-imagine, the things as humans. Said differently, unlike the remains that Wollstonecraft found so abhorrent and mistreated, bog bodies seem in some ways exempt from becoming "a monument of the instability of human greatness" because the kind of "poetics of detachment" in which they are couched gives license to examination of the corpse as the historical, scientific, even artistic artifact it has become.[55] As such, bog bodies are already imitations, objets d'art, or relics—or, as Seamus Heaney suggested, "these can now be classed as *objects,* if you like, because they seem to belong with heads made of clay or bronze or marble that we see in the art museum."[56] Although one might object that a comparison between artistically represented bodies and bog bodies is off-kilter, the point I hope to make is that the matter of naturalness and artificiality, authenticity and inauthenticity, in archaeological display culture becomes particularly complex when human remains are put on view. Even when museum displays use reconstructed faces, these reconstructions offer a roundabout way of "seeing" originals by way of copies. The underlying quest seems linked to a general thirst for authenticity.

The Problem with Authenticity

Swiss folklorist Regina Bendix has discussed this pursuit for the authentic as "a peculiar longing, at once modern and antimodern."[57] Looking for authenticity is, she points out, "oriented toward the recovery of an essence whose loss has been realized only through modernity, and whose recovery is feasible only through methods and sentiments created in modernity."[58] Our "craving for experiences" and for "unmediated genuineness" paradoxically points to a longing away from the modernity to which it is (ironically) wedded. Produced in and of the modern, authenticity thus defined is also decidedly anti-modern and nostalgic. It seems to promise transcendence, yet is elusive and contingent. It articulates our anxieties about losing ourselves by promising the real, the genuine, and the uncorrupted against the falsity or even faithlessness of an alienated present. It is this problem with authenticity that most poignantly shows how difficult it is to place the archaeological object, human remains particularly, as purely (and pure) visual portals to the past.

Authenticity is both seductive and deceptive. It is seductive in that it promises access—directly—to a sense of the real or a place of origin seemingly unspoiled by the messiness of later, and auxiliary, markings and demarcations. It is deceptive in that the authentic often needs to be certified by someone or something, and therefore is not immune to the projections and subjectivity promised by its very authenticity. Authenticity itself has served to guarantee a given object's worthiness of examination within academia; "authenticity" as a concept is therefore imbedded in the history of institutions and disciplines of knowledge. Because authenticity spells originality and originality in turn creates canons (or so convention holds), the scrutinizing of what is—or is not—canonical has brought about a questioning of authenticity as well.[59] Expert assessments (historians verifying historical facts, archaeologists attesting to the genuineness of an artifact, or literary scholars professing to the validity of canons, to name but three disciplinary fields) produce variations on the theme of authenticity, but also necessarily open the floodgates to interpretations.

No institution seems more fraught with questions of authenticity than are museums. This, as American historian Spencer Crew and museum curator James Sims point out, becomes all the more important because authenticity is not about "facticity or reality"; it is "about authority. Objects have no authority; people do." To implement the authority, the museumgoer must have confidence in the "voice of the exhibit."[60] The problem is that the "voice of the exhibit" at times obfuscates the lines between "morbid curiosity and scientific interest." Ethical questions surrounding the display of bog bodies in museums are similarly positioned between the "morbidity" of looking and the pursuit of scientific probing. Looking at *human* remains—no matter through which optic—is therefore always, I would argue, a matter of "appropriate" respect. But the question of how the parameters of such respect are drawn is complex.

Maintaining authenticity, an essential part of the foundation of ethnological and archaeological museums, is seen by some curators as an ethical choice and as a way to preserve the "realness" of the human body. Yet, as we have seen, insistence on authenticity is easily compromised.[61] Similarly, "the taint of artistry," to use American literary critic Stephen Greenblatt's terminology, is often unwelcome when we want to have authentic experiences or look at authentic material.[62] Yet the question is whether such a "taint of artistry" is different from the other kinds of glossing in which most display culture is embedded.[63] After all, unless the bog body remains in situ, authenticity has already been compromised; when bodies are moved into an unfamiliar realm, authenticity is always and necessarily at peril. British archaeologist David Crowther has a similar point when he notes that a museum object, of necessity, undergoes a shift: "From *in situ* and unknown, to removed and researched, before it has any impact on knowledge, the evidence has largely lost its original integrity. Its meaning rests only in the information attached to it in the form of the associated records."[64] Therefore museums, traditionally seen as harboring objects of authenticity, always submit to the process of representation and to the inherent inauthenticity of displays. Said differently, when a bog body has been excised from its natural environment it has been submitted, semantically speaking, to the art of the metonym. The absent whole (here the bog and its environment) is only suggested, and is left to be filled out by the power of the viewer's imagination.

In regard to bog bodies, the best way to finish our reflection on the ethics of display and the nature of authenticity is to investigate the art of face reconstructions. In today's world, face reconstructions—such as those of Lindow Man, Yde Girl, Grauballe Man, and others—may seem rather trite to the hard-core museumgoer when they are compared to the new technologies of visual display. Nevertheless, face reconstructions call for a particular kind of physio-emotional response in which the experience of authenticity (the emotive aspect of being face-to-face) is continually renegotiated. The visceral aspect, so fundamental in museum displays of human remains, has been bifurcated so that the "raw" body is buffered and glossed by the "processed" copy, in effect highlighting a vacillation between the bog body as a person and as a thing. As we shall see in the next chapter, when the archaeological specimen gains a "human" face in reconstruction, the place on the human body that is most intimately tied to individuality and personality can become a site for inauthenticity.

CHAPTER 7

Making Faces

I started this study with a face. Let me now finish with a question: Why are faces so important? "Face-to-face" is by far *the* most commonly used expression when it comes to describing bog bodies, in strong competition with the metaphor "frozen time." One would be hard pressed to find any study of bog bodies that does not use this figure of speech. Why is that? The answer seems almost too obvious—namely that the face is at the very core of our understanding of "identity," "personhood," "personality," and, not least, our "humanity." This understanding is rooted, it seems, in a quest for authenticity and in the assumption that no two faces are completely alike.[1] But identity and humanity are more complicated, less universal, and less straightforward than the nomenclatures "identity" and "humanity" seem to suggest.

Faces are grasped not just as material physiognomies, but also as intimate corporeal geographies with legends and keys. A face consists of various co-players: the eyes, the nose, the mouth, and so on. To have or not to have a face is perceived as having or not having a self; to have a face is to be individual, unique, and distinct from others. To be robbed of our face is to be robbed of our identity marker. The language we use when we talk about our personal characteristics is telling. As the metaphor "faceless" illustrates, without a face we are anonymous, lacking personality—or, as in the case of "losing face," we are vulnerable and weak, unable to hold on to status, honor, and pride. The grammar of the face is complex. We need to keep a "straight face" when we "face" the world and its demands, etc. Yet the face is an unreliable document. We can be "two-faced" like Janus—and unlike fingerprints, which we traditionally locate as a privileged site of identity, faces are malleable and changeable. As a result, identity and authenticity can be challenged.[2]

The importance of faces for anthropological or archaeological purposes can hardly be underestimated, and our fascinations with prehistorical and historical visages are difficult to separate from present-day projections. In scrutinizing faces we seek depth through surfaces; we invest faces with historical meaning. To read closely into faces is to read, or so we hope, some truth about who *they were* and who *we are*. We are inevitably caught in a field between the heterogeneous and the homogeneous, between wanting the face to be individual and unique, like no other, and wanting it to be inclusive, like ours. The use of the face for mass identification (history is full of such examples) and the apparent seductiveness of gazing into a visage that symbolizes one nation, for example, often causes the face to be condensed into symbol.[3] And as symbol, faces are easily hemmed in by prejudices and historical or political claims. The inherent danger in such projective representations pushes us to look beyond specific material representations and to open up to an examination of the face with respect to a broader understanding of humanness. With this in mind, any attempts to read faces from the past must be navigated with care.

The French philosopher Emmanuel Lévinas's notion of the *face,* for example, offers a way to explain the ethical demands that others make on us. The Levinasian *face* is "definitely not a plastic form like a portrait," but is, as he puts it, "the most basic mode of responsibility"; it is "the other who asks me not to let him die alone." And most importantly: "The face is what one cannot kill, or at least it is that whose meaning consists in saying, 'thou shalt not kill.'" To Levinas, then, the face is an "epiphany," and an "event."[4] The visual representation of a face, on the other hand, is the place "where humanization and dehumanization occur ceaselessly," so that the *face* in representation essentially becomes a paradox. The paradox is that the human face, when captured in "a plastic form," can lose the very humanness it proposes to show. The question for this chapter becomes, therefore, how face reconstructions of bog people fit into this paradox.

Beneath the Mask

Face plays a key part in archaeological lore. "I have gazed upon the face of Agamemnon," the German archaeologist Heinrich Schliemann is said to have boasted in a telegram to the king of Greece in a dramatic gesture to sensationalize the already sensational discovery of Agamemnon's mask. His moment of unfettered access to the real face, however, was brief. The face beneath the mask instantly and famously dissolved into dust.[5] Even if, as has been shown by many, there is good reason to believe that Schliemann's utterance was fabricated, the face-to-face encounter between the archaeologist and the ancient king serves well as a trope for archaeology's straddling of appearance vis-à-vis disappearance. In the archaeological act of digging, the necessary altering of the location of the find always curbs the moment of potential authenticity and free and open access to an elusive past. Like the countenance of Agamemnon—brought to light only to vanish—other archaeological objects are damaged not only by time but also by their excavation. To reestablish and restore is, then, an obligatory part of archaeology's didactic purpose; and with it comes a basic questioning of whether reconstructions of the past are, at least in part, fictitious. It also brings to the fore the phantasms of immortality that we assign to faces preserved.[6]

When we have the opportunity to observe faces from a distant, even ancient past, we seem to instinctually place ourselves in relation to these strangers, as if to make ourselves contemporaries with them. The same was the case for visitors to an exhibit entitled *Ancient Faces: Mummy Portraits from Roman Egypt* at the Metropolitan Museum of Art in New York in spring 2000. *New York Times* art critic Holland Cotter imagines a compressed time-travel when facing these mummy portraits: "Maybe you have met them. The faces certainly seem familiar. The prettiest girl from high school is here, and that smug 10-to-6'er she married. So is a distant, distant cousin, and a barely remembered college chum, and even an old flame you'd prefer to forget."[7] You could go over and talk to them, Cotter proposes, before he concedes that they are off in a world of their own, two thousand years ago. Such face-to-face encounters with ancient people, even if this takes place by proxy of replicas in the space of a museum, can be seen as a variant of what Italian semiotician Patrizia Magli in "The Face and the Soul" has called "face recognition practices." The face, she writes, has a "fluctuating and unstructured logic" but also a "sort of perceptual perseverance" that allows us to recognize each other.[8] Western thought, she shows, has not only been "fascinated by this paradox" but has also "from its origins, attempted to record and explain face recognition practices."[9]

Mummy portraits and bog body reconstructions do not recognize us, of course, so here face recognition is a one-way street and part of a broader concern with assigning meaning and place to things in the complexity of our humanness. Neither do bog bodies come neatly wrapped in linen with painted portraits on sarcophagi to help facilitate the realism of face reconstructions. In fact, bog bodies may be entirely without face, grossly disfigured, or simply in need of a face-lift. Nonetheless, faces matter greatly for our understanding of their stories. As you might expect, the few bog bodies with relatively pleasant and recognizable faces, Tollund Man first among them, have received more attention in bog body literature than those with absent or severely damaged faces. Despite the fact that plastic representations of faces, as Levinas argued, embody an inherent and paradoxical fluctuation between humanization and dehumanization, face reconstructions also supply voices and stories to articulate this paradox.

Contemporary Prometheans

With face comes voice. And with voice comes story. So when we see flattened or vacant and distorted skin envelopes, our curiosity is aroused ("I wonder what he or she *really* looked like"), and to satisfy this inquisitiveness the attention of museum curators have turned progressively to "face making."

The face reconstruction of Yde Girl will help illustrate this. According to the Dutch archaeologist Wijnand van der Sanden, who initiated the making of her face, her new appearance served as a mirror for some museum visitors who claimed to recognize their own features in hers.[10] But more conspicuously, after the face reconstruction, a number of poets, dramatists, and others have been moved and motivated to give her a life story. The humanity of her "new" face, and the face-to-face familiarity which museum visitors experience, transports her out of her anonymity as an archaeological specimen and into the familiarity of a recognizable fellow human being. The number of texts produced after her face reconstruction suggests how important these face-makings are for the literary imagination.[11] When Yde Girl's face was fleshed out, so was her (fictional) life story. The vulnerable and fragile-looking girl who came into view from the collaborative effort of British archaeologist John Prag and medical artist Richard Neave turned out to be different than they had imagined—more peculiar, "with small features, wide-set eyes and a very high, straight forehead."[12] But as they submit: "We have no alternative but to accept the ruling of the skull." The face, although "not common," was "by no means unique and can be seen on people today, just as one sees the features of the lower part of her face reflected in numerous modern Dutch faces. As if to prove the point, our colleague Bob Stoddart looked up as someone sat down opposite him on a train crossing Denmark some time later, and found himself gazing at the spitting image of the Yde-Girl."[13] In other words, "The metamorphosis that a reconstruction undergoes," they point out, "never ceases to amaze, and even those of us who have seen the process before still find it slightly bewildering to see one of these heads looking so incredibly lifelike. The Yde Girl was no exception. This reconstruction was intended both to illustrate the final report on the study, and to serve as the centerpiece in the museum at Drents"[14] (figs. 7.1–7.6).

The result, a "small, slightly peaky face with its blue eyes set well apart and wide mouth, with long blonde tresses falling down on either side of a remarkable high forehead, creates a powerful impression as soon as one meets her. The proportions of her skull have given her a face that one may not call beautiful, but it is certainly striking."[15]

Prag and Neave's explorations in *Making Faces* offer a kind of poetics of face reconstruction.[16] They describe it as a visual art-science, and they demonstrate a keen understanding of the ramifications and interdisciplinary aspects of their work. In contrast to the so-called Russian or American method (more on the schools later), Prag and Neave's so-called Manchester school relies primarily upon "the musculature of the face, but the soft-tissue data is also used in order to ensure consistency."[17] To demonstrate the scientific validity of their method they have conducted controlled (and therefore presumably repeatable) experiments on cadavers, which have shown that "a different shape of face would result from each skull," that it was "possible, through reconstruction, to link each skull with its undissected face," and finally that there was an obvious similarity between the final reconstructions and the original faces. Starting with the premise that "faces are fascinating and faces from the past hold a particular fascination," and subscribing to the credo that archaeology is about *people* not *things,* they attempt to bring visual archaeology (or *visage*-archaeology) one step further to mimetic perfection.[18] They contend that the scientific usefulness of the reconstruction goes "far beyond the mere thrill of seeing what our ancestors looked like," yet it is precisely this "thrill" that allows us, I think, to see the face reconstructions as more than archaeological props—as amalgams of artistic and scientific interpretation.[19]

Tradition holds that a death mask can be seen as the last authentic photograph of the deceased, but according to Prag and Neave, death masks may be "realistic and are certainly individual, [but] they are modeled upon the superficial features of the face and thus have more in common with a sculpture created from the outside inward than a reconstruction founded on the skull."[20] For them, the skull is the origin, the inner structure from which the true face

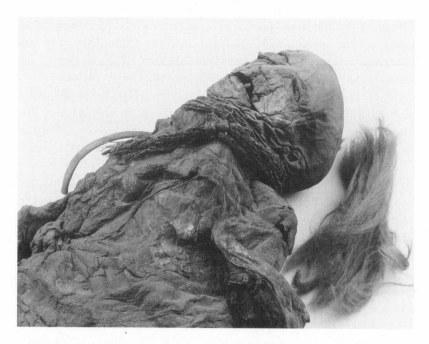

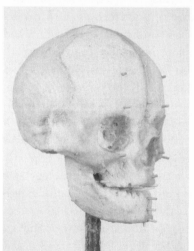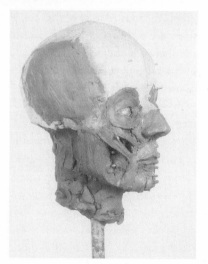

Figs. 7.1–7.6. The various stages of the reconstruction of Yde Girl read like a visual manual for "face making." Photographs © Drent Museum in Assen, Netherlands.

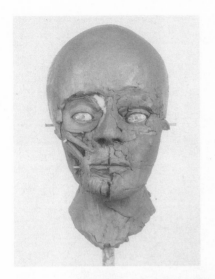

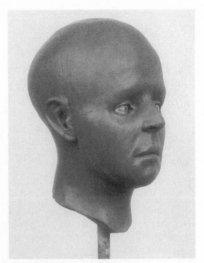

Fig. 7.7. The Dutch archaeologist Wijnand van der Sanden with the face reconstruction of Yde Girl. Photograph © Drent Museum in Assen, Netherlands.

is built. They explain in anatomical detail the ways in which the face gradually takes shape from the skull, and they maintain that their "methodological approach is the most logical and foolproof way of ensuring that the face grows from the surface of the skull outwards of its own accord and according to the rules of anatomy, and reduces to a minimum the possibility of subjective interference by the artist."[21] But they also allow that a face reconstruction "can never be regarded as a portrait. As a general rule what seems to happen is that one creates a face which is very similar to the kind of face which the individual had when alive. In terms of the forensic application such a reconstruction will become just one part of the story."[22]

Face reconstructions are to some degree an adventure in speculation, a fiction mapped onto and into the countenance of someone who once lived; they are a way of filling in and fleshing out what is no longer here. Similarly, a masked and concealed corporeal reality lies behind the face reconstruction of bog bodies. Prag and Neave acknowledge that the final phase, which is the phase in which individuality and personality are applied, is fraught with the subjective; a method cannot be followed in a ritualistic manner. And they concede that without this final projection, the results are often "wooden or lumpish and lack individuality."[23] Nevertheless, they maintain that it is important to start with no "preconceived ideas about how the face will look"[24] and they are also cognizant of the ethical issues involved in face reconstruction: "On the one hand, the academic detachment that goes with both medical and archaeological research tells one that any form of preservation is only an interruption in the natural process of total decay: 'earth to earth, ashes to ashes, dust to dust' is how the Prayer Book puts it. Yet as fellow human beings, one can readily feel intrusive as one seeks out someone's personal details. Is the making of a reconstruction a prying into a person's private life, almost a form of voyeurism?"[25]

In the end they offer two explanations in defense of face reconstructions. First, facial reconstruction is essentially no different than other kinds of historical or archaeological research which result in written reports. In this case the report takes the shape of a visual product. Second, they avoid the term "in the flesh," emphasizing that the reconstructions are not life and that one should be careful with the production of uncomfortable waxwork illusions. If the waxwork is too vivid and lifelike, they say, the secondary medium of photography can help flatten the effect by making it two-dimensional. Whatever form it takes, the "alchemical hotline" that reconstructed faces provide to the past is long and complicated and goes deeper that one might first expect.[26] Let us therefore stray from bog bodies for a moment to look into the tradition and the complexities of reconstruction history.

"Fossils Do Not Speak for Themselves"

The wish to know what our ancestors looked like was a subject of inquiry in nineteenth-century archaeology, and the tradition of reconstructing faces, now a rather commonplace practice, was in effect inaugurated to recreate our fossil ancestor, the Neanderthal.[27] In "Created in Our Own Image," American paleoanthropologists Erik Trinkaus and Pat Shipman point out how the malleability of the Neanderthal, with his or her "special physical features and inferred psychological status," allowed for this early human to serve as a test stone for various theories and constructions of our past.[28] They lament: "Infuriatingly, the fossils do not speak for themselves. It is the examining scientists who bring them to life, often endowing them with their own best or worst characteristics. Each generation projects onto the Neanderthals its own fears, culture, and sometimes even personal history. They are a mute repository for our own nature, though we flatter ourselves that we are uncovering theirs rather than displaying ours."[29] Such projections are manifest in the "full-flesh reconstructions, sometimes in the form of drawings or paintings and at other times in stone, plaster, or metal."[30] Citing a truism, "We do not see things the way they are; we see them the way we are," they point out how the discovery of the Neanderthal "provoked the painful growth of a new field, anthropology," and how these fossil humans "have led a tortured 'life after fossilization,'" not least as "one of the major battlegrounds of the war for acceptance of evolutionary theory."[31]

> These mute fossils shouldered the burden as anthropology [. . .] underwent its birth struggles. Neandertals were the fossil upon which so many new techniques and approaches were tried [. . .] Indeed, Neandertals were the mirrorlike fossils to which everyone looked for evidence on human nature—and found it. What they have revealed, more than their own lives, is the lives of those who have gazed at them and pronounced. It has been, for humans, a journey of revelation and self-discovery, an exercise in ogling the fun-house mirror.[32]

The burden Neanderthals had to shoulder, as British archaeologist Clive Gamble argues, is deeply rooted in Western thought (starting with Aristotle's theory of the natural slave) and in our understanding of human progress. "As a result, our ancestors were not so much enslaved as domesticated within a tradition that judged individuals, races and nations by their position on the ladder of civilization."[33] And the imagery chosen to depict such distant others would carry the characteristics of someone on the lowest end of this ladder, with excessively hairy, stooped, and naked bodies. The visual language used in each reconstruction was intensely influenced by deep-rooted preconceptions. During the rise of science in the nineteenth century, for example, the scientist did not discard "the classical and national claims on our visual language of the past."[34] In fact, archaeologists and paleontologists embraced already established visual paradigms and hereby added to the difficulty, which lies in "translation from evidence to image."[35]

In the case of Neanderthals, the classical iconography of primitiveness most certainly colored the images produced. An iconographical tradition of highlighting specific "uncivilized" characteristics of humanoids in the distant past to some extent still fabricates the visual language in popular archaeological reconstructions. As British archaeologist Stephanie Moser explains in *Ancestral Images,* scientific illustrations and reconstructions across the past few centuries have been used as a seductive didactic tool in spreading the knowledge about humans or humanoids from the deep past. The lure of the visual language and the potency and effect of reconstructions are so strong, she argues, that it produces a fundamental paradox: "Images appear to be creative and free of constraint, yet at the same time they serve to confirm and reinforce established ideas [. . . .] It is clear that once we had imagined the past and translated this into imagery, it effectively became what we pictured it to be."[36]

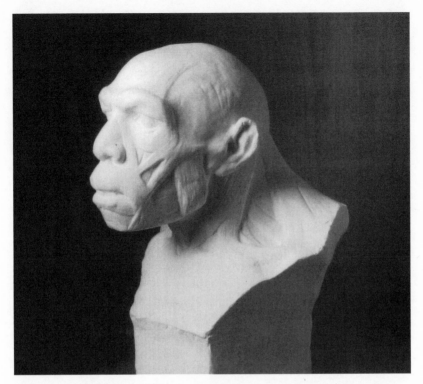

Fig. 7.8. Paleontologist Marcellin Boule's 1911 interpretation of *Homo Neanderthalensis* from Chapelle-aux-Saints resulted in a face reconstruction that emphasized the brutish features of early man. Musée de l'Homme de Neandertal, France © 1910 Marcelin Boule.

Other reconstructions, such as those of dinosaurs, are also connected to our concern for a human past and a human present. As W.J.T. Mitchell has shown in *The Last Dinosaur Book,* "One key to our fascination with dinosaurs is ambivalence. They are like us, yet unlike us. They are terrifying monsters, yet safely distinct."[37] The "archetypical fascination," to use American paleontologist Stephen Jay Gould's expression, is, as Mitchell points out, not only a craving for "spectacles of gigantic violence" potentially facing human beings, but a complex "series of crises in modern history, a linkage between this image and controversies in politics, science, and culture."[38] The fossilized bones speak about giants, but also about "monstrous double[s]" with "double face[s], both human and bestial."[39] As Mitchell rightly points out, they force us to think about the things themselves and ask "what 'real things' are, and what it means to see them firsthand or in representations."[40] The bones of dinosaurs are so familiar, he argues, that we forget "that fossil imprints and bones are the only real, natural traces of the dinosaur that anyone has ever seen. All dinosaurs are products of the creative imagination, assembled out of fragments and augmented with speculations about skin color, ornamentation, sounds, and movement."[41] The dinosaur as a constructed image, he goes on, is always accompanied by words: "a whole lexicon of names, descriptive terms, narratives, and statements. We never *see* a dinosaur without *saying* something about it, naming or describing it, or telling its story [....] We never see the 'real' dinosaur, but only an artifact, a visual-verbal-tactile construction based on its remains and an array of prototypes we use to make sense of those remains."[42] This composite is circumscribed by and imbedded in cultural meaning, and the scientific history of the dinosaur is driven by, or at least influenced by, the myths, metaphors, and images that surround them.[43]

If there is a similarity in the reconstructions of dinosaurs and of hominids, it is that both are part of our quest of self-discovery and both point to essential ways in which nature and culture operate vis-à-vis each other. Dinosaurs are not humans, but as Mitchell convincingly shows, they have been created as a "hybrid scientific/cultural object that gives us a glimpse of the total form of our modernity."[44]

From Dry Bones to Identities

The tradition of making faces in three-dimensional reconstructions harks back to 1895, when an anatomist from Leipzig, Wilhelm His, attempted with the help of a sculptor, Carl Seffner, to reconstruct the face and head of the composer Johann Sebastian Bach. Other faces—such as those of Schiller, Raphael, and Kant—had been remade in two-dimensional forms, but according to the Danish professor of anatomy, Frederik C.C. Hansen (1870–1934), who in 1921 wrote a lengthy study, *Identifikation og Rekonstruktion af Historiske Personers Udseende paa Grundlag af Skelettet* (Identification and Reconstruction of the Likeness of Historical Persons built on Skeletal Evidence), His's reconstruction of Bach was the first in a series of such plastic reconstructions of famous persons.[45] Bach's grave outside St. Thomas Church in Leipzig had been left unmarked after his death in 1750, and a church expansion in 1894 offered the opportunity to locate his remains. In 1895 His and Seffner performed a face reconstruction which confirmed that Bach's cranium had indeed been found. According to Hansen, they created a bust in which both the remains and known portraits of the composer were used. As a result, "the most significant of the [painted] portraits' characteristics were combined in the bust, which in regard to life and distinctive expression surpassed each and every one of the portraits."[46] Seffner, Hansen tells us, even experimented by "forming Händel's head over Bach's cranium; but although this bust outwardly resembled Händel quite well it was an anatomical impossibility [....] such busts are 'anatomical lies.'"[47]

Soon after this early attempt, the new technology (or science-art) was tested on, as Hansen puts it, "reconstructions of faces on crania of prehistoric or contemporary races."[48] Swiss anatomist Julius Kollmann produced the first three-dimensional plastic reconstructions of what Hansen calls a "race portrait," and took it as his thesis that the human races had maintained their distinct characteristics through time as "long as there was no race mixing," thus making it possible to get information about the appearance of prehistoric races by combining His's method with new measurements.[49] Kollmann used measurements taken from "certain key points in the forming of physiognomies in the present population,"[50] and the result of his work was a racialized portrait of an approximately twenty-five-year-old woman from neolithic Switzerland. Another German anatomist, Friederich Merkel, made a reconstruction of an approximately fifty-year old man, whose cranium was around twelve to fourteen hundred years old, from the area around Gottingen. This portrait, too, would resonate with features of what Hansen calls a "race-portrait." Both portraits depicted strong yet even-featured faces precariously close to those of stereotypical, idealized Aryans. In Hansen's assessment, the human eye is sensitive to nuances and peculiarities in the physiognomy of "the familiar type"—that is, of faces that belong to one's own race—"whereas the peculiarities in the physiognomy of other races to all intents and purposes work through their *typical foreignness,* so that the individual differences, for example, in Chinese or Negro faces, [...] in the beginning cannot be perceived as precisely by Europeans."[51] Even amongst Europeans, Hansen goes on, the sensitivity of the eye and its ability to read the delicate differences in various faces are somewhat hampered by the divergence between southern and northern European physiognomies. As such, he concludes: "It is not without merit to talk about *race-portraits* as an expression of a typical and characteristic standard."[52]

Hansen acknowledges that face reconstructions cannot avoid subjective projections, but says this does not make it an "empty search for knowledge or pointless curiosity, but part of a chain of studies that will explore past life and its traces before obliteration."[53] Plastic face reconstructions, then, are anatomy's effort to join forces with paleontology and archaeology in a reading of the body as inscribed "by life itself with thousands of signs like a kind of writing." Hansen concludes with an unattributed citation in Latin from "the old seer," who gazed out over a valley full of dry bones and saw them as living: *et dabo super vos nervos, et succrescere faciam super vos carnes, et superextendam in vobis cutem: et dabo vobis spiritum et vivetis!*"[54] Hansen assumes that his reader is familiar with the Latin quote from the Old Testament prophet Ezekiel and his "vision of the valley of dry bones" (Ezekiel 37:6), with its allusion to Genesis and the biblical description of creation of man.[55] The use of Ezekiel, then, is a both traditional and rather bold gesture that implicitly casts the "science" of face reconstruction as a creation that can ignite life in the remains from graves and hence implement the words of the prophet: "And I will lay sinews upon you, and will cause flesh to come upon you, and cover you with skin, and put breath in you, and you shall live."

Not all scientists would look as favorably on face reconstructions as did Hansen. The uneven tests made on the so-called type specimen of the European Neanderthal skull found at La Chapelle-aux-Saints in France in 1908, for example, and the experiments made under supervision of a professor from Jena University, Heinrich von Eggeling, would give rise to scathing criticism. Still, attempts at reconstructing faces would continue, and different methodologies for that work have been developed in the last several decades—such as the "Russian method," in which the "development of the musculature on the skull and neck is regarded as being of fundamental importance," and the "American method," "which relies primarily upon the measurements of the soft tissues that lie over the bone." The failure of earlier face reconstruction gave rise to skepticism. Critics objected that the results were "probably best left to the ample literature of detective fiction."[56] Others thought that "recognizable reconstruction would be impossible in the majority of Whites."[57]

There are other ethical questions to consider, however, if we investigate what brings the faces of ancient peoples into direct contact with present-day expectations and projections—not least if they offer us images of humanity, or inhumanity. Therefore we are forced to confront a number of questions about our assumptions and interest in faces.

Face Controversy

One case in point is the so-called Kennewick Man. His story made headlines for a few years in newspapers and on television in the United States after his discovery in 1996 by two college students on the bank of the Columbia River near the town of Kennewick, Washington. The human remains turned out to be some 9,200 to 9,500 years old, and they soon prompted speculations in the press, in such articles as "Was Someone Here before the Native Americans?" (*The New Yorker*), "Europeans Invade America: 20,000 BC" (*Discover*), and "America before the Indians" (*U.S. News and World Report*). The controversy started soon after the discovery of the remains but took hold in the press when American archaeologist James Chatters, who had been called by the coroner to assess the date of the find, explained that the remains were probably from an early settler and that the skull resembled someone with Caucasoid-like features. In Chatters's words: "He was simply not who we expected to see. I had been searching for a modern face in order to answer the question 'what did he look like?' [The actor Patrick] Stewart—an unusual (even odd) looking European himself—has some similar features—projecting face, large nose, high sloping forehead. In doing the reconstruction, we were guided some by features of photos of the Japanese Ainu, who many if not most of the early American skulls most closely resemble morphometrically. In the face, I found some of the features of Kennewick Man in Patrick Stewart, not the other way around."[58]

Nevertheless, this observation and the face reconstruction made by Chatters and sculptor Tom McClelland brought about a frenzied debate about ancestry and the ownership of ancient remains. In the press Kennewick Man took on a life of his own. In the *New York Times* of November 9, 1999, for example, American science journalist John Noble Wilford reported how "the uneasy relationship between archaeologists and the people they study has erupted into a bitter court battle over the skeleton [of Kennewick Man]."[59] Wilford wrote that the controversy had "opened new wounds over how archaeologists and other scientists treated Indians in the past." Who owned the body?[60] Chatters responded: "When confronted with a decision that would bias toward one modern group or another, we chose the human norm. That is, for eyes, we did not include an epicanthic or Nordic fold; for lips, we stuck with average turn-out. The color of the clay is gray to stay out of the color issue. Bone and conventions dictated the facial form, the nose width and length. Put a hat on him and he ceases to resemble Patrick Stewart, or for an older generation, Yul Brynner."[61]

In the case of Kennewick Man, science was pitched against religion, the right to examine archaeological remains against the taboos and funeral practices of native tribes. While the press was feeding on this "good" story, it also incited the Odin-worshiping Astura Folk Assembly to lay claim to the remains and to the past it represented. "If study shows this skeleton to be more closely related to Europeans than to Native Americans . . . we think it should be turned over to us for proper study and for the proper religious rites."[62]

Why do face reconstructions ignite such controversy? Do faces, more than any other part of our anatomy, test the limits of how to read bodies as historical evidence? From Plato to Montaigne—from Lavater, for whom the face and body were a manifestation of an inner reality, to Gall, whose famous cranial interpretations laid the foundation for nineteenth-century phrenology (and

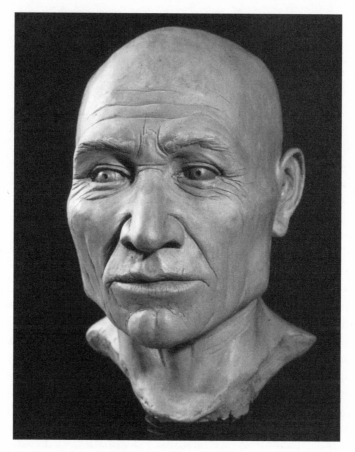

Fig. 7.9. The face reconstruction of Kennewick Man caused a controversy. Who owned the body? Reconstruction © James Chatters and Tom McClelland.

the complex issues connected with that "science")—reading faces has been potentially perilous.[63] Historically, the work of decoding them has been influenced by our circumstances: time, nations, identity, race, and gender. Phrenology and ethnology are twins, or as American literary scholar Samuel Otter articulates it: "To draw ethnological lines on the human body was to interpret them."[64]

Skeptics would say that the final "life-giving" act in face reconstructions, this reanimation of people from the past, is mere fiction-making. But how do these visual representations differ from the scenarios produced by most archeological writing? Is our skepticism about the accuracy of facial reconstruction rooted in a general skepticism about visual representation in archaeology? Are we less suspicious toward written texts, specifically when their topics and concerns are related to archaeology and history? If so, perhaps that is because language can and does make use of qualifiers—self-applied limits that guarantee a distance between what we *know* and what we *think* we know about people of the past. While we may have no reservations about being face-to-face with past peoples in the metaphorical sense, we may cringe when we are brought face-to-face with physical reconstructions. Should our skepticism about (and our fascination with) face reconstructions be tied, at least in some sense, to the battle between words and images? This is a battle that harks back to Plato, Horace, and others, and which has been revived often and vigorously through the centuries until today. If suspicions about face reconstruction are tied to general suspicions about imagery, then it is problematic simply to dismiss the archaeological face reconstructions as fictions and/or lies.

Interestingly, the controversies over disturbance of the dead, and the various archaeological transgressions that have been perceived as insensitive (often Eurocentric) claims to the rights of science and the right to knowledge about the past, seem to have been less pressing in the case of bog bodies. In fact, the kinds of cultural, religious, and ethical controversies that have risen in ethnographical display culture regarding the propriety of appropriating indigenous peoples' remains are strangely absent in bog body discourse. While this may be due to the bog bodies' status as pre-Christians ("primitive," prehistoric, strange, raw, and ethnographically interesting) we have, as seen in chapter 3, located a particular moment in time when race and ethnicity also played a role.

Can one avoid subjective projections? And if not, what do these projections tell us? Is the last phase described by Prag and Neave necessarily an artistic process and not a scientific one? Are the face makers, as Hansen implied, on a par with Ezekiel? Or are they akin to Pygmalion or Prometheus? It is worth remembering that Mary Shelley's novel *Frankenstein* was subtitled "The Modern Prometheus." Today's face reconstructionists cannot lay claim to being latter-day Promethean scientists and artists because their intentions differ fundamentally from those we find in the myth. Nevertheless there are similarities. Prometheus sought fire for human betterment; Mary Shelley's scientist declared that he had "benevolent intentions." Postmodern interest in giving face also carries with it questions of human accountability. If face reconstructions, at least in some sense, are a recasting of the Pygmalion scenario, some (if not all) of the baggage that the myths and their modern articulations carry with them should be considered.

Poetological Archaeologists

Let us return briefly to the matter of humans and things. The objective of face reproduction is in many ways the exact opposite of thingness, in that its aim is to give the illusion of an unbroken and lifelike face. It is a paradox, nonetheless, that in plastic reconstructions the illusion of *life-giving* can only be created by making *lifelessness* manifest. The "fiction" is that reconstruction quite literally enlivens the vacancy we encounter in bare bones or mummified skin. It adds story and voice where there is relative silence. When the clay is formed into flesh and the envelope of skin yields the familiarity of a human countenance, it is not only a scientific forensic enterprise we see, but also the making of a particular kind of physiognomic poetry: a poetics of face in which the history and uses of reconstruction calls ethics, politics, and ideology into view. In faces from the past, "something new has emerged," says American anthropologist Michael Taussig: "A mystery has been reinvigorated, not dissipated, and this new face has the properties of an allegorical emblem, complete with its recent history of death and shock, which gives it this strange property of 'opening out.'"[65] If we stay within this logic, the ways in which face-makers operate (be they poets or scientists) could well resemble the rhetorical employment of prosopopeia, a figure of speech in which an absent or imaginary person is made to speak or act. Similarly, "giving face" is a way to decode the remnants from the grave and to preserve, flesh out, and give voice.

If we return to Prag and Neave's "muscle-by-muscle" and "feature-by-feature" approach, I have already suggested that their bid to translate skeletal remains from "things" to recognizable human beings can be compared to a poetological project. Not surprisingly, they point directly to Heaney as someone who, like them, desires to unveil, dig out, the living face: "He [Heaney] is able to stand before them [the bog people] as they are exhibited to public view and yet understand them for the people they once were, to explain the way in which their bodies can change from expressing a living, personal identity to being an object which shares much with the sculpted heads and statues that are displayed in museums, and which thus speaks across the divide directly to the viewer."[66] They are correct, of course, in placing Heaney's poetic universe vis-à-vis forensic archaeological efforts in making something inanimate come alive. Heaney's bog body poems, they infer, give voice to the museum objects and allow them to speak "across the divide directly to the viewer" on par with other museum objects like sculpted heads and statues. Thomas Docherty similarly describes Heaney's poems, as noted earlier, as being akin to the reconstruction of faces; Heaney's poem on Grauballe Man, Docherty argues, displays less of a "moment of knowledge of the past, but rather an actual recreation of the past,

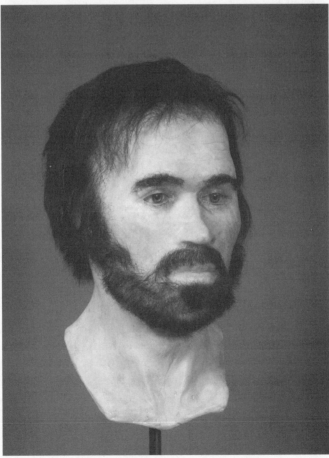

Figs. 7.10 and 7.11. The relationship between the raw and rebuilt faces of Lindow Man makes visible the complexity of dehumanized and humanized faces. Photographs © The Trustees of the British Museum.

now present fully: that is, as it were, the actualization of the virtual."[67] If this process of "actualization of the virtual" is central to Heaney's poems, then face reconstructions seem to be the ultimate actualization in visual form.

Although they do not use the idiom, Prag and Neave essentially point to Heaney's employment of ekphrasis—and in their own way, they could be said to have become "ekphrastic archaeologists" who create history as story and offer the spectator a mirror (not unlike the photograph) in which to search for some kind of kinship. Archaeologists, artists, and poets might well wish to see the past as more than an immobile artifact; it has to be reanimated. Perhaps one might call this archaeological and poetical maneuvering a re-facement of the past, in which words and images come together in the shape of a face, or as a face in the text of the past.

Missing Bodies

To conclude, let me put forward one final point on the relationship between an original face and its reconstruction. American literary scholar Mark Sandberg has shown how wax mannequins in various museum practices allow spectators to imagine "living images" even when the images in question are essentially immobile. This kind of revivification is fraught, he argues, with "the scandalous physicality of the figure, leaving behind a vague discomfort, a whiff of seediness that clings to wax figures to this day."[68] This discomfort is tied to the fact that "the corpse is the hidden secret in the wax museum."[69] Sandberg's point is that wax effigies become corporeal traces of missing persons, and as such perform a pantomime of absence and presence that is both anxiety-filled and reassuring. A wax figure, he argues, relies on "a combination of iconicity (its power of resemblance) and indexicality (its physical connection to the source) for its realistic effect."[70]

A similar kind of iconicity is at stake in face reconstructions of bog bodies, but in this case the linkage to the source body is slightly different from the one that found its genesis in Madame Tussaud's freshly minted masks of the French Revolution's guillotined heads. Her wax portraits relied on a direct correlation between face and imprint, an assumption that inside the wax model there was a manifest trace of the now missing face. This kind of straightforward indexical suturing, however, is complicated in the case of bog bodies. This is not because reconstructions of bog body faces can never be absolutely verifiable since the faces of bog people do not belong to persons known in historical time, nor is it because there are no imprints to capture the last moments before physical corruption set in. Rather, it is because bog body face reconstructions offer a *parallel,* not a *replacement,* to the bog bodies. There is no corpse hidden *inside* the replica. There are no "missing bodies." Indeed, the body is quite present in the museum space, providing stiff competition to its own reconstruction. In other words, the slight indexical shift in the reconstruction's physical relation to the source body is of such importance that most bog body exhibits, as we have seen, negotiate this problem by allowing for ample physical space between source body and replica. This separation indicates a sense of discomfort on the part of curators in allowing the new face to claim superiority and exercise authority over the original, albeit ruined, remains.

As shown through the present study, there are always ethical issues to consider when the intent of face reconstructions is to put a human face on the remains. A character in Sri Lankan-Canadian Michael Ondaatje's novel *Anil's Ghost* goes so far as to say that face reconstructions "look like historical cartoons to us. Dioramas, that sort of thing."[71] But there is a wide gap between seeing face reconstructions as cartoons and seeing them as giving the past a human face. What one *can* say is that these visual reconstructions place the face (of *this* man or *this* woman) squarely in a field where humanization *and* dehumanization occur. This means that if, as I argued at the beginning of this study, bog bodies are liminal, unstable, uncanny, and potentially threatening to social, sexual, and historical order, then face reconstructions can be seen as the ultimate attempt at stabilization, taking control of peoples of the past by making their faces clear and present—and recognizable. In doing so, as the fraught history of reconstruction demonstrates, face makers inevitably become part of a thorny and complex practice and are unable to fully escape the liminality that began in the bog. Nonetheless, by completing what the bog started—preserving, recording, archiving—they also offer us the ultimate prosthetic memory.

Frozen Time and Material Metaphors

If we think of a bog body as an envelope of frozen time, a corporeal time capsule that rests in quietude for centuries, it is undeniable that this envelope of silence becomes quite eloquent after its excavation. Its silence is always broken, or "thawed," in one way or another. It is broken not only by the spade or hand that uncovers it, nor by the scholars or scientists who examine it, but also by the many voices projected or injected into it. These voices, whether they are meant to be scientific or fictive, all have stories to tell. If frozenness in time denotes a certain kind of cessation in the production of meaning, reentry into time also represents reentry into ever-changing meanings. This is not to say that bog bodies in archaeological discourse are treated in a way that conflicts with their irrefutable status as material specimens. But it is to say that fantasizing about the lives they once lived can be, and has been, powerfully articulated both within and outside the discipline of archaeology proper. Indeed, when wrapped in a more generalized definition of archaeological imagination, bog bodies often resemble actors dressed in costumes to play assigned roles directed by inquisitive imaginations. Like dressmaker's dummies, they are given voices by inventive ventriloquists to tell tales of sacrificial victimization, nationality, power, and sexuality.

The materiality of human physical remains, in all its forms, has been subject to trends and to ever-evolving acts of registration, contextualization, and interpretation. Mute, the remains speak "only" in material metaphors. Yet the "image" of mummies, the semiotics of the kind of representations and articulations they undergo, is often pried away from the actual materiality of their remains. Many bog bodies, as I have shown in this study, are known chiefly for the hermeneutic attention they have attracted or for the poetic and artistic ventriloquism to which they have been submitted. They have been eroticized or positioned as ideological and political bodies, as emblems of or for national identity. They have been used as stand-ins for all kinds of historical atrocities—as Holocaust victims, freedom fighters, victims of sexual discrimination, or people punished for other transgressions. They have been seen as noble or horrid, as mnemonic emblems or as symbols of that which has been repressed or forgotten. They have become a strangely personal part of the archaeological record, even if—or perhaps because—we do not fully understand them and their significance.

British archaeologist Rick Turner, one of the leading bog body archaeologists, suggests that the "miraculous survival" of the bog bodies, the fact that they have cheated death, speaks to an almost existential perceptivity in the modern observer: "These few individuals seem to have escaped the fate which awaits us all. They also personalize prehistory, providing a snapshot rather than a panorama of a culture, a human interest in a subject otherwise dominated by material culture. The emaciated, often tortured and monotone appearances of the famous bog bodies have resonances with some of the most disturbing images of modern times, the victims of the concentration camps and of the war in Vietnam, and the countless refugees and starving people who stare out of our television screens."[1] Along the same lines, archaeologists Bryony Coles, John Coles, and Mogens Schou Jørgensen suggest that bog bodies "shock us and shake our preconceptions, and open our eyes to the spectrum of beliefs and relationships that we are intrigued to glimpse, even if frustrated at the present limitations to our understanding."[2] Simply by virtue of the astonishment they raise in us, they seem to test the boundaries we draw around archaeological artifacts and prompt us to clarify our assumptions about the past. As they come to us with an air of fiction, indubitably imaginable even as we address them as actual physical objects, they call attention to the precariousness of "iced up" dichotomies between space and time, between past and present, and between text and image. They demonstrate how such binaries *are, can,* and even *should* be challenged.

The fact that they "look as if they died yesterday" and their uncanny ability to overstep fixed dichotomies allow for time travel in countless ways. This kind of unique temporality is routinely described via the metaphor "frozen time."

Yet "unfrozen" bog bodies seem to live more than twice and they leave us with a long and complicated trail to follow, both in so-called high and low culture.[3] Sometimes they appear robust and resilient, other times vulnerable and fragile when submitted to the gaze and curiosity of a modern (or postmodern) world. Whatever the case may be, as tropes for poetry and prose, or as objects in museum displays, bog bodies test the rather porous line separating aesthetics and ethics. Therefore, when they are persistently re-embalmed in a range of political, aesthetic or cultural wrappings, the ethics of "making art" (in all its forms) from dead bodies is also tested.

In the last few decades archaeology as a concept has had a renaissance, and we have become attuned to the elasticity of material-temporal imagination particularly when it comes to preservation. David Lowenthal's *The Past is a Foreign Country,* from 1985, provided a plethora of examples of how the strangeness of the past has been domesticated in strategies of preservation. He starts quite apropos for the present study by citing Nigel Dennis: "We moderns have so devoted the resources of our science to taxidermy that there is now virtually nothing that is not considerably more alive after death than it was before."[4] Although bog bodies are not subject to taxidermy per se, since they are naturally preserved in the first instance, the point made by Dennis and Lowenthal is that taxidermy can be seen as a metaphor for the kind of preservation that aims to enliven dead material in an extended sense of the word. Actual scientific paradoxes and philosophical and theoretical possibilities may or may not factor into the imagination's work on the bog bodies; in fact, when the precision of archaeology comes under pressure from fabulations, novelizations, and other ways to imagine past lives, imagination occasionally froths with the absurd and uncanny. These are spatio-temporal quandaries that we find not only in the historical manifestations of and interests in archaeological bodies, but also at the very core of both modernity and postmodernity; the impossibility of grasping contemporary life in its totality has given archaeological digging—in the widest sense of the word—a new pitch and valence.

Mummies, of course, have been "modern" at various times. The nineteenth century was absorbed in Egyptomania and indulged in absurd preoccupations with mummies. Tomb looting and semi-public unwrapping of mummies served as quasi-academic "edifying" entertainment. So-called Mummia (pulverized mummies) was digested as medication and mummies were used to produce paint pigment. In the twentieth century, systematic scientific examinations guarded the remains against such destruction, but at the same time mummies found their way onto the movie screen with Boris Karloff's appearance in *The Mummy* (1932), as the reanimated mummy of an Egyptian prince, as prime example. At the beginning of the twenty-first century forensics (and with it, mummies) seem to have had an almost explosive renaissance in the entertainment industry.[5] Bog bodies, too, partake in the entertainment industry, although their appearance in television entertainment is so far limited to the *National Geographic* and *Discovery* channels, or programs like PBS's *Nova*. They have been used in novels, theater plays, poems, and even ballet,[6] in music ranging from jazz to rock[7] and pop ballads, and in photography, painting, sculpture, and Internet art. They have also been placed in absurd and anachronistic settings: as floating objects inside pens,[8] as decorations on ties, T-shirts, posters, cups, and pencil-holders (fig. p.1), and on a number of other curiosa. In the museum world the borderline between entertainment industry and museum strategies has become increasingly blurred, and virtual unwrappings of mummies are now commonplace as part of many didactic museum entertainment packages.[9]

There is a particular slipperiness in humans "frozen in time": a challenge to the concept of embodied time that makes it difficult to hold on to as something (or *someone*) fixed and static.[10] Their frozen-in-time-ness may be used as metaphor, but because they are human beings and look like us (even when

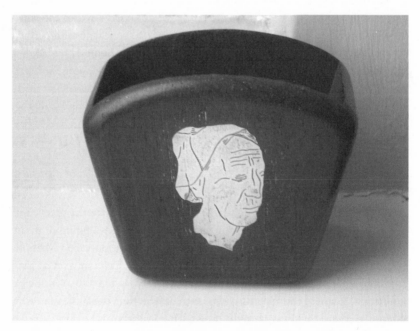

Fig. p.1. A silvery depiction of Tollund Man's face on a wooden pencil holder from the late 1960s. Courtesy of the Paaby family. Photograph by Karin Sanders.

they are unrecognizable), we often experience a heightened appreciation of the nearness they bring, nearness with other humans far removed in time. As the inanimate remains they have become, they do not possess a consciousness of time but they do exist in time and they do have a past(ness) which is projected into our present (and future) and somehow fused with our ways of thinking and of using language about them. Because they have lived human lifetimes— have experienced infancy, childhood, adolescences, and perhaps adulthood before their time was stopped—their temporality mirrors our own experience of personal time.

But personal time aside, we are also tied to these ancient bodies on a larger temporal-corporeal scale. That is to say the human body has not differed genetically for centuries (perhaps for as many as fifty million years), which also means that bog bodies, in such an expanded understanding of humans in time, are not in any physical sense different from us, but are virtual contemporaries removed by a mere two thousand years or so. As the British archaeologist Colin Renfrew has argued (in the context of art and archaeology), the hardware of our prehistoric ancestors is the same as that of modern man, but the software has changed: while our anatomy and genetic makeup is shared, the context in which we live is radically different. Renfrew calls it the *sapient paradox*. If we are the same as our ancient relatives, he asks, how come we are so different? Although bog bodies are much closer to us in time than Renfrew's examples of prehistoric man, they engage a comparable kind of paradox. While their physical remains suggest proximity and commonality with us, they are, or rather *were,* eyewitnesses to a past we do not know.

It is this paradox that makes bog bodies and other human remains different from archaeological objects such as pieces of pottery.[11] We may project all sorts of human traits into material artifacts—in fact we habitually do—but in bog bodies, which are already inhabited by a shared sense of humanity, we are confronted more directly with the push and pull of identification and alienation, our closeness with and distance from the past and its history. In this meeting we can find ourselves in a state of historical flux, a flux that provides the archaeological imagination (in its many forms) with spatio-temporal poignancy.[12] While we are accustomed to seeing artifacts as metaphorical bodies and mapping human attributes onto them, such designation of human attributes to an archaeological object is particularly (and ironically) tested when the object *is* a human body, and therefore *was* a living being who experienced time and traveled through stages in life—someone who lived and died. Their having-been speaks to our being. The artifact that the bog body has become is therefore strangely rehumanized in our engagement with it; indeed it is anthropomophized, made human (again).

Bog bodies challenge the *pastness of the past,* its presumed stillness, by their *presence* in *the present.* We might call them anamnesis receptacles or, more mundanely, corporeal time-vessels. Such presence of the past does not necessarily call forth precise knowledge about the past, but merely suggests the possibility of actualizations in the present. These actualizations, which take place in reception stories, museum displays, and writings of various sorts, call into question not only how a dead body interacts with living persons but also how these bodies engage—in a larger sense of the word—with history. What then does it mean when history is personalized in "frozen" bodies, which have seemingly defied normal time and space confines—bodies that have resisted the standard material temporality of decomposition? As I have shown, in bog body representations the prehistoric oftentimes becomes present-day, and the border between past and present becomes porous.

Time, it goes without saying, is never frozen in a literal sense of the word. Since the beginning of the twentieth century, when Albert Einstein and French philosopher Henri Bergson, each his own way and not always in agreement, made clear that time was relative, we have been accustomed to (or have at least learned to acknowledge) a sense of time's elasticity.[13] It is therefore erroneous in many ways to say that things are frozen in time. But as Bergson pointed out, there is a difference between the concept "time" and the experience or perception of time. Therefore it is not impossible to *imagine* frozen time—which helps explain why the metaphor clings so stubbornly to our experience of bog people. Augustine's legendary question in *Confessions* ("For what is time?") and his lamentation over how resistant time is to definition ("Who can easily and briefly explain it?") tells us that time is hard to grasp because of the persistent conundrum: that the present was once the future and will be the past.[14] Consequently, as Augustine goes on, time essentially exists only as "it tends toward nonbeing." Metaphors like "frozen time" are fundamentally tied to visions of overcoming such resistance to definition. But to imagine that time can come to a standstill is to imagine the opposite: that time can accelerate and expand beyond normal spatio-temporal confines and allow us to travel through it. Einstein tried to demonstrate that time travel is indeed theoretically possible, but as American astronomer Carl Sagan has pointed out, there is always a deep paradox in traveling back in time because of the risk of "interfering with the scheme of causality [that] has led to our own time and to ourselves."[15] What is important here, however, is that what may or may not be doable in the real world, what may or may not be possible in theory only, has always been possible for the imagination. In British science fiction writer H.G. Wells's *The Time Machine,* most famously, the time traveler comes up with an elaborate way to manipulate the relation of time and space that allows for travel. In fictions such as this, the crux of the matter is not whether time travel is *probable*, but rather that it is *possible* as an imagined probability.[16] Similarly, the imagination surrounding bog bodies as frozen-in-time allows us to fantasize that we, with them, can travel through time—for most part to visit the past, but also in some instances to see the future. Although this is not real travel, of course, it can be a joyride of considerable range and scope.

Elasticity of time—taken as a loosely defined model of temporality, and not as a strictly Einsteinian model—has bearing on bog bodies in a number of ways. As they are pulled from their diachronic trajectory *in* time and *through* time and made synchronically present, bog bodies essentially embody two moments in time: the moment of death (the end of *their* life) and the moment, or rather shifting moments, in which they are viewed, understood, and described. The way we approach them is inevitably couched in such double temporality. It is the mystery of the first moment—the fact that we know so little about the deaths and lives of these prehistoric human beings—that fuels the second moment, spurring the imagination. In each case, the second moment or (shifting) moments are typically, but not always, located at the time of the discovery. These shifting moments and the interpretations they produce reflect the times—the zeitgeist—and, as I have shown in the previous chapters, the views and beliefs of the interpreters.[17]

As we acknowledge the obvious, that "frozen time" is a metaphor, we need also allow, as American linguist George Lakoff and philosopher Mark Johnson have shown, that metaphors are based in physical experience and *can* be material and solid, and that we regularly come across other forms of "frozen time" in nature.[18] Consider, for example, the traces of bygone centuries lodged into the memory of a tree, such as in the famous slice of ancient redwood displayed at the entry to the cathedral-like grove of giant trees in Muir Woods, California.[19] In this cross-section, layers of time appear in the form of concentric circles, each denoting a permanent record of times past. Arrows have been attached to point out tree rings that represent years of particular historical interest: the landing of Columbus in America, the Napoleonic Wars, and so forth. For the modern eye, the tree rings serve as a symbolic historical archive that allows park visitors to wonder at how the span of time and the events of history are captured in measurable inches.[20] Trees are often endowed with anthropomorphic features and resonance, not unlike X-rays of human bodies. The various thicknesses and sizes of the tree rings, aptly known by arborists as "sensitivities," show conditions and changes in climate at various points—just as in the case of human bones, where lean years or sickness can result in readable markings known as Harris lines.[21] In his analysis of objects in modernity, French philosopher Jean Baudrillard, for example, has noted that wood has a particular kind of atmosphere, which means that even when manufactured into furniture it retains "latent warmth." Wood, he writes, "draws its substance from the earth, it lives and breathes and 'labours.'"[22] This anthropomorphic feature is further elaborated when Baudrillard notes that wood has odors; that it ages, and indeed has "*being*." "Time is embedded in its very fibres," he argues, "which makes it a perfect container, because every content is something we want to rescue from time."[23] Although the naturalness of wood, in Baudrillard's terms, makes it "warm," its ability to "rescue something from time" is structurally similar to what I here call the freezing of time. Like the natural and organic "time capsules" found in woods or on seashores, or like the ones created by the artistic visions of artists like Smithson, bog bodies seem "naturally" sutured to nature and earth. Many of these examples are about nostalgia, others are of a more traumatic kind. And like the "time capsules" teased out in DNA laboratories, they are linked to the anatomic and molecular history shared by all mankind.

Other natural phenomena provide material metaphors for "frozen time." Amber, for example, is an ancient captor of time with a translucent window that allows us to view objects such as insects fortuitously "frozen" within. Amber insects—"enshrined in their own nectar"—offer the most obvious parallelism to bodies captured in bogs.[24] As early as the first century AD, Pliny the Elder noticed the magic of the "discharge of a pine-like tree" that "originated in the north and often contained small insects."[25] Wedged inside these time capsules of fossil resin, amber insects have endowed poetry with tropes for remembrance and immortality.[26] Hans Christian Andersen, for instance, reflected on amber's mnemonic property as connected to smell. "Memory is like amber," he writes in a letter to a friend; "if you rub it some of the fragrance returns." And Alexander Pope marveled:

> Pretty! in Amber to observe the forms
> Of hairs, or straws, or dirt, or grubs, or worms;
> The things, we know, are neither rich nor rare,
> But wonder how the Devil they got there.[27]

In a less metaphorical sense of frozen-in-time-ness, the drillings of so-called ice cores in Greenland and Antarctica provide tangible and readable traces of sediments in time. In ice cores we can penetrate deep geological time and pull into view "slices of frozen time" containing lines of light and dark snow that represent seasonal changes and temporal layers much like those in tree rings.

Other combinations can be made with ingredients different from these. But each in its own way, amber, wood slices, and ice cores offer material metaphors for frozen time. Each provides the opportunity for imaginary time travel, and each conjures compressed images of the "deep" past humans all share.[28] The prehistoric groves of Muir Woods and the dark and light layers in ice cores cause most visitors to feel dwarfed and experience a heightened sense of wonder about their own existence in time, their own mortality. Peat bogs, of course, do not have the illustrious and transparent color of amber, nor do they have the didactic obviousness of a slice of wood or an ice core. But in their own (albeit opaque) way they are related to these other captors of human history. Whatever our reaction may be when confronted with such visual displays of time, we rarely remain untouched or unaware of the fact that time, even in its most abstract sense, has taken form as graspable and attainable material traces. These traces are both tangible objects-in-the-world and metaphors of time. In them time is represented as duration *and* as immobility. Sometimes, as in the Muir Woods display where visitors are allowed to touch the sliced section of wood, material time is offered as an *experience,* a palpable sense of pastness, a possibility to *feel* a literal relationship to time.

Prehistoric and futuristic imaginings are often blended, and the magic of conserved time in the sophisticated natural archives of DNA can serve as example. These minuscule building blocks form their own unique time capsules, and as molecular travel clocks they are used both as forensic crime solvers and as a way to give us a "deep" view into the fabric of the past. But unlike the Jurassic vision explored in a number of Hollywood movies, DNA in real mummies often falls apart, as Arthur Aufderheide, an American founder of modern paleopathology, has pointed out.[29] This means that "compared to modern bodies, dissecting mummies is salvage pathology" and it means that scientists often have to work "with an alphabet soup of broken-down proteins where there used to be organs."[30] But this reality does not prevent the imagination from extracting imaginary DNA. In fact, bog mummies provide a particularly rich "alphabet soup" in form of untold stories ready to be "cooked up."[31]

Some humans of course are quite literally frozen and they, more than any of the other examples I have used, offer tangible proof of frozen time. I have already mentioned the Ice Man, known as Ötzi, who has been frozen for five thousand years. During that time, as American children's book author David Getz explains: "His fellow Europeans moved on from stone and copper tools to plastic, atomic weapons, and superconductors. They engaged in innumerable wars. They explored the mountains, other continents, the bottom of the ocean, other planets."[32] Another famous discovery of a body literally frozen in time happened in 1990, when the so-called Inca Ice Maiden, freeze-dried and mummified on the mountains in Peru, was found and soon nicknamed "Juanita." She soon made it into popular imagination as a "frozen girl" sacrificed some five hundred years ago as an offering to the Inca gods. The jacket of Getz's book on her discovery reads like a gruesome fairy tale: "More than five hundred years ago, a young Incan girl lived in the mountains of Peru. Because of her perfect beauty, she was given the greatest honor—to be the official sacrifice to the god of Mount Ampato. Before the ceremony, feasting went on for days and priests gave the girl elaborate costumes and jewelry to wear. Then the time came for her to be given to the mountain."[33]

The most recent example of a "frozen man," discovered by hikers in the melting snow of the Sierra Nevada in 2005, was described as a "mummified human time capsule" in the *New York Times*.[34] He was found with a number of temporal markers on him, such as a coin in his pocket from 1942 and a Sheaffer fountain pen. Soon it became clear that he might have been a missing airman from World War II, but his name tag was too corroded to be legible and his face was gone. Only long strands of blond hair seemed to give the "frozen airman" any kind of individuality, and *Times* readers were held in suspense until forensic experts finally confirmed that he had been the twenty-two-year-old son of Finnish immigrants from Minnesota.

Yet while the frozen man from the Alps, the frozen girl from the Andes, and the frozen airman from World War II are preserved *because* they are iced up, time itself has not come to a standstill. Both the bog bodies and the ice people may seem to be stationary *in* time, but they are essentially *out of time*—and at least in the case of the bog bodies, they are also *out of place*. That is to say, the place of their time has changed. To be precise, bogs have their own circuitous time line: while still here *now*, they are different from what they were *then*. As they emerged after the Ice Age, they took around ten thousand years to create, only to be altered and destroyed by human activities of various kinds in the past thousand years. Today, bogs are no longer spaces of sacred or ritual burial, and only rarely are they peat-digging sites or excavation sites. For the most part they have been turned into protected nature areas and have changed radically, both physically and functionally, from what they were. As protected nature areas, many northern European bogs today can be seen as ecological museums for still undiscovered treasures and unseen people. So when bog bodies resurface in a place that is no longer *their* place; when they resurface in a time that is not *their* time, it seems inevitable that moments of *anachronism* take place. Anachronism, "something which does not fit in with its context chronologically," in turn calls forth moments of the *incongruous*.[35] Out of time and out of place, bog bodies pull our focus not only to anamnesis and anachronism, but also to the bizarre.

To sum up, bog bodies are not and should not be seen as pure presences, resting in their timeless essence. They are unique anamnestic time vessels understood differently at different times, challenging our assumptions of an absolute knowable past. As the context shifts, the meaning of what bog bodies are made to represent also shifts. This means that the diachronic and synchronic lenses through which we view the remains give us the chance to understand them both as physical testimony about the time that was *their* time, and as a kind of souvenir from the past, making traces *through* time. One way to grasp such multilayered historicity in bog bodies is to accept the absurdity of their unique temporality. We know of course, that they are *here, now,* in front of us. We can see them, even touch them if permitted to do so by the institutions and curators who house and preserve them. We know that they are remains of human beings from the past. We can date them with uncanny precision. But we are bound by limits. We do not know their names; we do not know their identity; we do not know with certainty why they were placed in the bog; we have to fill in the blanks.

I started this book by conjuring a face and later suggested the image of a handshake between people across time. I want to finish by invoking the famous footprints left by two hominids that once walked the earth across a muddy field of ash left from a volcanic eruption in Laetoli in Tanzania millennia ago. The hardened ash has "frozen" their footprints in time and allowed anthropologists and archaeologists to speculate about our common ancestry. But the footprints also permit the mind's eye to fantasize about the story of a couple walking together in time, side-by-side or so it seems. This imagined couple, germinating upwards into full-bodied beings from the indentations in the ash, has materialized as a diorama at the American Museum of Natural History in New York. It depicts a life-size re-creation of a male and a female, two hairy and chimp-like figures. In American author Bill Bryson's popular "travel guide" through the world of science, *A Short History of Nearly Everything* (itself a testimony to the bridge between the kinds of scientific-academic and popular-story discourse discussed earlier in this study), he points out how the tableau figures have been made human-like by the placement of the male's arm "protectively" around the female's shoulder. "The tableau," Bryson states, "is presented with such conviction that it is easy to overlook the consideration that virtually everything above the footprints is imaginary."[36] The fact that the tableau is "suppositional" ("we can't even say that they were a couple") and the fact that Bryson had heard that the figures "were posed like that because during the building of the diorama the female figure kept toppling over" makes him question the curator, American paleoanthropologist Ian Tattersall, about the museum's intention and display strategy. Tattersall "insists with a laugh that the story [about the toppling female] is untrue" and explains: "'Obviously we don't know whether the male had his arm around the female or not, but we do know from the stride measurements that they were walking side by side and close together—close enough to be touching. It was quite an exposed area, so they were probably feeling vulnerable. That's why we tried to give them slightly worried expressions."[37]

I want to bring in Bryson's narrative and Tattersall's diorama because they so clearly conjure the image of two prehumans "close enough to be touching" and allow the "suppositional" to feed our imagination (and make emotive connections) as we unfreeze the hitherto "frozen" footprints. Colin Renfrew has called the Laetoli footprints "the most arresting, most breathtaking" of archaeological traces, and says they can be seen as art-in-time, or art-from-time "more worthy of our contemplation" than much of the twentieth century art he has compared with archaeological artifacts and traces.[38] I would like to add that our archaeological imagination easily connects these prehistoric footprints to footprints from the futuristic world of space travel. That is to say, the Laetoli footprints are eminently comparable to the equally famed footprints left by the first humans on the moon. Likewise, both the handshake between the conservator and the bog man and the double set of footprints made by prehumans and astronauts suggest a great deal of flexibility in our ability to bridge temporal chasms. Frozen and unfrozen in time, transformed and "re-embalmed" bog bodies should not, as I have shown throughout this book, be seen exclusively as antiquated remnants from a bygone time, but also as part of an ongoing activity performed by the archaeological imagination and by ethics and aesthetics in culture. It follows that bog bodies are small but evocative threads in the larger fabric of memory culture.

An example of the flexibility of temporal-archaeological imagination came into sharp focus at the brink of the new millennium. Leading up to this event, the *New York Times* published two seemingly different articles on human history, reconstruction, and archaeology. One was in the Living Arts section; the other was printed in the weekly Science Times section.[39] While the first discussed a "Times Capsule" to be sealed in the new millennium and reopened in the year 3000, the other dealt with the time capsule offered by ancient human remains and its significance for understanding the past. The "Times Capsule" was to be displayed in an exhibition at the American Museum of Natural History called "Capturing Time." Here the public could view "samplings of the items considered essential for understanding the 20th century" and "explore concepts like biological time, time and the universe, and time and civilization." According to Ellen V. Futter, president of the museum, its "vast collection represents a history of Life on Earth" and the Times Capsule "offers a snapshot of civilization today intended to last 1,000 years."

In an intriguing if unintentional exchange of the archaeological imaginary, the Science section in the same issue offered the reader a more traditional archaeological account from John Noble Wilford. He describes how the loss of a paradigm had "plunged American archaeology into a new period of tumult and uncertainty over its oldest mystery, one critical to understanding how modern humans spread out through the world." In spite of the apparent differences, the irony in printing both pieces in the same edition could hardly escape the readers. Somehow the imagination seems to leap eagerly and easily from the possibility of picturing our descendants one thousand years from now, reading us and our time from the artifacts chosen for the capsule, to our own continued fascination with our ancestors and with the remarkable remains of a past preserved in a less strategized form, such as the accidental sustentation provided by natural phenomena, like bogs.[40]

While I may be accused of evoking a by now overused image, there is a particular point brought out by American art historian and Museum of Modern Art director Philippe de Montebello in his analysis of the archaeological nature of the collapsed twin towers of the World Trade Center which speaks to my study of bog people. The "searing fragment of ruin is already an icon," he writes in the *New York Times* of September 25, 2001, just two weeks after the collapse; "it should stand forever as a sculptural memorial, incorporated into whatever other structures or landscapes are chosen as fitting for this site." The surviving remnant of the disaster "already constitutes a solemn and authoritative statement"; it is "a relic of destruction" and a "testament to renewal." Montebello ties the experience and emotion of the loss and the trauma of September 11, 2001, to the "inexplicably durable" fragment. The trauma, it seems, cannot be separated from the artifact, which was produced and left by the calamity. In fact, it was "authenticated"—over and over again—before our very eyes as television repeatedly showed the surreal yet real moment. The buildings have been seen as "the newest Laocoon,"[41] as "artifacts of anguish" and "raw material for museums."[42] The pain of experience again authenticates the pain in the object, and its authenticity renders it worthy of museal interests both institutional and in situ, in the form of the preserved footprints of the now missing buildings.[43]

Bog bodies are not victims of modern wars, and I do not want to claim that their dead bodies can be seen straightforwardly as representatives of our concern with the kind of terror and politics that haunt modern life. Neither are they prehumans who have left only footprints behind, giving us no more than a faint likelihood of conjuring "correctly" from those prints the shapes that once stood tall. But I can say with some confidence that the representations in which bog bodies have been couched have created a place for them in literature and art as cultural interlocutors that allow us to examine them not only as anonymous "faces," "voices." and "bodies" of human beings in time, but as points of pressure that stress that the traces of human beings force us to articulate ethical and other inter-human concerns of the present.[44] That is why bog bodies matter in contexts outside of archaeology proper. Likewise, the archaic footprints in the desert sand, if seen in sync with the footprints of DNA, hold a particular poetics of time. From the first hominids who lived two hundred million years ago, and from the change in the "fabric" of these ancestors some fifty to sixty million years ago (which gave us the first "human" as we know the term today), the remains and traces of humans tell stories with infinite possibilities.

Two field trips marked the end of my investigation. The first took place in October 2005 when I walked Smithson's *Spiral Jetty* on Salt Lake in Utah. Smithson interested me not only because some of the bog artists (mentioned in chapter 5) were influenced by earth art, nor only because he in turn had been influenced by William Carlos Williams (who had written one of the earliest poems on Tollund Man), but also and more importantly because for Smithson time, to borrow American art historian Jack Flam's succinct formulation, "is never only a disembodied abstraction but always a tangible and material reality. Time must have coordinates in space, must even be made manifest in a quite specific kind of place and thus inhabit as well as contain the material world."[45]

Less than two months later I found my way into the bogs of Jutland, the place where Tollund Man was unearthed. By then I had visited the museum in Silkeborg, where his remains lie, a number of times. On this last trip I did not want to contemplate the bog body itself; instead I drove around the city to find the place where Heaney (in his poem "Tollund," written twenty-two years after his "Tollund Man") had stood and observed how modernity mingled with the ancient landscape. In his "pilgrimage" to the excavation site he depicts how the "hallucinatory and familiar" "Tollund Moss" (notice the word choice: moss, not bog) was marked by the present:

> A path through Jutland fields. Light traffic sound.
> Willow bushes; rushes; bog-fir grags
> In a swept and gated farmyard; dormant quags.
> And silage under wraps in its silent mound.
>
> It could have been a still out of the bright
> 'Townland of Peace', that poem of dream farms
> Outside all contention. The scarecrow's arms
> Stood open opposite the satellite

Dish in the paddock, where a standing stone
Had been resituated and landscaped,
With tourist signs in *futhark* runic script
In Danish and in English. Things had moved on.[46]

With Heaney's poem in hand, I had always imagined that modernity had taken over Tollund Bog, that "things had moved on" and that sounds and signs and lights and satellites and phony *futhark* for tourists had infused bogland with so much presentness so as to make nostalgia and an emotive response to "being there" utterly improbable. I was unprepared to experience any sense of being frozen in time. Yet when I finally stood there, after having had a great deal of difficulty in finding the place, and after walking for more than half an hour without seeing any other visitor to the bog, the first lines that sprang to mind were not Heaney's but those in Glob's first paragraph of *The Bog People:* "Evening was gathering over Tollund Fen in Bjaeldskov Dal. Momentarily the sun burst in, bright and yet subdued, through a gate in blue thunder-clouds in the west, bringing everything mysteriously to life. The evening stillness was only broken, now and again, by the grating love-call of the snipe. The dead man, too, deep down in the umber-brown peat, seemed to have come alive."[47]

I also remembered a passage I had read from one of the earliest descriptions of the bog phenomenon, by William King, archbishop of Dublin, in 1685: "I know not if it will be worth the observing, that a *Turf-Bog* preserves things strangely, a Corps will ly entire in one, for several years; I have seen a piece of leather pretty fresh dug out of a *Turf-Bog,* that had never in the memory of man been dug before."[48] And I recalled the photographs of landscape in Glob's book as my own digital camera took in snapshot after snapshot of scenery very much like the ones used by him. And while there *were* faint sounds from the highway, it was all too easy to imagine that I had stepped over a threshold in time and that it was well "worth the observing that a *Turf-Bog* preserves things strangely." At the same time I thought of the scale of the *Spiral Jetty,* which in Smithson's own words "tends to fluctuate depending on where the viewer happens to be"—how he saw the Jetty as an echo and reflection of "Brancusi's sketch of James Joyce as a 'spiral ear,'" which as such presented a "sense of scale that resonates in the eye and the ear at the same time [. . .] a reinforcement and prolongation of spirals that reverberates up and down space and time."[49]

Somehow the two fieldtrips and experiences came together. Different as they were, both landscapes—one in Denmark, the other in America—suggested prehistory, and each in its way was marked by modernity. It made sense, to me, to see the two as symbolic of how time and matter, time and place, coalesce.

PREFACE

1. P. V. Glob, *The Bog People; Iron Age Man Preserved.*, trans. Rupert Bruce-Mittford (Ithaca, NY: Cornell University Press, 1969), 82.

2. Seamus Heaney, "The Man and the Bog" in *Bog Bodies, Sacred Sites and Wetland Archaeology.* ed. Bryony Coles and John M. Coles (Exeter, UK: University of Exeter and National Museum of Denmark, WARP, 1999), 4.

3. Anne Ross and Don Robins, *Life and Death of a Druid Prince. How the Discovery of Lindow Man Revealed the Secrets of a Lost Civilization* (New York: Touchstone, 1991), 15.

4. Wijnand van der Sanden, *Through Nature to Eternity: The Bog Bodies of Northwest Europe,* trans. Susan J. Mellor (Amsterdam: Batavian Lion International, 1996), 7.

5. Marc Bloch, *The Historian's Craft,* trans. Peter Putnam (New York: Vintage, 1964), 8.

6. Mieke Bal, *Travelling Concepts in the Humanities: A Rough Guide* (Toronto: University of Toronto Press, 2002), 4.

7. Glob, *The Bog People,* 17 (my emphasis).

8. Gaston Bachelard, *Earth and the Reveries of Will* (Dallas: Dallas Institute of Humanities and Culture, 2002), 19.

INTRODUCTION

1. See Michael Kimmelman, *The Accidental Masterpiece: On the Art of Life and Vice Versa* (New York: Penguin, 2006) for a description of how a variety of objects can become unintended art.

2. Heaney, "The Man and the Bog," 5.

3. See Glob, *The Bog People,* 33. For the most complete and detailed description of the northern European bog bodies, see Wijnand van der Sanden, *Through Nature to Eternity.* Children are rare finds. It is unclear how many bog bodies have surfaced through time. In Denmark, the number has been estimated as high as from 441 to 500 (according to the German scholar Alfred Dieck). A more conservative calculation by van der Sanden estimates that of forty-one women and forty-five men, many of which are bog skeletons; only around twenty are bog mummies proper. Ibid., 84. The oldest bog body in Denmark is the Koelbjerg Woman from Funen, whose partial skeletal remains are more than ten thousand years old; the earliest recordings of the discovery of bog bodies date from the seventeenth (England, Wales) and eighteenth (Denmark) centuries.

4. Some bog bodies are presumably the remains of accident victims. Bog bodies are not, as previously thought, exclusive to northern Europe; they have been found on Crete and in Russia. Nor are they exclusive to the Iron Age; some are as much as eleven thousand years old, while others are but a century old.

5. Although bog bodies encompass all sorts of human remains from bogs—skeletal remains included—the most common association is with those that are mummified. From now on I will refer to bog mummies simply as bog bodies.

6. Here cited from R.C. Turner and C.S. Briggs, "The Bog Burials of Britain and Ireland," in *Lindow Man: The Body in the Bog* (London: British Museum Publications, 1986), 144. Turner and Briggs do not tell us who Charles Leigh was, but they most likely refer to the British physician and naturalist who lived from 1662 to approximately 1701.

7. [Den 4de Junii sidsteleden blev i Raunsholdts Tørve-Long, Lindkier-faldet, funden et dødt Menneske een og en halv Alen under i Tørvegrunden, da en Karl, som skar Tørv, stak den ene Fod af med Tørvespaden; og] da de fornam saadant et Kiendetegn af et Menneske, skar 4 Karle Tørve-Jorden oven fra ud, hvorved et heelt fuldkommen Mands Menneske blev funden, som af mig underskrevne tilligemed tvende Mænd blev synet og forefundet liggen-des, som meldt, een og en halv alen i Tørven lige udstrakt paa sin Ryg, med begge Armene over Kors under Ryggen, ligesom de kunne have været sammenbundne, hvortil dog intet Tegn af Baand eller paa Armene var; Legemet var ellers ganske nøgent, uden alene om Ho-vedet var svøbt en Faareskinds Pels, ved hvis Fratagelse kiendelig kunne sees, at Mennesket har havt rødagtig Skæg og gandske kort Haar som han kunne have baaret Paryk; Huden over det gandske Legeme var, førend det blev rørt eller omveltet, gandske heel, undtagen under Halsen, hvor man kunne see lige ind til Beenet, og de forreste Tænder ligesom indstødte i Munden; Legemet var ellers gandske heelt og holdent, og alle Legemets Lemmer kiendelig at see, undtagen den ene Fod, som ved Tørvespaden blev afstødt; Legemet var ellers meget mørt og raadent, og efter al Skiønsomhed der i mange Aar havde ligget, eftersom der var svær sammengroet Tørve-Jord over ham halvanden Alen, og foran ved Balken [Valken?] hvor han blev funden ligeledes, hvor der saavel i aar som i Fior var bleven saaet Tørv. Ellers kan ikke skøinnes anderledes, end at dette Menneske med Forsæt maa være kommen der; da, foruden anførte, befandtes, at der var lagt nogle smaae Riis, og derudover igjen stukket nogle smaae Pinde korsviis over ham, ligesom at hindre Legemet at skulde flyde eller sætte sig op; men desuagtet fandtes Tørve-Jorden over ham ligesom compagt, tæt og sammengroet, som den øvrige Tørve-Jord, og ei nogen Kiendetegn til at det har været en gammel Grav, hvoraf man slutter, han maa have lagt der i mange Aar. Thi bekendtgiøres dette; om nogen kan give

nogen Oplysning i denne Sag, om noget Menneske kunne været bleven borte eller saa videre, da at melde sig hos mig. Ravnsholdt den 4de Juni 1773. Hans Christian Fogh. Birkedommer og Skriver ved Ravnsholdt Birk." Transcribed from facsimile in van der Sanden, *Through Nature to Eternity*, 40. The English translation I use is partly the one offered by Rupert Bruce-Mitford in Glob, *The Bog People*, 66, and partly my own.

8. Conversation with Wijnand van der Sanden, Assen, May 2003. See also his *Through Nature to Eternity*, 40.

9. Clive Gamble, *Archaeology: The Basics* (London: Routledge, 2001), 74. R.C. Turner and C.S. Briggs also reflect on the nature of bog bodies vis-à-vis written sources: "In common with other antiquarian finds, records of their discovery relate closely to the availability of printing and the spread of an intellectual awareness of the past. Consequently, during the period of greatest dependence upon peat fuel, when both peasant agriculture and more sophisticated drainage projects were in hand, bodies were only recorded when they came to the notice of men (or women) of learning." Turner and Briggs, "The Bog Burials of Britain and Ireland," 153.

10. Cornelius Tacitus, *The Agricola and the Germania,* trans. Harold B. Mattingly and S. A. Handford (Harmondsworth: Penguin, 1970), 111.

11. For a detailed description of the various translations of *corpores infames* and their impact, see Allan A. Lund, *Moselig* (Højbjerg, Denmark: Wormianum, 1976). Also see his *Hitlers håndlangere: Heinrich Himmler og den nazistiske raceideologi* (Copenhagen: Samleren, 2001). According to Wijnand van der Sanden, the first time Tacitus was connected to bog bodies was in 1924 by the German F. Arends. But as we shall see later, earlier connections between Tacitus and bog bodies are to be found. For an excellent summary of the history of bog body interpretations, see van der Sanden, 166–77. Also note that the earliest known written interpretation by an Irish countess from Moira, who describes a bog body found in Drumkeragh in 1783, speculates that the body stems from a druid ceremony. For more on this, see van der Sanden, 166. In Anne Ross and Don Robins, *The Life and Death of a Druid Prince: How the Discovery of Lindow Man Revealed the Secrets of a Lost Civilization,* this thesis is taken to the extreme. Van der Sanden also points out the importance of German archaeologists in the first half of the twentieth century—including the work of Alfred Dieck, who concentrated his efforts on written accounts and most likely fabricated or at least exaggerated them, including some that detailed female bog bodies with intact hymens.

12. Tacitus, 117.

13. Glob, *The Bog People,* 151. Glob's theory was in part built on a thesis by a fellow Danish archaeologist, Else Thorvildsen.

14. The Danish archaeologist Christian Fischer writes: "When we excavate graves from that time [the Iron Age] we feel that we are approaching the spiritual world." Christian Fischer, "Face to Face with Your Past," in *Bog Bodies, Sacred Sites and Wetland Archeology,* ed. Bryony Coles, John Coles, and Mogens Schou Jørgensen (Jutland, Denmark: National Museum of Denmark, Silkeborg Museum, 1996), 7. Another Danish archaeologist, Flemming Kaul, writes: "In all cases one can presume that through the bogs people thought that they had a link with something much more than the spirits of human beings, that this was a gateway to another world beyond the world of men [. . . .] The bogs and the forces found in them or beyond them were perhaps more integrated in a wider religious system where the

bog was regarded as an especially sacred place." "The Bog: The Gateway to Another World," in *The Spoils of Victory: The North in the Shadow of the Roman Empire,* ed. Lars Jørgensen, Birger Storgaard, and Lone Gebauer Thomsen (Copenhagen: Danish National Museum, 2003), 21. See also J.R. Beuker, "The Bog: A Lost Landscape," in *The Mysterious Bog People,* ed. C. Bergen, M.J.L.Th. Niekus, and V.T. van Vilsteren (Zwolle: Waanders Publishers, 2002), 12–17. And in V.T. van Vilsteren's "Discoveries in the Bog: History and Interpretation" we read: "The lowering of fog over the water only contributed to the ominous character of the bogs. In historic times, this often led to superstitions about bog ghosts and mythical creatures." Ibid., 22.

15. Indeed, as some have argued, it might have been a privilege to be hung and then placed in a bog grave. Hanging was not seen as degrading until Christianity made it so in an effort to stamp out pagan beliefs and practices. See Kaul, 39.

16. Ibid.

17. As the Danish historian Allan A. Lund succinctly puts it, it is "the circumstances surrounding each individual find which determines the characteristic of each find." *Moselig,* 70. A great deal of effort went into ensuring that the bodies remained in the bog, unable to return from the dead and harm the still living. Therefore stakes and poles, used to secure the dead, are the most frequent auxiliary artifacts found on, in, above, or under the bodies. Many bog bodies show evidence of torture or "overkill," often in form of a *triple-death;* hanging combined with drowning, stabbing, or other forms of death were not uncommon. On "overkill," see Allan A. Lund, *Mummificerede moselig* (Copenhagen: Høst & Søn), 82–85. See also Timothy Taylor: *The Buried Soul. How Humans Invented Death* (London: Fourth Estate, 2002), and Miranda Aldhouse Green, *Dying for the Gods: Human Sacrifice in Iron Age & Roman Europe* (Gloucestershire, UK: Tempus Publishing Limited, 2001).

18. The irony that bog bodies found their deaths around the same time as Christ's crucifixion, and that their sacrificed bodies would in time become relics of a sort, is not lost on many poets and visual artists, as we shall see later.

19. Anthony Vidler, The *Architecturally Uncanny: Essays in the Modern Unhomely* (Cambridge, MA: MIT Press, 1992), 11.

20. Ibid.

21. Tacitus' expression is sometimes translated to: "unhealthy marshes." I follow here Simon Schama's suggested translation in *Landscape and Memory* (New York: Vintage, 1996), 583 n. 4.

22. For a fleshed-out description of the bog and its properties and myths, see Kaul, 18–43.

23. The lures of water nymphs are manifold. Consider Homer's description of Odysseus's fight to withstand the temptation of sirens by tying himself up; or Hans Christian Andersen, who gave this motif a famously different spin in "The Little Mermaid." Here the mermaid siren saves her love from the watery death and finally sacrifices her own sexual-physical desire in anticipation of immortality. Andersen also explores the liminality of bogs in fairy tales like "The Bog-King's Daughter" and "The Girl Who Stepped on Bread."

24. R.C. Turner: "Boggarts, Bogles and Sir Gawain and the Green Knight: Lindow Man and the Oral Tradition," in *Lindow Man: The Body in the Bog,* ed. I. M. Stead, J. B. Bourke, and Don Brothwell (London: British Museum Publications, 1986), 170.

25. Plato *Phaedo,* trans. Harold North Fowler (Cambridge, MA: Harvard University Press, 1913), 281. See also Davide Stimilli, *The Face of Immortality: Physiognomy and Criticism* (Albany, NY: State University of New York Press, 2005), 38–39.

26. Nicholas Royle, *The Uncanny: An Introduction* (New York: Routledge, 2003), 1. In Anne Ross and Don Robins's *The Life and Death of a Druid Prince,* the significance of naming in Celtic cultural tradition is seen as "imprinted on the Celtic subconscious" and the authors go so far as to assign the name Lovernio—meaning "fox," a name of great import and status—to the bog body known as Lindow Man. They speculate that the fox armband found on him indicates this name and with it the status of a highborn, or a Druid prince.

27. Vidler, 10–11. In this quote Vidler cites Homi Bhahba: "[A] boundary that secures the cohesive limits of the western nation may imperceptibly turn into a contentious internal liminality that provides a place from which to speak both of, and as, the minority, the exilic, the marginal and emergent." I shall return to questions of the uncanny in chapters two and three.

28. Similarly, as Jennifer Wallace has argued: "the excavation of perfectly preserved bodies disturbs the boundaries we maintain between what is natural and what is unnatural." Jennifer Wallace, *Digging the Dirt: The Archaeological Imagination* (London: Duckworth, 2004), 55.

29. I am thinking here of Fernand Braudel and others. See, for example, his *On History,* trans. Sarah Mathews (Chicago: University of Chicago Press, 1980).

30. Georges Bataille, *Erotism: Death & Sensuality,* trans. Mary Dalwood (San Francisco: City Lights Books, 1986), 44.

31. I use "contact zones" as a variation of what Mary Louise Pratt has described by that term in *Imperial Eyes: Travel Writing and Transculturation* (London: Routledge, 1992).

32. See Franco Moretti, *Atlas of the European Novel, 1800–1900* (London: Verso, 1998).

33. Schama, 6–7. Interestingly, Schama describes his book in archaeological terms as "constructed as an excavation" and intended to dig "below our conventional sight-level to recover the veins of myth and memory that lie beneath the surface." Ibid., 14.

34. Ibid., 7.

35. Ibid., 10. Landscapes, Schama goes on, are "culture before they are nature; constructs of the imagination projected onto wood and water and rock." Ibid., 61.

36. Pierre Nora, "Between Memory and History," in *Realms of Memory: The Construction of the French Past,* ed. Lawrence D. Kritzman, trans. Arthur Goldhammer (New York: Columbia University Press, 1996), 3.

37. Ibid. For the importance of memory, see also Paul Ricœur, *Memory, History, Forgetting,* trans. Kathleen Blamey and David Pellauer (Chicago: University of Chicago Press, 2004). For a critical note on the debate about history and memory, see Kerwin Lee Klein: "On the Emergence of Memory in Historical Discourse," *Representations* 69 (Berkeley and Los Angeles: University of California Press, Winter, 2000), 127–50.

38. Gaston Bachelard, *Earth and Reveries of Will: An Essay on the Imagination of Matter,* trans. Kenneth Haltman (Dallas: Dallas Institute of Humanities and Culture, 2002), 309.

39. Ibid., 101.

40. See also Vidler, 64.

41. These ruptures challenge a number of conventions. We should recall, for example,

that archaeological periodization, at least since the beginning of the nineteenth century, has been wedded to a breakdown of pre-historical time into chronologies (stone, bronze and iron ages) that helped facilitate the conviction of European superiority and suggested development over time in which the people lifted themselves, through physical and mental struggles, to ever-higher stages of civilization. Furthermore, the chronologies implied that unlike other peoples, Europeans had fulfilled this historically important mission in that they had conquered and developed the world and brought it to a presumably higher plateau than had other peoples.

42. Hayden White's seminal study *Metahistory: The Historical Imagination in Nineteenth-Century Europe* (Baltimore: Johns Hopkins University Press, 1975) is, of course, a key player in this regard.

43. Clive Gamble goes on: "The past is always in front of you." Gamble, *Archaeology: The Basics,* 1.

44. Julian Thomas, *Time, Culture, and Identity: An Interpretative Archaeology* (London: Routledge, 1996), 63.

45. Christopher Tilley, *Metaphor and Material Culture* (Oxford: Blackwell, 1999), 265. While Tilley stays close to a nomination of the objects as something that exist in immediacy to be taken in by a glance, he allows, fortunately, that the density of the object "need[s] to be understood temporally in [its] actional and biographical contexts." Ibid., 264, 261. There is a rich body of work on archaeological narratives. Cultural assumptions and biases, it implies, are built into—indeed, invented—by language and its users. Empiricist and positivistic understandings of the past have yielded to interpretative models adopted from critical theory, hermeneutics, phenomenology, post-structuralism, and gender studies. Disbelief in facts and skepticism toward truth claims serve instead to acknowledge the imbedded story telling and the "artistic" license in archaeological discourse. Indeed, it has become convention in some (albeit far from all) archaeologies to see interpretation as provisional; to assign agency to the personality and background of the interpreter, to see his or her special interests and alliances as unavoidably interacting with interpretations and choice of tropes. Paul Atkinson talks about "the poetics of authoritative accounts" as a way to articulate the reliance within so-called factual genres on rhetorical methods and textual conventions. Paul Atkinson, *The Ethnographic Imagination: Textual Constructions of Reality* (London: Routledge, 1990), 35. Archaeology too can be said to offer a poetics of authoritative accounts. To link archaeology with narrative (or poetic) models of telling seems rather obvious, perhaps. David W. Anthony goes as far as to say that, "all really good science *is* poetry, in a way." David W. Anthony. "Nazi and Eco-Feminist Prehistories," in *Nationalism, Politics and the Practice of Archaeology,* ed. Philip L. Kohl and Clare P. Fawcett (Cambridge: Cambridge University Press, 1995), 84. Neil Asher Silberman points to the basic narrative structures in any kind of archaeological utterance: "Whether spoken, written, or visually depicted, these interpretations usually take the form of narratives: sequences of archetypal story elements, didactically arranged with clear beginnings, middles, and ends." Neil Asher Silberman, "The Politics and Poetics of Archaeological Narrative," in ibid., 250. Or as the archaeologist Mark Pluciennik argues: "Typically, archaeologies are presented in the form of narratives understood as sequential stories." Mark Pluciennik, "Archaeological Narratives and Other Ways of Telling," *Current Anthropology* 40, no. 5 (December 1999), 1. See also Christopher Evans, "Digging

With the Pen: Novel Archaeologies and Literary Tradition," in *Interpretative Archaeology,* ed. Christopher Tilley (Providence, RI: Berg, 1993). Evans makes this comment: "Archaeology often envisages its potential (non-academic) audience as a unified whole, uniformly appreciative of its practices. The many uses of, and the diversity of attitudes directed towards, archaeology in twentieth century literature should warn us against such simplistic, and essentially derogatory, interpretations. It should lead us to question whether *the public* is ever 'knowable'; are their responses to the past predictable, and how is time variously appreciated? The past is so inextricably linked to the cultural present that it is absurd to think that academia has any prerogative over it." Ibid., 439.

46. M.M. Bakhtin, "Forms of Time and Chronotype in the Novel," in *The Dialogic Imagination: Four Essays,* ed. Michael Holquist, trans. Caryl Emerson and Michael Holquist (Austin: University of Texas Press, 1981), 84.

47. Ibid.

48. Ibid., 85.

49. Anthony Purdy goes on to say: "Of course, the mnemotrope might come in many guises and be inflected by attitudinal values ranging from nostalgia and desire through obsession to horror and denial. Its preferred genre might be pastoral, but it might equally well be ghost story or fantasy, historical romance, satire or memoir." "Unearthing the Past: The Archaeology of Bog Bodies in Glob, Atwood, Hébert and Drabble," *Textual Practice* 16, no. 3 (2002), 447–8.

50. Sigmund Freud's use of archaeological depth metaphors will receive particular attention in chapter 2, but let me offer a few references for the others here. To see archaeological objects as mnemonic, to connect archaeology, memory, and language, is apparent in Walter Benjamin's thoughts on excavation. While he directs his meditation in large measure on the subject of urban ethnography, his comments are useful in their linking of language and archaeology, memory and archaeology, and human psychology and archaeology. Yet in spite of the implied archaeological nature of Benjamin's work as a whole, he has but one brief reflection on archaeology, called "Excavation and Memory" from around 1932, in which he links the tasks and intentions of the archaeologist with memory and language. See Walter Benjamin, "Excavation and Memory," in *Selected Writings, vol. 2,* trans. Marcus Paul Bullock, Michael William Jennings, Howard Eiland, and Gary Smith, ed. Marcus Paul Bullock and Michael William Jennings. (Cambridge, MA: Belknap Press, 1996), 548. Foucault's archaeology tries to untie the knots tied by historians, and provides a principle for the articulation of discourses which are moored both to materiality and to a temporality marked by discontinuities and ruptures. See his *Archaeology of Knowledge,* trans. A. M. Sheridan (London: Routledge, 2002). A Heideggerian archaeological model of "being-in-the world" calls for an understanding of human existence, according to Julian Thomas, as "thoroughly invested in the world"—not a world in which we can make distinctions between external (the body and material objects) and internal (the mind), but a world in which "human beings are always already enmeshed in a structure of meaning, and the interpretation of their bodies as one thing or another is constitutive of what those bodies *are*." Julian Thomas, *Time, Culture, and Identity: An Interpretative Archaeology,* 17. See also Heidegger's "Building Dwelling Thinking" and "The Thing" in *Poetry, Language, Thought,* trans. Albert Hofstadter (New York: Harper and Row, 2001). For a more full description of these theories' relevance for archaeol-

ogy, see my article "The Archaeological Object in Word and Image," *Edda, 03* (2002). See also Julian Thomas, *Archaeology and Modernity* (London: Routledge, 2004) for an excellent investigation of archaeology and critical theory.

51. Or, as Mark Pluciennik says, if "we benefit from thinking of artifacts as people, we can also think of people as artifacts." "Art, Artefact, Metaphor," in *Thinking through the Body: Archaeologies of Corporeality*, ed. Yannis Hamilakis, Mark Pluciennik, and Sarah Tarlow (London: Kluwer/Academic Press, 2002), 227.

52. Wallace, 50.

53. Gamble, *Archaeology: The Basics*, 194.

54. Kristian Kristiansen, ed., *Archaeological Formation Processes: The Representativity of Archaeological Remains from Danish Prehistory* (Copenhagen: Danish National Museum, 1985), 28.

55. I interviewed a number of bog body specialists for this book, and it became clear that in spite of Glob's, in Kristensen's words, "slightly piquant and macabre appeal," almost everyone professed to not only a continued fascination with, but also respect for, Glob's study.

56. Purdy "Unearthing the Past," 443.

57. Ian Hodder, *The Archaeological Process: An Introduction* (Oxford: Blackwell, 1999), 56. The influence of Hayden White's model of historical imagination (he posits that the historical imagination, like literary representations, follows grammatical elements such as scene, agent, act, agency, and purpose in combinations that reflect the intent or position of the author) has resonated across several fields of study, archaeology included. One of White's key contributions is to wed rhetorical tropes to historical discourse, to see the *kind* of language, which is used to describe a historical moment as an active component in the *kind* of history told. White sustains in part Kenneth Burke's critical vocabulary from *A Grammar of Motives* (Berkeley and Los Angeles: University of California Press, 1969), but augments this grammatical positioning and points to a set of linguistic rules in which the style used by a given historiographer resonates with the various "'explanatory' strategies [he] used to fashion a 'story' out of the 'chronicle' of events contained in the historical record." History is, then, indentured to language, or to the "types of linguistic protocols" used. See White, 14. n.8, 426. White also draws on Northrop Frye's *Anatomy of Criticism. Four Essays* (Harmondsworth: Penguin, 1990).

58. But, as Hodder goes on to say: "The trouble with the term 'story' is that it implies a lack of concern with links to data and this is not advocated here. Rather a term is needed which both takes account of the need to link interpretations to data and of the wider structure of the account being told. Perhaps 'narrative interpretation' fits this dual role best. Perhaps most important, the emphasis on narrative or story-line implies narrating or telling *to* someone. Archaeologists have to produce different accounts for different audiences. The notion of narrative takes us beyond the idea of pure knowledge to recognition of archaeology as a practice within society." Hodder, 56.

59. "Arkæologerne graver eventyr frem." Martin A. Hansen. *Orm og Tyr*, 6th ed. (Copenhagen: Gyldendal, 1963), 82. Martin A. Hansen was a Danish author with a strong interest in the archaeological world. Amongst other archaeologically suggestive works he authored a short story entitled "The Man from Earth," an existential fable about being of earth and returning to it. "Manden fra Jorden" in *Agerhønen*, 1947.

60. See also Kristian Kristensen, "Fortids kraft og kæmpestyrke: om national og politisk brug af fortiden," in *Brugte historier. Ti essay om brug og misbrug af historien,* ed. Lotte Hedeager and Karen Schusboe (Copenhagen, Akademisk Forlag: 1989), 187–218.

61. Glob, *The Bog People,* 18.

62. Margaret Conkey addresses how our consciousness inevitably becomes entangled with earlier presumptions: "Yes, the viewer is an interpreter, and yes, interpretation changes as the work changes and thus we cannot claim final or absolute knowledge. Many of the core concepts that we invoke and depend upon [. . .] are not just cultural constructions, but are cultural concepts that have changed over time: 'art,' 'symbolism,' 'meaning,' 'consciousness,' to name a few." "Context in the Interpretative Process," in *Beyond Art: Pleistocene Image and Symbol,* ed. Margaret Conkey et al. (San Francisco: California Academy of Sciences, 1997), 358. She warns against any potential blindness that prevents acknowledging the difference between the meanings of contexts in which cultural objects were imbedded in the past and the meanings we attach to them in the present.

63. Marc Bloch, 8.

64. One archaeological bog body study, Anne Ross and Don Robins's *The Life and Death of a Druid Prince,* shows how easily authors can be seduced by their own narrative ploy. In their departure from a "traditional" model of objective scientific writing, they describe how "a possible scenario" of Lindow Man's life and death "gradually turned into a historical and archaeological detective story." In *The Life and Death of a Druid Prince,* the very title of which suggests poetic license, Ross and Robins deliberately place themselves as third-person characters and call themselves by their proper names—"Anne Ross' interest in the bog body was now profoundly aroused," etc.—thus intentionally splitting themselves into narrators *and* characters. This narrative strategy illustrates by example how the gradual archaeological process of compilation of facts and clues inevitably generates a pliable character. The archaeological object, here Lindow Man, cannot by virtue of such a strategy hold just one identity—after all, or so the logic would go, his is an imagined identity. As Ross and Robins sift through evidence and clues, the bog body in turn assumes different identities. Lindow Man becomes "the silent witness" and is renamed Lovernios. Critics of such an "author-saturated" discourse would argue, no doubt, that Ross and Robins sacrifice scientific prudence in favor of archaeological imagination. Ibid., 7. A similar kind of "pressure to outrun the evidence," as David Papineau has pointed out, is easily detectable in the interpretations of the imagined fate of the Iceman from the Alpine Glacier. Konrad Spindler's version of the man's last days, for example, is in the words of Papineau, a fable. Konrad Spindler, *The Man in the Ice: The Discovery of a 5,000-Year-Old Body Reveals the Secrets of the Stone Age* (New York: Random House, 1996).

65. Heaney visited Aarhus in 1973 and the museum display of the bog man, but as we know, he had originally been introduced to him via photographs and text in Glob's *The Bog People.*

66. Heaney, "The Man and the Bog" (italics in parenthesis are mine), 4.

67. See also Jefferson Hunter, *Image and Word: The Interaction of Twentieth-century Photographs and Texts* (Cambridge, MA: Harvard University Press, 1987), 185.

68. Heaney's use of the visual representation of the bog man as a partner in crime for poetry does not, however, spill over into an embracing of the word-image relationship *per*

se. On the contrary, Heaney confesses to a reluctance to engage in such an undertaking and claims to harbor "unease about the problematic relationship between image and text . . . . I think I felt that in these cases (to be extreme about it) the photograph was absolute and the text a pretext. This was more of a hesitation than a conviction: all I can say is that I tended to foresee misalliance of some sort between the impersonal, instantaneous thereness of the picture on one side of the page, and the personal, time-stretching pleas of the verse on the other." Seamus Heaney, preface to *Sweeney's Flight: Based on the Revised Text of "Sweeney Astray": With the Complete Revised Text of Sweeney Astray* (London: Faber and Faber, 1992).

CHAPTER 1

1. Glob, *The Bog People,* 41 (my emphasis). Heaney's line is from his poem "Grauballe Man."

2. Siegfried Kracauer, "Photography," in *Classic Essays on Photography,* ed. Alan Trachtenberg (New Haven, CT: Leete's Island Books, 1980), 246.

3. Jefferson Hunter has also pointed out this analogy between bog and photography. I became aware of his article after I had formed my own concept of the bog as photographic laboratory, and I agree with him on most points. Hunter, 8.

4. Douwe Draaisma, *Metaphors of Memory: A History of Ideas about the Mind* (Cambridge: Cambridge University Press, 2000), 110. Freud's so-called "mystic writing pad" is yet another example of the common interconnection of imprint and memory and prosthetic memory metaphors, as a variation of the archaeological metaphor that Freud used frequently. Sigmund Freud, "A Note Upon the 'Mystic Writing-Pad,'" in *The Standard Edition of the Complete Psychological Works of Sigmund Freud* (London: Hogarth Press, 1961).

5. The importance of the wax tablet metaphor is that memory is seen not only as imprints made by the senses, but also as writing. Draaisma, 2.

6. André Bazin, "On the Ontology of the Photographic Image," in *Classic Essays on Photography,* ed. Alan Trachtenberg (New Haven, CT: Leete's Island Books, 1980), 241.

7. Roland Barthes, *Camera Lucida: Reflections on Photography,* trans. Richard Howard (New York: Hill and Wang, 1981), 110.

8. Christian Fischer, "The Tollund Man and the Elling Woman and Other Bog Bodies from Central Jutland," in Coles, Coles and Jørgensen, 96.

9. Glob, *The Bog People,* 36. The italics are mine.

10. André Malraux, *The Voices of Silence,* trans. Stuart Gilbert (Princeton, NJ: Princeton University Press, 1978), 14.

11. The original body parts are kept in a safe at Silkeborg Museum.

12. See Jean Baudrillard's theories of simulacrum in *The System of Objects,* trans. James Benedict (London: Verso, 1968). For a useful description of the relationship between original and copy, see also Hillel Schwartz, *The Culture of the Copy: Striking Likenesses, Unreasonable Facsimiles* (New York: Zone Books, 1996).

13. Lionel Trilling, *Sincerity and Authenticity* (Cambridge, MA: Harvard University Press, 1972), 93.

14. Ibid., 94.

15. Walter Benjamin, "The Work of Art in the Age of Mechanical Reproduction," in *Illuminations,* ed. Hannah Arendt, trans. Harry Zohn (New York: Schocken Books, 1986), 220.

16. Barthes, *Camera Lucida,* 89. While Barthes is concerned with the meaning of visual representation, his narrative also grapples with the *loss* inherent in representation.

17. See Michael Charlesworth "Fox Talbot and the 'White' Mythology of Photography," *Word & Image* 11, no. 3 (1995).

18. Both postcards were purchased from the museum shop at Moesgaard Museum and the photographs are attributed to P.V. Glob.

19. Glob, *The Bog People,* 49. The emphasis is mine. Glob's description of Grauballe Man's embryonic nature might well have inspired Seamus Heaney's line "forceps baby" in the poem "The Grauballe Man."

20. Walter Benjamin, "The Work of Art in the Age of Mechanical Reproduction," 226. See also Anthony Purdy "Unearthing the Past," 446.

21. Ibid.

22. Susan Sontag, *On Photography* (New York: Anchor Books, 1990), 15.

23. Barthes, *Camera Lucida,* 78–9.

24. Roland Barthes, "The Photographic Message," in *Image, Music, Text,* trans. Stephen Heath (New York: Hill and Wang, 1977), 42.

25. See Sontag, *On Photography,* 187; See also Hillel Schwartz, 94. Elizabeth Bronfen also suggests that there is a kind of vacillation between animation and de-animation in death photographs: "Even as the dead return in any photographic reproduction of what is past and lost, the dialectic of animation and de-animation is, such, however, that the photograph always mortifies the subject it reproduces by transforming it into an object imitating the subject." Elisabeth Bronfen, *The Knotted Subject: Hysteria and its Discontents* (Princeton, NJ: Princeton University Press, 1998), 88–89.

26. The use of photography to record bog bodies started in 1873 when the so-called Rendswürden Man, found in 1871, was shot in a studio in a rather peculiar, theatrical pose. Standing erect on his two legs, partially draped in a shawl and posed in "ballet-posture," as Winjand van der Sanden puts it, this early photograph treats the bog body as a curiosity. The earliest bog body photograph made in situ was shot a few years later and shows the man found in Nederfrederiksmose in 1898. Here the body is still lodged in the peat, and the photograph is in accordance with "proper" archaeological documentation. In other cases, photography serves as the equivalent of the so-called paper bodies: a bog body from Kreepen, for example, found in 1903 and photographed shortly after while resting in a wooden coffin, was lost during the Second World War. But the photograph survives and testifies to his having-been.

27. Susan Sontag, *Regarding the Pain of Others* (New York: Farrar, Straus and Giroux, 2003), 26.

28. I have asked a host of students, colleagues, and friends about their responses to photographs of bog people compared to photographs of other dead bodies, and the responses echo the sentiment expressed above.

29. Ibid., 42.

30. Ibid.,76.

31. W.J.T. Mitchell, "The Photographic Essay: Four Case Studies," in *Picture Theory: Essays on Verbal and Visual Representation* (Chicago: University of Chicago Press, 1994), 288.

32. See also Mark Sandberg's analysis of how life and death are blurred when wax models are photographed. He writes: "In the idea of recording, the frozen image of the wax tableau

and the frozen frame of the film converge, and one loses track of a clear conceptual path back to the body." Mark B. Sandberg, *Living Pictures, Missing Persons* (Princeton, NJ: Princeton University Press, 2003), 57.

33. I want to point out here that each and every one of these points would be moot, of course, if the onlooker were not cognizant of the fact that what he or she is seeing is in fact an *archaeological* body and not the physical remains of a victim from more recent times.

34. Barthes, *Camera Lucida,* 15.

35. Bataille, "Musée." Here cited from Vidler, 86.

36. Maurice Blanchot, *The Space of Literature,* trans. Ann Smock. (Lincoln, NE: University of Nebraska Press, 1982), 256–58.

37. Sontag, *On Photography,* 15.

38. Glob, *The Bog People,* 31

39. Hunter, 10.

40. See Michael Wood's foreword in *The World Atlas of Archaelogy,* ed. Christine Flon (London:. Mitchell Beazley, 1985), 8. Glob's frequent appearances in the press—not only when bog bodies were in the news, but also when he went on his many archaeological adventures to Bahrain and the Arabian Gulf (in fact it was one of these excursions that delayed his publication of *The Bog People* by several years)—mplicitly positioned him within the kind of popular perception that archaeologist Neil Asher Silberman has called the "fable of the Archaeologist as hero."

41. Glob, *The Bog People,* 22.

42. Glob had been submitted to what he calls a "bloodless battle" between science and local lore. Newspaper cartoonists drew satiric pictures of him vis-à-vis his famous archaeological find and made fun of his propensity for snaps, the Danish national alcoholic drink. In the end, however, his archaeological interpretation prevailed. For a full description of this story see my article, "Anachronistic Encounters: A Reception Story," in *Nordic Naturecultures. Eco-critical Approaches to Film, Art and Literature,* ed. C. Claire Thomson and Christopher Oscarson (forthcoming).

43. Glob, *The Bog People,* 39.

44. "Han har været en smuk og statelig Skikkelse, omtrent af Garderhøjde og meget kraftig bygget, og eventuelle Formodninger om en vis Plumhed hos vores germanske Forfædre i Jernalderen kuldkastes af den Kendsgerning, at han har haft smukke Hænder og mandelformede Negle. Hans kraftige, røde Haar har efter alt at dømme været bundet i en saakaldt Germaner-Knude ved venstre Øre, en Mode hos Germanerne fra Perioden omkring Kristi Fødsel." *Jyllandsposten,* May 2, 1952.

45. "Ansigtets lidt mongolske præg behøver ikke være noget medfødt, men kan maaske skyldes en voldsom Kvæstelse af Hovedet." *Jyllandsposten,* May 2, 1952.

46. "Rene og regelmæssige Ansigtstræk med de tænksomme Rynker i Panden" *Nationaltidende,* November 8, 1953.

47. "Om Grauballe-Manden i sidste Øjeblik er blevet grebet af Panik, og Offerpræsten har maattet strække ham til Jorden med et Kølleslag, er naturligvis Gisning. Men mon det er det rene Gætteri?" "Verdens to Ældste Mænd dansk Verdens-Sensation." *Nationaltidende,* November 8, 1953.

48. Miranda Aldhouse-Green, *An Archaeology of Images: Iconology and Cosmology in Iron Age and Roman Europe* (New York: Routledge, 2004), 89–90.

49. See also Susan Stewart, *On Longing: Narratives of the Miniature, the Gigantic, the Souvenir, the Collection* (Durham, NC: Duke University Press, 1993), 153.

50. For an excellent discussion of this, see Linda Haverty Rugg, *Picturing Ourselves: Photography & Autobiography* (Chicago: University of Chicago Press, 1997).

51. See Walter Benjamin, *One-way Street and Other Writings,* trans. Edmund Jephcott and Kingsley Shorter (London: NLB, 1979).

CHAPTER 2

1. Sigmund Freud, *The Uncanny,* trans. David McLintock (New York: Penguin Books, 2003). See also Anthony Vidler, 22.

2. Ibid., 151.

3. Freud, "Fragments of an Analysis of a Case of Hysteria ('Dora')," in *The Freud Reader,* ed. Peter Gay (New York: W.W. Norton and Company), 176. See also Sandra Bowdler, "Freud and Archaeology," *Anthropological Forum* 7 (1996), 419–38.

4. Freud. *The Ego and the Id,* trans. Joan Riviere (London: L & V Woolf; Institute of Psycho-Analysis, 1927), 53.

5. Sigmund Freud, *Civilization and Its Discontents,* trans. James Strachey (New York: W. W. Norton & Company, 1961), 17.

6. Ibid.

7. It is worth repeating the definition proposed by Freud: "The German word 'unheimlich' is clearly the opposite of 'heimlich' ['homely']"; "And it seems obvious that something should be frightening precisely because it is unknown and unfamiliar. But of course the converse is not true: not everything new and unfamiliar is frightening. All one can say is that what is novel may well prove frightening and uncanny; some things that are novel are indeed frightening, but by all no means all. Something must be added to the novel and the unfamiliar if it is to me uncanny." Freud, *The Uncanny,* 124–25.

8. Although we should also acknowledge, as does Whitney Davis, that: "In archaeology, an object is never outside the *physical* assemblage in which it is found. In psychoanalysis, an object is never outside the *psychic* assemblage in which it is found." Whitney Davis, *Replications: Archaeology, Art History, Psychoanalysis,* ed. Richard W. Quinn (University Park, PA: Pennsylvania State University Press, 1996), 3.

9. Freud, *The Uncanny,* 148.

10. Ibid.

11. Ibid., 135.

12. C. G. Jung, *Memories, Dreams, Reflections,* ed. Aniela Jaffé, trans. Richard Winston and Clara Winston (New York: Vintage Books, 1989), 156.

13. Ibid., 152–3.

14. Ibid., 152. Jung uses the "black tide of mud" expression earlier when he speaks of Freud's suspicion against occultism. In his memoirs, Jung repeatedly refers to Freud's "black tide of mud" phobia as if Jung wanted to flood the "father" in his own revealing vocabulary. A few years after the first fainting episode, during a discussion brimming with other archaeological resonances, Freud fainted once again. This time the archaeological topic was

concerned with the interpretation of the Egyptian Pharaoh Amenophis IV's destruction of his father's stele cartouches. While Freud once again saw this as connected to negativity toward the father, Jung had no patience for Freud's obsession with father-murder theories and saw the destruction as an expression of creativity. Ibid., 157.

15. "After your departure I determined to make some observations, and here are the results. In my front room there are continual creaking noises, from where the two heavy Egyptian steles rest on the oak boards of the bookcase, so that is obvious." Jung, [Appendix: "Letter from Freud to Jung"]. Ibid., 361. Years later, in his article on the uncanny, Freud returns to the ways in which serendipitously reoccurring numbers can be seen as examples of uncanniness.

16. Ibid., 155 (my emphasis).

17. Ibid., 160.

18. Ibid., 159.

19. Ibid., 159–60.

20. Ibid., 161.

21. Ibid., 287.

22. Ibid., 288.

23. On Pompeii as an uncanny location for the fear of being buried alive, see Vidler, 45–55.

24. Jung, 288.

25. Ibid. (my emphasis).

26. As Sarah Kofman has pointed out, Freud overlooks an interesting detail in the text: When the young man has opportunity to touch the hand of the young woman, he refrains from doing so because he fears that the statue-hand will dissolve into air. See Sarah Kofman, *Freud and Fiction,* trans. Sarah Wykes (Cambridge: Polity Press, 1991), 107.

27. I have also made this point in "'Upon the Bedrock of Material Things': The Journey to the Past in Danish Archaeological Imagination," *Northbound,* ed. Karen Klitgaard Povlsen (Denmark: Aarhus University Press, 2007), 147–66. The connection between art and archaeology has a long history and it is important to remind ourselves of the time-honored link between the two fields, particularly as it took hold in the eighteenth and nineteenth century. Art history tells us how painters and sculptors through time have paid attention to the past, either in the form of depicting historical scenes and mythological or religious motifs, or through mimicry, such as in the case of neoclassical sculpture's attempt to compete with ancient Greek and Roman art. This, naturally, does not make art archaeological per se. But if archaeology can be (and certainly has been) seen as a cousin to history (as discipline), its genealogical tree is equally tied to the field of art history. The eighteenth century's rediscovery of the ancient Greek and Roman past, in particular, facilitated intimate bonds between the two fields. Or more correctly, art and archaeology had not yet been separated as they would be in the nineteenth century. Art history and archaeology continue to share a premise; namely, that the material artifact *itself* is the key object of analysis. The rebirth of a Hellenistic past was a combined art-archaeology; Winckelmann's influential studies of Greco-Roman sculpture, for example, were written as part of his archaeological art studies and so forth. The importance of his and other's treatises on art and archaeology meant that for northern European antiquarians, interest in local and national prehistory was considered

secondary to interest in the classical birthplace of Western civilization. Also, in the Scandinavian countries national archaeology did not find solid footing until the nineteenth century, when archaeological sites were meshed with national romantic visions. Freud's interest in the classical past, in other words, is part of a conventional optic.

28. Gaston Bachelard, *Earth and Reveries of Will,* 98. For a description of uncanny dark spaces, see Vidler, 167–75.

29. In 1909, the same year as Freud's bog body encounter with Jung, he published the famous account of the so-called Rat Man's analysis and points out to his patient, and to us, that "the destruction of Pompeii was only beginning now that it had been dug up." Quoted from Lowenthal, 254, n. 402.

30. Gordon C.F. Beam. "Wittgenstein and the Uncanny," in *Soundings: An Interdisciplinary Journal* 76, no. 1 (1993), 33. Here cited from Royle, 88. I have omitted Beam's emphasis and added my own.

31. In fact, as Jennifer Wallace puts it, it was "the psychoanalytic equivalent of imagining the victims of Pompeii from witnessing Fiorelli's plaster casts." Wallace, 95.

32. Colin Renfrew, *Figuring It Out: What Are We? Where Do We Come From? The Parallel Vision of Artists and Archaeologists* (London: Thames & Hudson, 2003), 205, n.2.

33. Bog bodies are *abject* in the definition brought to bear by Julia Kristeva: that which disturbs the identity, disturbs order, as an "in-between." See Julia Kristeva, *Powers of Horror: An Essay on Abjection,* trans. Leon S. Roudiez (New York: Columbia University Press, 1982).

34. Vidler, 46.

35. Ibid., 48.

36. Wallace, 55. For an interesting combination of corpse and marble, see Mark Sandberg's description of "marble-ization" in Sandberg, 42.

37. Collectors of such erotica included Lord Hamilton, who in turn helped direct his wife Emma Hamilton's classical yet suggestive attitudes. Other aficionados of archaeological erotica included Lord Elgin, whose pillaging of the Parthenon on the Acropolis between 1800 and 1802 continues to this day to be a hotbed for discussions of archaeological appropriation and misappropriation. Elgin found himself in a midst of a sex scandal when he returned to London, and in 1816 sold "his" marbles to the British Museum. During the divorce proceedings when Elgin accused his wife of adultery, her defense was built in part on the disease that had eaten away Elgin's nose. The public trial was essentially an emasculization of Elgin. Lord Byron, who in *Childe Harold* voiced his abhorrence over Elgin's "rape" of the Greek past, turned his pen against Elgin and described him as a pathetic noseless statue on a "pedestal of Scorn." If Byron accused Elgin of archaeological rape, a young Keats in turn fell into enraptured silence, stunned at the sight of the marbles' visual power. He proceeded to write two poems on the Elgin Marbles, in which words and images dueled. Also the death of Winckelmann, who had celebrated the Greek body and was keen on locating a moral codex in antique statuary, has been tied to illicit sexuality: he was killed in what is believed to have been a homosexual encounter. Hawthorne's novel *Marble Faun* was filled with love and "buried" desire amidst Roman antiquities. In H.C. Andersen's first major novel, *The Improvisatore,* volcanic eruptions from Pompeii's subjugator, Vesuvius, are described as orgiastic outbreaks which needed to be cooled into ethereal immortality. The list is endless, but the connection between archaeology and erotica is clear and present.

38. As pointed out by Winfried Menninghaus: "aesthetic and anatomy, which boomed simultaneously in the second half of the eighteenth century, are thus opposing disciplines." *Disgust: The Theory and History of a Strong Sensation* (Albany, NY: State University of New York Press, 2003), 55. See also my study of sculpture and death imagery, *Konturer: skulptur- og dødsbilleder fra Guldalderlitteraturen* (Copenhagen: Museum Tusculanums Press, 1997).

39. David Lowenthal, *The Past Is a Foreign Country* (Cambridge: Cambridge University Press, 1985), 253.

40. Freud's consulting room did not resemble "a doctor's office but rather . . . an archaeologist's study. Here were all kinds of statuettes and other unusual objects, which even the layman recognized as archaeological finds . . . " The words are from Freud's famous patient known as the Wolf Man. Here cited by Peter Gay in *Freud: A Life for Our Time* (New York: W. W. Norton, 1989). This is not to say that bogs did not yield such objects. Museums brim with artifacts found in bogs.

41. Though the question lies outside the scope of this study, it nonetheless seems appropriate to ask: Do the sacrifices that Freud discusses in *Totem and Taboo* and places at the root of Western civilization hark back to his uncomfortable encounter with Jung's "primitive" death wish? Sigmund Freud, *Totem and Taboo: Resemblances Between the Psychic Lives of Savages and Neurotics* (Harmondsworth: Penguin Books, 1938)

42. In a footnote in *Totem and Taboo,* Freud reminds us that it "should not be forgotten that primitive races are not young races but are in fact as old as civilized races. . . . The determination of the original state of things thus invariably remains a matter of construction." Freud, *Totem and Taboo,* 128. Sandra Bowdler suggests that this implies that "Freud believed that there was a very fundamental difference between 'primitive' people and 'civilized' people, a mental difference, fundamental to all his work." In the same place where he inserts the last quoted disclaimer, he mentions "primitive modes of thinking." See Bowdler, 419–38.

43. Gay, *A Life for Our Time,* 170.

CHAPTER 3

1. "Wir haben es leider nicht mehr so einfach wie unsere Vorfahren. Bei denen waren diese einigen Wenigen Einzelfälle so abnormer Art. Der Homosexuelle, den man Urning nannte, wurde im Sumpf versenkt. Die Herren Professoren, die diese Leichen im Moor finden, sind sich bestimmt nicht dessen bewusst, dass sie jeweils in neunzig von hundert Fällen einem Homosexuellen vor sich haben, der mit seinem Gewand und allem im Sumpf versenkt wurde. Das war nicht eine Strafe, sondern das war einfach das Auslöschen dieses anormalen Lebens. Das musste entfernt werden, wie wir Brennesseln ausziehen, auf einen Haufen werfen und verbrennen. Das war kein Gefühl der Rache, sondern der Betreffende musste einfach weg. So war es bei unseren Vorfahren." Heinrich Himmler, *Geheimreden 1933 bis 1945 und andere Ansprachen,* ed. B.F. Smith and A.F. Peterson (Berlin and Wien, 1974), 93–104. Here cited from Lund, *Hitlers håndlangere,* 300n312.

2. Ibid., 221–2.

3. See Schama, 76.

4. Tacitus, 104. He continues with this: "They are less able to endure toil or fatiguing tasks and cannot bear thirst or heat, though their climate has inured them to cold spells and the poverty of their soil to hunger."

5. "Blut und Boten" was the motto of the "Nationalsozialitische Kulturgemeinde," which was housed in the foreign ministry and often called "Das Amt Rosenberg" after its ideological head Alfred Rosenberg. Its aim was to develop the nationalistic traits in Germany. Himmler's Ahnenerbe was a pan-Germanistic organization, formed in part to create ideological attributes for grouping all Germanic races in one nation. For more on this, see Lise Nordenborg Myhre, "Nationalism, Politics, and the Practice of Archaeology," in *Myter om det nordiske,* ed. Catherina Raudvere, Anders Andrén, and Kristina Jennbert (Lund: Nordic Academic Press, 2001), 159–90.

6. Klaus Theweleit, *Männerphantasien: Vol. 1.* (Berlin: Rowohlt, 1977), 388.

7. The body disappeared during the war, but a photograph was kept as record. See van der Sanden, *Through Nature to Eternity,* 89.

8. Tacitus, 117.

9. Besides Tacitus, Bonifitius in his letter 73 and Herodot in *Historia* also describe the practice of haircutting as punishment. On this see Lund, *Mumificerede Moselig,* 62–63.

10. See, for example, Philip L. Kohl and Clare Fawcett, "Archaeology in the Service of the State: Theoretical Considerations," in *Nationalism, Politics, and the Practice of Archaeology,* ed. Philip L Kohl and Clare Fawcett (Cambridge: Cambridge University Press, 1995), 3.

11. Ibid., 6.

12. But to see archaeology as a form of narrative does not exclude a plausible reconstruction of the past built on archaeological data or on "scholarly standards of logic and evidence." Neither an extreme positivism nor an exaggerated relativism will help us to steer clear of the many pitfalls in interpreting the past, but an awareness of how and what is at issue in each case is key. See Neil Asher Silberman, "Promised Lands and Chosen Peoples: The Politics and Poetics of Archaeological Narrative," in Kohl and Fawcett, *Nationalism, Politics and the Practice of Archaeology,* 249.

13. Ibid., 261. See also Bruce Trigger, *Alternative Archaeologies: Nationalist, Colonialist, Imperialist,* ed. Lotte Hedeager and Karen Schusboe (Copenhagen: Akademisk Forlag, 1989), and also Hedeager and Schusboe, *Brugte historier.*

14. David W. Anthony points out that "having lost its former objectivist guideposts, prehistoric archaeology has opened itself to innumerable popular reinterpretations of the past, ranging from nationalist bigotry to fantasies of spiritual roots-seeking." In "Nazi and Eco-Feministic Prehistories," in Kohl and Fawcett, *Nationalism, Politics and the Practice of Archaeology,* 84–85. Racist archaeology was not exclusive to Nazi ideology; see Kristiansen, "Fortids kraft og kæmpestyrke."

15. Bernard Wailes and Amy L. Zoll, "Civilization, Barbarism, and Nationalism in European Archaeology," in Kohl and Fawcett, *Nationalism, Politics and the Practice of Archaeology,* 34, n. 3.

16. Bettina Arnold and Henning Hassmann, "Archaeology in Nazi Germany: The Legacy of the Faustian Bargain," in Kohl and Fawcett, *Nationalism, Politics and the Practice of Archaeology.*

17. See Allan A. Lund's study of the role of archaeologists in Nazi Germany, *Hitlers Håndlangere.* Lund has written extensively on bog bodies and on the Nazi interpretations. See also, Lund, *Mumificerede Moselig.*

18. Schama, 36.

19. In doing so, as Julian Thomas points out, "archaeology requires us to *make the world static,* to freeze it in order to interpret it. That is, the past is in some sense seen as a "thing" which ignores temporal relations and seeks general rules that can be uniformly validated." *Time, Culture, and Identity,* 63.

20. See Avishai Margalit, *The Ethics of Memory* (Cambridge, MA: Harvard University Press, 2002).

21. Ricœur, Paul. *Oneself As Another,* trans. Kathleen Blamey (Chicago: University of Chicago Press, 1992), 164.

22. Margaret Drabble: "Millennium Lecture. 'Runes and Bones.'" Quoted from abstract posted on *www.abroad-crwf.com/abroadwritingworkshop.html.* Another author, Michael Ondaatje, reverses this point in his novel *Anil's Ghost* (New York: Alfred A. Knopf, 2000), 151, when he writes: "A good archeologist can read a bucket of soil as if it were a complex historical novel."

23. See W.J.T. Mitchell, "Ekphrasis and the Other," in *Picture Theory: Essays on Verbal and Visual Representation* (Chicago: University of Chicago Press, 1994).

24. Michael Ann Holly, *Past Looking: The Historical Imagination and the Rhetoric of the Image* (Ithaca, NY: Cornell University Press, 1996), 6.

25. Tournier does not credit Glob for the information on the bog bodies, but it seems more than likely that he has obtained his information from the Danish archaeologist's book. Volker Schlondorf made the novel into a film titled *The Ogre,* featuring John Malkovich.

26. Tournier, Michel. *The Ogre,* trans. Barbara Bray (New York: Doubleday & Company, Inc., 1972), 186.

27. Ibid., 188.

28. Ibid., 367.

29. Ibid., 190.

30. Ibid., 370.

31. For an excellent and more detailed analysis of this novel see Anthony Purdy, "The Bog Body as Mnemotype: Nationalist Archaeologies in Heaney and Tournier." *Style,* Spring 2002.

32. In a fictional and factual introduction we read: "During the Second World War, countless manuscripts—diaries, memoirs, eyewitness accounts—were lost or destroyed. Some of these narratives were deliberately hidden—buried in back gardens, tucked into walls and under floors—by those who did not live to retrieve them. Other stories are concealed in memory, neither written nor spoken. Still others are recovered, by circumstance alone." Page not paginated. Anne Michaels, *Fugitive Pieces* (New York: Vintage, 1998). I wish to thank Daria Fireman for pointing out the relevance of Michaels's book to my project.

33. Ibid., 8.

34. Ibid., 5.

35. The reflection in the novel is retrospective also in the sense that Tollund Man and Grauballe Man were not found until well after the Second World War, and therefore could not have been part of the young boy's consciousness.

36. Michaels, 49.

37. Ibid.

38. Ibid., 12–13.

39. Ibid., 143. See also Elaine Scarry's observations in *A Body in Pain: The Making and Unmaking of the World* (New York: Oxford University Press, 1987) on the relation of pain and silence.

40. Michaels, 165.

41. Ibid., 166.

42. Ibid.,240–41. Michaels makes use here of a historical fact. The so-called "Borger-moorlied" was written and performed in 1933. See Shoshana Kalisch and Barbara Meister, *Yes, We Sang! Songs of the Ghettos and the Concentration Camps* (New York: Harper and Row, 1985). See also Purdy. "The Bog Body as Mnemotype." He writes in a footnote, "According to Robert Merle's biographical novel, *La mort est mon metier,* the future Kommandant of Auschwitz was given the task of draining a swamp in Pomerania in 1929. It was Hoss who would later refer to Auschwitz as Anus mundi (Arschloch der Welt), an appellation that takes on a particular significance in Tournier's novel." [n. 9]

43. Michaels, 111.

44. Ibid., 104, 32.

45. Ibid., 21–22.

46. Ibid., 76.

47. Ibid., 161.

48. Ibid.,138 (my emphasis).

49. Pierre Nora: "Between Memory and History: Les Lieux de Mémoire," *Representations* 26, University of California Press (Spring 1989), 10.

50. Ibid., 12.

51. Michaels, 139.

52. Ibid., 221.

53. Ibid., 279.

54. Ibid., 233.

55. Ibid., 261.

56. Ibid., 265.

57. The writings by the three generations of survivors (Athos's book *Bearing False Witness* is finished by Jakob; Jakob's memoir in turn is discovered by Ben) are nested one inside the other, like Russian dolls. More importantly, the archaeologically layered narrative structure and the extension of various (auto)biographies across time is compressed into the metaphor of bog bodies. Although we read that "it's no metaphor to feel the influence of the dead in the world, just as it's no metaphor to hear the radiocarbon chronometer, the Geiger counter amplifying the faint breathing of rock, fifty thousand years old," metaphors can, as Michaels demonstrates in her use of bog bodies, capture the presence of a past. Ibid., 53.

58. Ibid., 274–75.

59. Ibid., 252.

60. Ibid., 253.

61. Ibid., 222.

62. In Cormac McCarthy's *The Road,* (New York: Vintage, 2006) bog bodies are used as the ultimate image of a more generalized collapse of civilization. A father and his son are walking through the ruins of some unnamed part of America after an apparent worldwide holocaust. "By dusk of the day following they were at the city. The long concrete sweeps

of the interstate exchanges like the ruins of a vast funhouse against the distant murk. He carried the revolver in his belt at the front and wore his parka unzipped. The mummified dead everywhere. The flesh cloven along the bones, the ligaments dried to tug and taut as wires. Shriveled and drawn like latterday bogfolk, their faces of boiled sheeting, the yellowed palings of their teeth. They were discalced to a man like pilgrims of some common order for all their shoes were long since stolen." Ibid., 24. I wish to thank Linda Rugg for providing me with this reference.

63. Stegner, Wallace. *The Spectator Bird* (London: Penguin, 1976), 26 (my emphasis).

64. Ibid., 24.

65. Ibid., 106.

66. Ibid., 26–27.

67. Ibid., 27.

68. Ibid., 147.

69. Ibid., 149–50.

70. Ibid.,179.

71. This point is also made by Martin Zerlang, "I historiens pariserhjul: Ebbe Kløvedal Reich," in *Dansk litteraturs historie. 1960–2000,* ed. Klaus P. Mortensen and May Schack (Copenhagen: Gyldendal, 2007), 32–33.

72. "Om betydningen af beretningen om begyndelsen"; "intet er så velkendt som glemsel." Ebbe Kløvedal Reich, *Fæ Og frænde: syv en halv nats fortællinger om vejene til Rom og Danmark* (Copenhagen: Gyldendal, 1977), 11.

73. Glob, *The Bog People,* 28.

74. "De gamle, seje og vandslunkne pletter på mosen fik menneskene til tørv, som de byggede af og kogte mad ved. Til gengæld smed menneskene forskellige slags interessante kroppe og andre ofringer ud i den nye mose, som Hundekød kunne øve sin udødeligheds-kunst på." Reich, 30.

75. "Hundekød er én eneste stor idé, fordelt i alle de små moseplanter, den består af: Ideen om at nedsænke og udødeliggøre alverden i ferskvand." Ibid., 29.

76. "Af vand er du kommet, til vand skal du blive, af vand skal du igen opstå." Ibid., 30.

77. "Han drømmer en behagelig drøm, for hans læber er kruset i et lille smil, der fortæller om fred, men også om en viden, der har gjort kål på evnen til at glædes dybt og længe. Det er en drøm. Og han ser ud, som om han kun smiler i drømme." Ibid., 31.

78. Ibid., 33.

79. "Odtor strammede til og tog tøjet af Geppu sammen med posen og slæbte ham ned til det sted, lige inden søen med åbent vand, hvor Hundekød stod lys og ung, og der var for vådt og sugende til at noget andet end den kunne vokse. Odtor tog Geppu i ventre arm og ben og svingede ham rundt et par gange, indtil der var kraft nok på—og han gav slip. / Der lød et svurp, og der gik en bølgen gennem Hundekød. Så var Geppu væk . . . " Ibid.,150–51.

80. Kløvedal Reich's use of Tollund Man in his vision of Danishness is shown in a news-paper article on April 9, 1978. Here the author locates three "typical Danes": one from year 0, Tollund Man; one from year 1700, the fictional character Jeppe from Ludvig Hol-berg's comedy *Jeppe paa Bjerget;,* and finally the Danish Social-Democratic prime minister, anno 1978, Anker Jørgensen.

81. There are other examples of how bog bodies become foils for national quandaries. In

A Natural Curiosity, from 1989, for example—Margaret Drabble's middle novel in a trilogy (sandwiched by *The Radiant Way,* 1987 and *The Gates of Ivory,* 1991)—bog body remains are used to give a satirical and critical view of present-day Britain. In the novel, Margaret Drabble exposes the pathology of Thatcher's Britain. Margaret Drabble, *A Natural Curiosity* (New York: Viking, 1989).

82. Reich, 485.

83. "[Den virkelige fjende er den magt, du sætter fri i Rom, efter at du selv er død. Den skal brede sig over alle de folk, du kender og over hele jorden.] Og skønt du er død, skal du få hele den hæslige historie at se med dine egne øjne som totusinde år." Ibid., 35.

84. After his initial attempts (*Death of a Naturalist,* 1966; *Door into the Dark,* 1969), Heaney had been encouraged by fellow poet Ted Hughes to write more bog poems, and with *Wintering Out* (1972) and *North* (1975) the Nobel laureate made his archaeology of bodies from the bog speak about the personal, the national, aesthetics, and ethics. He became, in the words of Helen Vendler, a "spectator to a renewed archaic violence, symbolized by bodies long nameless." Helen Vendler, *Seamus Heaney* (Cambridge, MA: Harvard University Press, 1998), 59.

85. Seamus Heaney, *The Government of the Tongue: Selected Prose 1978–1987* (New York: Farrar, Straus and Giroux, 1988), 107. See also Vendler, 11.

86. "Kinship" in *North,* here cited from *Opened Ground, Selected Poems, 1966–1996* (New York: Farrar, Straus and Giroux, 1998), 115–16.

87. Furthermore, when he imagines that "The bogholes might be Atlantic seepage. / The wet centre is bottomless" ("Bogland"), we may well associate *bogholes* with the equally bottomless *wormholes.* See the analysis of Queen Gunhild in the next chapter.

88. "The Tollund Man," from *Wintering Out.* Here cited from *Opened Ground,* 63.

89. Seamus Heaney, "The Man and the Bog," 3–4. The italics are mine.

90. But it was also connected to a persistence in the psyche that continues even after the "*frisson* of the poem itself had passed, and indeed after I had fulfilled the vow and gone to Jutland, 'the holy martyr for to seek.'" Seamus Heaney, "Feeling into Words" in *Preoccupations: Selected Prose, 1968–1978* (New York: Farrar, Straus, Giroux, 1980), 59–60.

91. Ibid., 57.

92. Heaney, "The Man and the Bog," 3.

93. Ibid., 4. A host of Heaney critics have commented on his use of history. Let me mention but a few. Conor Cruise O'Brian points to Heaney's ability to let us "listen to the thing itself, the actual substance of historical agony and dissolution." Conor Cruise O'Brian, *The Listener* (September, 25, 1975), 404–5. See also Edna Longley, "'Inner Emigré' or 'Artful Voyeur?': Seamus Heaney's *North,*" in *Seamus Heaney,* ed. Michael Allen (New York: St. Martin's Press, 1997), 31.

94. Heaney, "Tollund Man," in *Wintering Out,* from *Open Ground,* 63.

95. Ibid., 57.

96. See Holly, *Past Looking,* 4.

97. See also W.J.T. Mitchell: "Going Too Far with the Sister Arts," in *Space, Time, Image, Sign: Essays on Literature and the Visual Arts,* ed. James A. W. Heffernan. (New York: Peter Lang, 1987), 2.

98. David Lloyd writes: "Place, identity and language mesh in Heaney, as in the tradition

of cultural nationalism, since language is seen primarily as *naming,* and because naming performs a cultural reterritorialization[. . .] The name always serves likeness, never difference." In "'Pap for the Dispossessed': Seamus Heaney and the Poetics of Identity," *Boundary* 2, vol. 13, no. 2/3 (Winter/Spring 1985), 328. Thus, the Tollund Man is at its very core an emblem of cultural nationalism. I am not sure, however, that Lloyd is correct in defining naming solely as a restriction, and as a lack of awareness of difference.

99. Thomas Docherty, "Ana- or Postmodernism, Landscape, Seamus Heaney," in *Seamus Heaney,* ed. Michael Allen (New York: St. Martin's Press, 1997), 208.

100. Ibid., 212.

101. In *Preoccupations,* Heaney tells us that his poems come "sometimes like bodies come out of a bog, almost complete, seeming to have been laid down a long time ago, surfacing with a touch of mystery." *Preoccupations,* 34. Docherty points out: "But it is a poetry which lies uncertainly between image (the photograph which prompted the poem) and memory (where 'now he lies / perfected'), between history and its representation. If anything, then this is a poem which is about poetry, as mediation or about a specific act of reading. Heaney is confronting the bog as 'the memory of landscape,' the palimpsest record of history which is now conceived as 'a manuscript which we have lost the skill to read.'" Most importantly, "The Grauballe Man" is what we should think as a kind of "interstitial" event, a writing halfway between image and text, figure and discourse. The self-referentiality in the poem makes Thomas Docherty go so far as to see "The Grauballe Man" as a postmodern "event," a poem about a man "back up again," which eventually releases "the interior historicity" of anachronistic "time spots." Docherty's analysis brings attention to the anachronism which the use of bog bodies in modern poetry brings about. To Docherty, Heaney turns the poem into a temporal and spatial happening that eventually turns on itself. See also Docherty, 208.

102. James Randall, "An Interview with Seamus Heaney," *Ploughshares,* 5:3 (1979), 18. Here cited from Longley, 37.

103. Many have pointed out that the simile ennobles the dead. See for example Maurice Harmon, "'We Pine for Ceremony': Ritual and Reality in the Poetry of Seamus Heaney, 1965–75," in *Seamus Heaney: A Collection of Critical Essays,* ed. Elmer Andrews. (New York: St. Martin's Press, 1992), 79–80. In fact one of Heaney's critics, Edna Longley, suggests that violence here has been outweighed by beauty: "the poem almost proclaims the victory of metaphor over 'actuality.'" She goes on to say that "Heaney seems to regard a symbol or myth as sufficiently emblematic in itself: 'beauty' pleading with 'rage' within the icon of 'The Grauballe Man'—Man and poem synonymous—rather than through any kind of dialectic." Longley, 47.

104. Docherty, 209.

105. Ibid., 210. In times of postmodernism, Docherty goes on, the relation between "the Subject and History, or between the 'real' and its 'representation,'" is no longer easily accessible, if accessible at all. This is because "reality"—as Docherty argues, true to the post-structuralist credo—"is supposed to ground our representations, be it in the presence of History as exterior fact or the presence-to-self of the supposed transcendental Subject, has itself become an image." Ibid.

106. Longley, 31, 35.

107. Elmer Andrews "The Spirit's Protest," in Andrews, *Seamus Heaney: A Collection of Critical Essays,* 215.

108. Docherty, 216. When Docherty sees Grauballe Man as a mirror image of Heaney himself ("it is his Imaginary"), and calls the poem a "therapeutic act of recovering what […] had been repressed and facing it," such post-structural imagination seems rather too creative.

109. Ibid., 208–9.

CHAPTER 4

1. "Neppe tilhörende en Person af den arbeidende Klasse." J.F. Christens, "Oplysninger om et i en Mose nær Haraldskjær fundet kvindeligt Lig," a forensic report in *Annaler for Nordisk Oldkyndighed* (Denmark: Det Kongelige Nordiske Oldskrift-Selskab, 1836–37), 166.

2. "Sandsynligen levende nedpælet i Mudderet"; "Ansigtstrækkene, hvilke strax efter Optagelsen næsten tydelig kunde kjendes at være fortvivlede." Ibid., 165.

3. "Maaden, paa hvilken Liget saa omhyggelig og möisommelig var nedpakket, synes ligeledes at vidne om noget Overordentligt, og har neppe været eet eller to Menneskers Foretagende i en kort Tid." Ibid., 168.

"Bemærkninger om et Fund af et Mumieagtigt Kvinde-Lig i en Mose ved Haraldskjær i Jylland" by Oldsags-Committeen.

4. Glob, *The Bog People,* 70.

5. In 1846, the dramatist Danish Christian Hostrup gave Queen Gunhild a dramatic part in his comedy *En Spurv i Tranedans.* His use of the bog body was an underhanded attack on Petersen's Queen Gunhild theory. A friend of Worsaae, Hostrup initially included direct references to a hapless Petersen, but revised the play and cut out the offending lines before the premiere at the Royal Theater. For a full description of this and Queen Gunhild's reception story, see my article "A Portal Through Time," *Scandinavian Studies,* 2009.

6. Fordum / Da varst Du klædt udi Zobel og Maar, / Og pyntet med prægtige Smykker, / Juveler og Perler i / guldfagert Haar, / I Sindet de slemmeste Nykker. / /Nu / Da ligger Du nögen og skrumpen og fæl, / Alt med / Dit skallede Hoved, / Langt sortere end den Egepæl, / Hvormed Du blev Sumpen troloved. / Fordum / Du / glimred som Stjernen af förste Rang, / Og Tusinder adlöd Din Villie, / Forelskede / Sukke ledsaged Din Gang, / Naar yndigt Du dansed paa Tillie / Nu / Da ligger Du stille paa / Dödningeliin, / Med haarde, visnede Hænder; / Med dette stive og isnende Griin / Du ingen Hjerter antænder. St. St. Blicher, "Dronning Gunnild," in *Steen Steensen Blichers Samlede Skrifter* (Copenhagen: Det Danske Sprog og Litteraturselskab, 1928), 274.

7. Fordum /Da satte Du gjerne Blodöxen i Gang, / Kong Erik Du lettede Byrden, / Du deelte paa Fjeld og paa Hav og paa Vang / Din Tid mellem Leflen og Myrden. / Nu / Nu har Du det temmelig tört og svalt, / Men hæsligt med Munden du vrænger: / Dödsvrælet, da du blev I Mosen kvalt, / Neddukket af Bödlernes Stænger. Ibid., 275.

8. Hvor kunde Du troe Harald Blaatand saa? / Selv brödst Du jo over dig Staven; / Nu ligger Du blaatandet selv paa Straa, / Mens han taer en Luur sig i Graven. / Der er endda gledet en heel Deel Aar, / Mens Du haver ligget i Sölen; / Gad vide, hvordan det dernede staaer: / Hvad nyt, Dronning Gunnild, fra Pölen? / Blevst du udi Helvede saa forbrændt […] I tusinde Aar snart—det falder mig paa— / Min naadige haver sig fjælet; / De vil til

Dagen paa Ny opstaae; / De længe med Döden har hælet. / Maaske det skal være en Over-retsdom, / Som De nu paa den Maade lider. / De laa udi Mulmet et Tusind aar om, / Og atter til Lyset nu skrider. / De viser dem ret som en Antiqvitæt / Fra gamle hedenske Tider; / God Nat! og sov vel, deres Majestæt! / Det sorte Laag over Dem skrider. Ibid., 275–76.

9. Menninghaus, 54.

10. Ibid., 129.

11. Ludwig Tieck, *Das Alte Buch*. Here cited from Ibid., 129 (my emphasis).

12. Menninghaus, 130.

13. Høje Gladsaxe is a satellite city complex outside Copenhagen.

14. Camilla Christensen, *Jorden Under Høje Gladsaxe: Roman* (Copenhagen: Samleren, 2002), 94–95.

15. Hun har medfølelse med Lenin, der i trekvart århundrede har ligget på lit de parade i mausoleet på Den røde Plads, mens hans hjerne ligger i en skuffe på det medicinske institut tre gader derfra. Hun har medfølelse med faraoerne og deres hustruer, der trækkes ud af pyramider og kister og kan beses på alverdens museers ægyptiske samlinger, og med Egtved-pigen og sin egen navnesøster, dronning Gunhild, moseliget i Vejle, men mest af alt har hun medfølelse med Tollundmanden, som han ligger i sin montre på Silkeborg museum og kan beskues fra alle sider, han er smuk, han er betagende, han er fredfyldt (trods fremvisningen), en sovende skønhed, der på det grusomste udstilles for alverdens nyfigne blikke, og nu skal den langt mindre tiltrækkende Grauballemand (og 'gudskelov da, at det er ham og ikke Tol-lundmanden') udsættes for det danske sygehusvæsen, han skal ikke længere kun beses udefra, men også indefra. Ibid., 91.

16. A nod to the most famous line in the Danish Romanticist Adam Oehlenschläger's poem "The Golden Horns," from 1802.

17. " . . . hun er opslugt, selvforglemmende opslugt lader hun det ene link tage det andet og higer og søger efter flere oplysninger, først om Grauballemanden, men snart om jernal-derens øvrige europæiske moselig, og i løbet af aftenen bevæger hun sig fra lokale Jubii ud i globale Google og lærer sig stort set alt, hvad der for tiden kan vides om velbevarede lig, og ikke bare moselig [. . .]." Christensen, 92 (my emphasis).

18. Ibid., 144.

19. "Arthur ser lige ind i hende og danner sig et billede af hendes skelet, som det var dengang; han genopbygger hende, rekonstruerer hende ud fra de mange lag knogler, indtil hun ikke bare [. . .] er harmonisk bygget, men lydefri, fuldendt, og udenpå lægger han sener, vener, muskler, kød og fedt, og frem af forfaldet dukker en kvinde på sit livs højde i stedet for et oldtidslevn med frakturer efter utallige hoftebrud." Ibid., 201.

20. For the latest interpretation of Queen Gunhild as a depraved and oversexed woman, see Terry Gilliam's 2006 film *Tideland*, adapted from Mitch Cullin's 2000 novel. In a surreal and drug-infested world, a young child tries to overcome the death of her depraved mother Gunhild, played by Jennifer Tilly. After her bog-body-obsessed father, played by Jeff Bridges, dies of a drug overdose and slowly turns into a mummified Tollund Man lookalike, the child continues to live in a liminal world where the borders between living and dead, humans and dolls (and bog bodies), are broken down.

21. Jennifer Wallace, for example, points to Michael Shanks's defense of archaeology as striptease and "a pleasure existing in the interplay of performer and audience." She also

argues that Freud's obsession with collecting can be seen as "a substitute for sex," and goes on: "Freud once confessed to his physician that acquiring ancient sculptures was an 'addiction second in intensity only to his nicotine addition.' Since he believed that all addictions were just substitutes for the primal addiction, masturbation, both cigar-smoking and looking at his phalanx of figurines could be understood as displacements for less acceptable desires." *Wallace,* 81.

22. Renfrew, 42.

23. See Joan M. Gero and Margaret Wright Conkey, eds., *Engendering Archaeology: Women and Prehistory* (Oxford: B. Blackwell, 1991).

24. Again, we should recall how it has been pointed out that sacrifice and violence walk hand in hand and that dead bodies disrupt and restore order. See René Girard, *Violence and the Sacred,* trans. Patrick Gregory (Baltimore: Johns Hopkins University Press, 1977), 254–56.

25. Besides science fiction and horror, Michael Talbot (1953–1992) also wrote books such as *The Holographic Universe* (1991) in attempts to mix science and spirituality.

26. Michael Talbot, *The Bog* (New York: Jove Books, 1986), 21.

27. Ibid., 22.

28. Ibid., 54.

29. Ibid., 18–19 (my emphasis).

30. Ibid., 62.

31. Ibid., 66.

32. Ibid., 54.

33. Ibid., 129.

34. Ibid., 134–35.

35. Ira Levin's horror novel *Rosemary's Baby* seems to have been an inspiration here.

36. Talbot, 147.

37. Ibid., 308.

38. Ibid., 193.

39. See Margaret Atwood, "The Bog Man," *Playboy* 38.1 (January, 1991).

40. Margaret Atwood, "The Bog Man," in *Wilderness Tips* (Toronto: M & S, 1991), 84–85.

41. Ibid., 88.

42. Ibid., 89.

43. Ibid., 85–90.

44. Ibid., 91.

45. Ibid., 88.

46. Ibid.

47. Ibid., 79.

48. For a different reading, see Anthony Purdy, "Unearthing the Past," 450. Purdy's own excellent reading makes the case, I think, that Atwood's story is indeed about gender.

49. Sylvia Kantaris, "Couple, Probably Adulterous (Assen, Holland, circa Roman Times)," in *Dirty Washing: New & Selected Poems* (Newcastle upon Tyne: Bloodaxe, 1989), 128.

50. Glob, *The Bog People,* 110.

51. Here quoted from Jonathan Dollimore, *Death, Desire and Loss in Western Culture* (New York: Routledge, 1998), 64.

52. According to conversations I had with Wijnand van der Sanden, no postcards were ever produced of the Weerdinge Couple in the museum in Assen.

53. Whenever bog bodies of different sexes have been found in close vicinity of each other, they have been coupled in the popular imagination. The so-called Elling Woman, for instance, found in Denmark in 1938 and displayed at the same museum as Tollund Man, has time and again been imagined to be his lover. One of the more bizarre instances of such an imagined love affair was published as late as July 17, 1982, in a local Danish newspaper with the heading: "Tragic Love Story in Tollund 2000 years ago." The story is told in a facetious manner, drawing on all possible stereotypes of gender and race. Two thousand years ago, we read, a Negro wandered around Tollund town and "smiled roguishly to the Elling Girl." The young blond Danish girl, bored by the local men, was drawn toward the "beautiful muscular negro." He in turn fell deeply in love with her but the color of his skin, we read, was "apparently an obstacle." The (Tollund) man consequently throws himself into the bog in an act of suicide and the girl in turn is punished by hanging and death in the bog. The story, as the author-journalist concedes, is "not entirely authentic," but the fact that the bog man was a black man can, he assures the reader, be verified by a museum anecdote: "The other day a colored American couple visited the museum. The man bent over and looked the little dark cranium deep into its eye-pits and then exclaimed in genuine American: 'Look Ma, this fellow is a negro.'" The American tourist and Tollund man, who we are told "looked alike like twins," thus meet across time in a Danish museum. Obviously the story was not meant to be taken seriously, and the ethnographic stereotypes may seem rather innocuous were it not for the history of archaeological misuse; a misuse which oftentimes, as we have already seen, takes place on behalf of ethnic or racial claims. The absurdity of Tollund Man as lovestruck "Negro" lies not as much in the fact that the reality of the bog body's skin color is used (it is rather surprising that so few representations, visual or verbal, engage the irony of bog bodies' skin color changing over time); the preposterousness is connected to the way in which the self-sacrifice is fabricated to spin a tall tale of ethnic misfortune in love. Sexuality and violence are interwoven, as René Girard has taught us in his anthropological studies; and sacrifice is supposed to restore order to society. Girard has pointed out how sacrifice and violence walk hand in hand, and how dead bodies disrupt and restore order. The self-sacrifice of Tollund Man-as-"negro" to all intents and purposes (albeit in a rather whimsical way) thus restores (racial) order to the provincial village of Tollund—and to the Danish past. See Girard, 254–56. Along the same lines, one of the most famous examples of an erotizing of the past is tied to the so-called Egtved Girl from Denmark, found in 1921. While she is not a bog body proper, but received a "proper" burial in the Bronze tradition in an oak casket placed within a mound, her blond hair and skimpy clothing—in the form of a short see-through string skirt on which pubic hairs were still lodged—would send many an archeologist and many a layman into amorous daydreams for decades to come. Also, Skydstrup Girl, found in a coffin in 1935, was seen as the quintessential blond Danish girl, and the poet Piet Hein wrote a poem that captured the erotic fantasy of a rendezvous across time: "Was it her I should have loved / a breaking Stone Age spring / with light in her white teeth / and leaves in the wet hair / A girl who smiles to me / through three thousand years." Hein's fantasiz-

ing of being a prehistoric lover was ironically echoed when President Bill Clinton in 1996 viewed the remains of the so-called Inca Ice Maiden, freeze-dried and mummified on the mountains in Peru and nick-named "Juanita." His infamous comment after seeing her in a National Geographic exhibit in Washington—"If I were single, I'd ask the mummy out. It's one good looking mummy"—has become part of Clinton lore, but it also shows how easily the imagination can fill in, and fill out, female mummies with sexual fantasies.

54. Heaney, "Feeling into Words," 41. See also Christine Finn's *Past Poetic,* in which she suggests that both archaeologists and poets "are transformers with the skills of 'seers' to project into places, or sites, to retrieve 'things' and illuminate a general past, or a past which is personal and specific." *Past Poetic: Archaeology in the Poetry of W.B. Yeats and Seamus Heaney* (London: Duckworth, 2004), 147.

55. See Maurice Merleau-Ponty, *The Visible and the Invisible: Followed by Working Notes* (Evanston, IL: Northwestern University Press, 1968), 139–45.

56. Seamus Heaney, "Digging" from *Death of A Naturalist* (London: Faber & Faber, 1966). Here cited from *Opened Ground,* 4.

57. Johann Wolfgang Goethe: Römische Elegien," in *Gedenkausgabe der Werke, Briefe und Gespräche, Bd. 1.* (Zürich und Stuttgart: Artemis, 1795), 167. I use a translation by Linda Haverty Rugg.

58. Seamus Heaney, "Bog Queen," in *North* (London: Faber & Faber, 1975). Here cited from *Opened Ground,* 108.

59. Vendler, 45.

60. Ibid., 46–47.

61. Ibid., 47. On the story of Lady Moira, see also van der Sanden, *Through Nature to Eternity,* 46–48.

62. Again, for a fuller description of the Queen Gunhild story and its political implications, see my article, "A Portal Through Time."

63. Seamus Heaney, "Strange Fruit," from *North.* Here cited from *Opened Ground,* 114.

64. Helen Vendler sees this poem as one of Heaney's less successful bog body poems because it relies "too heavily on lavish but conventional adjectives." Vendler, 48. But the very reliance on such conventional adjectives ("murdered, forgotten, nameless, terrible / Beheaded girl") could also, I would argue, be part and parcel of the conventions of expressing *pathos;* a pathos, which gains particular poignancy when it is transferred to museum displays, as suggested in the poem.

65. There are a number of interesting contextual references in the poem. One is to the Greek historian Diodorus Siculus from the first century AD, and we are told that he "confessed / his gradual ease" at murdered sacrifices. The reference seems to point to two moments in Diodorus Siculus's oeuvre: first, his description of his own unease at witnessing an angry mob in Egypt, his demand for the death of a Roman citizen (lynching!), and his gradual desensitizing to such atrocity; second, his descriptions of the mythological figure Osiris, who had made mankind give up cannibalism and instead cultivate the fruits of the land.

66. The lyrics and music were written in 1939 by a Jewish schoolteacher, Abel Meeropol, and released under his pen name Lewis Allan. Meeropol wrote "Strange Fruit" after seeing a photograph of a lynching of a black man, hanging "like fruit" from a tree while the lynch-

ers and their children look toward the camera, seemingly unaffected. Lewis Allen (Abel Meeropol), "Strange Fruit," in *Scanning the Century: The Penguin Book of the Twentieth Century in Poetry*, by Peter Forbes (New York: Viking, 1999).

67. See the documentary film by Joel Katz from 2004.

68. Vendler, 48.

69. Seamus Heaney, "Bone Dreams," from *North*. Here cited from *Opened Ground*, 104.

70. The re-placement of the cord over her eyes instead of over her mouth is, according to Allan A. Lund, a possible misinterpretation. The string tied around her mouth would have prevented her from screaming, and "[...] the garrote (or blindfold) might have served irrational motives: the intention would have been to prevent the doomed in spewing her particularly effective curses. The dying girl's last words or gaze would have been specially dangerous and harmful." See Lund, *Moselig*, 28–37. Gaze or scream, the archaeological interpretations seem to suggest that the girl has been silenced or blinded to avert retribution.

71. Jefferson Hunter has pointed out that Heaney's attentiveness to the political power of photography was sharpened through photographs in the *Guardian* or the *Irish Times*. Hunter argues that newspaper photographs like one in the *Times* of London which "shows the fifteen-year-old Elizabeth Hyland with her hair cut off," and another "widely distributed Associated Press photo [which] shows nineteen-year-old Marta Doherty shaven, tarred, and tied to a Londonderry lamppost," as well as the well-known images of shaven-headed girls from the Second World War, function very much like the photograph of Windeby Girl in Glob's book. Hunter, 192. Indeed, if we read Heaney's poems with Glob's bog photographs in hand, if we "go back and forth between photograph and poem," it brings into view how Heaney's "world," as Hunter argues, is "photographically distinct, empty of poeticizing, and full of peculiar beauty." Hunter, 189.

72. Seamus Heaney, "Punishment," from *North*. Here cited from *Opened Ground*, 112.

73. The poem's narrator, the argument goes, admits to cowardice and "would have cast, I know / the stones of silence." Heaney has been much criticized for this poem. John Wilson Foster, for example, argues that: "From the sidelines of that procession of defeats [the Ulster conflict] Heaney felt helpless, unmanned, a voyeur capably male but rendered passive, in some sense womanly." John Wilson Foster, *The Achievements of Seamus Heaney* (Dublin: The Lilliput Press, 1995), 33. Also, David Lloyd locates "an explicit affirmation of a sexual structure in [...] the writer's relation to a land or place already given as feminine." And we find, he says, a "Romantic schema of a return to origins which restores" that which is lost. *Critical Essays on Seamus Heaney*, 121. Helen Vendler dismisses the condemnation of Heaney and argues that "if Heaney had no ambivalences about the fraternizing women and their abusers, he would not have been moved to write the poem [....] In the self-indictment of the end [in the poem "Punishment"] the poet has passed beyond 'veneration' and beyond 'atrocity': he has replicated himself in the very posture of the silent onlooker." Vendler, 49–50. I agree with Vendler on most points.

74. Lori Anderson Moseman, *Persona* (Jamaica Plain, MA: Swank Books, 2003), 35.

75. Ibid., 37.

76. Ibid., 38. After learning about the DNA results of the Bog Girl, Anderson Moseman has recently produced new poems in which the gender change has been addressed.

77. Ibid., 46. There is an interesting visual similarity between Anderson Moseman's poem

and Robert Smithson's "From the Center of the Spiral Jetty," which starts: "North—Mud, salt, crystal, rocks, water." I will return to Smithson and his relevance for bog body art. Smithson's poem is part of his article, "The Spiral Jetty," in *Robert Smithson: The Collected Writings,* ed. Jack Flam (Berkeley and Los Angeles: University of California Press, 1996), 149.

78. Tacitus, 28.

79. Moseman, 70.

80. Ibid., 36.

81. Anderson Moseman's use of cyberspace as an analogy to the bog—as a place for symbolically burying and excavating bog bodies—echoes Camilla Christensen's use of cyberspace as a contemporary bog realm.

82. Geoffrey Grigson, *Collected Poems 1963–1980* (London: Allison & Busby, 1982), 56.

83. William Carlos Williams, "The Smiling Dane," in *Journey to Love* (New York: Random House, 1955), 34. See also Thomas Tranströmer's "Elegi" (1954) in which the Swedish bog body known as Bocksten Man is located at a center of silence that "resonates" from the immobility of a stilled landscape. In the poem the dead human is fused with nature and shares a kind of mental property; the hand of an orchid sticks out of the soil. "Immobile wood, immobile surface of water / and the hand of the orchid being pulled from earth." I wish to thank Andreas Lombnæs for calling my attention to Tranströmer's poem.

84. The obsession with forensics in prime-time television also homes in on the time clock of stomach contents. Any *CSI* program, for example, will tell and retell this scenario. For a more detailed description of the last meal as a forensic/poetic trope, see my article about the Grauballe Man reception story: "Anachronistic Encounters: A Reception Story."

CHAPTER 5

1. For more on ecovention, see Sue Spaid's *Ecovention: Current Art to Transform Ecologies.* Published by Ecoartspace on the occasion of the exhibition "Ecovention," held at the Contemporary Arts Center, Cincinnati, Ohio, June 9–August 18, 2002.

2. Caroline Tisdale, *Joseph Beuys,* (London: Thames and Hudson, 1979), 39 (my emphasis).

3. I wish to thank Jens Toft for bringing my attention to the relevance of Beuys's Grauballe Man installation to my study.

4. According to Lone Hvass, *Dronning Gunhild: et moselig fra jernalderen* (Copenhagen: Sesam, 1998), a search of the Danish National Museum in Copenhagen has not been able to locate the painting.

5. Mario Vargas Llosa, "Botero: A Sumptuous Abundance," in *Making Waves* (New York: Farrar, Straus & Giroux, 1996), 264. Here quoted from Wendy Steiner, *Venus in Exile: The Rejection of Beauty in Twentieth-Century Art* (Chicago: University of Chicago Press, 2001), xv.

6. Tisdall, 34. A video recording of Beuys's installation practice also describes his work with Grauballe Man. Since Tisdall quotes Beuys's remarks about Grauballe Man from this recording, I use her transcription rather than my own notes.

7. Ibid., 34.

8. Ibid.

9. Ibid.

10. Ibid.

11. Christopher Lyon, "Beuys. Thinking is Form. The Drawings of Joseph Beuys," in catalogue to exhibit at Museum of Modern Art in New York, February 21–May 4, 1993.

12. Ibid.

13. Julie Luckenbach, *Beuys/Logos,* a hyper-essay, *http://www.walkerart.org/beuys/hyper.*

14. Tisdall, 36.

15. Luckenbach, "Beuys/Logos."

16. See also Lyon, "Beuys. Thinking is Form."

17. The expression is from Schama, 123.

18. In a letter found in the archives at Silkeborg Museum (built to house the work of Asger Jorn, and a neighbor to the museum that holds Tollund Man), the curators were asked for a picture of Tollund Man, which the letter writer wanted to give to Pablo Picasso, who was working on ceramics at the time. It seems likely that Picasso's interest in primitivism extended to archaic artifacts of the North. According to my correspondence with museum director Troels Andersen, who informed me of the letter, it is uncertain whether the picture ever reached Picasso.

19. Jorn would in time publish a volume called *Guldhorn and Lykkehjul* alone. See *Asger Jorn og 10.000 års nordisk folkekunst,* ed. Troels Andersen and Tove Nyholm (Silkeborg, Denmark: Silkeborg Kunstmuseums Forlag, 1995). Before Glob's collaboration with Jorn ended, the archaeologist wrote for *Helhesten,* the Danish pendant to the French surrealists' periodical *Minotaur.* Glob's article "Kurve og Keramik" was published in *Helhesten* vol. 1., nr. 1; and his "Helleristninger og Magi" in vol. 1, no. 2. I wish to thank Mette Mia Krabbe Meyer for locating these articles for me.

20. For a thorough introduction to the Cobraists, see Willemijn Stokvis, *Cobra: spontanitetens veje* (Copenhagen: Aschehoug, 2003).

21. Ibid., 410.

22. Bachelard, *Earth and Reveries of Will,* 8.

23. Ibid., 20.

24. Ibid., 69.

25. Girard, 37. Girard argues that Bachelard does not understand the paradoxical nature of violence and therefore does "not even realize how real the sacrificial process can be and how appropriate the major metaphors and symbols through which it is expressed." Ibid.

26. The cross-pollination between archaeologists, visual artists, and poets is remarkable: while Glob inspired Jorn, Jorn inspired Vandercam, and Vandercam inspired Claus. An avant-garde jazz trio based in the United States, *The William Bhati Trio,* has produced an instrumental piece called "The Man from Tollund" inspired by Hugo Claus's poem *and* Vandercam's pictures. Claus's poem is published in Gedichten (Amsterdam: De Bezige Bij, 2004). Here quoted from William J. Goegebeur's English translation at *http://cobrafineartgallery.com/media/psfs/TheManFromTollund.pdf* (accessed August 2008).

27. Bachelard, *Earth and Reveries of Will,* 93.

28. Ibid., 7.

29. Ibid., 13.

30. Ibid., 60.

31. Ibid., 89.

32. Aloïs Riegl, *Historical Grammar of the Visual Arts,* trans. Jacqueline E. Jung (New York: Zone Books, 2004), 51.

33. My description of Vaughan's paintings is built on a visit to the exhibit called "Kathleen Vaughan & the Bog Series" in November/December, 2004. Assen, Holland. Vaughan's work can be seen in color at http://www.akaredhanded.com/kv3cbogseries.html.

34. Kathleen Vaughan, "Artist Statement—The Bog Series," 1 (unpublished, n.p., 1995–1996). I am grateful to the artist for providing me with copies of her written work.

35. Ibid.

36. Ibid.

37. Ibid.

38. My use of the term "overlay" here and elsewhere is inspired by Lucy R Lippard, *Overlay: Contemporary Art and the Art of Prehistory* (New York: The New Press, 1983).

39. Mieke Bal has described a similar kind of overlay in her analysis of "contemporary art, preposterous history" with help of Maurice Merleau-Ponty's phenomenological framework when she points out that "the eye, like the skin, is another site where culturally constructed opposites turn out to be inseparable, where mind and body collaborate." See Mieke Bal, *Quoting Caravaggio: Contemporary Art, Preposterous History* (Chicago: University of Chicago Press, 1999), 234.

40. See Kathleen Vaughan, "Modes of Knowing and Artistic Practice: Beauty, Bog Bodies, and Brain Science," (unpublished article, n.p., n.d.), 28, 19–20.

41. Ibid., 23.

42. Robert Emmett Mueller: "Mnemesthics: Art as the Revivification of Significant Consciousness Events," *Leonardo,* 21, 2 (1988), 193. Here cited from Vaughan, "Modes of Knowing," 9.

43. Vaughan, "Artist Statement," 3.

44. Ibid.

45. Ibid., 1.

46. I am grateful to the artist for providing me with a video of her working process, entitled *Leven in het licht van het verstrijken van de tijd. De beelden van Désirée Tonnaer.*

47. Mitchell defines paleoart as follows: "It articulates the past-present contrast central to modernity in its most extreme form, fusing remote scenes of 'deep time' with the immediate present. Far from evoking nostalgia for a primitive past, paleoart is engaged with technology, environmental devastation, and questions of entropy, catastrophe, and extinction. It is characterized by a corrosive, mordant irony about pretensions to human greatness." W.J.T. Mitchell, *The Last Dinosaur Book: The Life and Times of a Cultural Icon* (Chicago: University Of Chicago Press, 1998), 272–73.

48. Although Hal Foster is critical of what he calls pseudo-anthropological and pseudo-ethnographical art, he concedes that: "All such strategies—a parody of primitivisms, a reversal of ethnographic roles, a preemptive playing-dead, a plurality of practices—disturb a dominant culture that depends on strict stereotypes, stable lines of authority, and humanist reanimations and museological reconstructions of many sorts." Hal Foster, *The Return of the Real: The Avant-garde at the End of the Century* (Cambridge, MA: MIT Press, 1996), 199.

49. Ibid., 168.

50. Ibid., 274, n. 82.

51. See Lippard, *Overlay*. W.J.T. Mitchell also engages similar concepts but points to the difference between his formulation and Lippard's: "we must distinguish it [paleoart] sharply from what Lucy Lippard has called 'prehistoric art,' which attempts to revive primitive and traditional man-made art forms (Stone Age sculpture, mound building, ritualistic art) and to evoke a mystical, premodern era when, as Lippard puts it, 'art and life were one.'" Mitchell, *Last Dinosaur Book*, 272.

52. Lippard, 8.

53. *Bogland Symposium Exhibtion* (Ireland: Crescent Art Center, 1990).

54. Another symposium of bog art was held in Oldenburg, Germany, in 1999. It too yielded a plethora of examples of bog art. See Ewald Gässler: "Wenn Künstler heute ins Moor gehen: Das Moor in der Kunst nach 1945 bis in die Gegenwart," in *Moor: Eine verlorene Landschaft.* (Oldenburg, Germany: Isensee Verlag, 2001).

55. *Bogland Symposium.*

56. Here quoted from Suzaan Boettger, *Earthworks: Art and the Landscape of the Sixties* (Berkeley and Los Angeles: University of California Press, 2002), 43.

57. Peter Schjeldahl "What on Earth," *The New Yorker,* September 5, 2005, 159. Schjeldahl's comment refers to Smithson, not to the Bogland Earth Art pieces.

58. Interestingly, although earth art is often contingent on the preservation powers of photography, Smithson argued: "Photography steals away the spirit of the work." See Robert Smithson, *The Collected Writings,* 251.

59. *Bogland Symposium,* 12.

60. Ibid., 22. See also Ross and Robins, *Life and Death of a Druid Prince.*

61. *Bogland Symposium,* 20.

62. Catherine Harper, *A Beginning* (Derry: Orchard Gallery, 1991), 4.

63. *Bogland Symposium,* 6.

64. Glob, *The Bog People,* 31; Mieke Bal, *Quoting Caravaggio,* 9.

65. Ibid., 233.

66. Svetlana Boym, *The Future of Nostalgia* (New York: Basic Books, 2001).

67. One artist, Yvonne Struys, to name yet another bog body artist, traces the signs and marks of the past as they materialize in the bog and "raises that history above its temporality," in the words of Jana Loose. In a piece called "Sacrifice to the Gods," from 1996, we see the outline of a figure formed like a cross, the large hands hanging over what might look like the bark of a tree or fingerprint patterns? The intimate connection to the earth is also highlighted in a sculptural piece by Bennie Klazinga of a bog body surfacing in large fragments, head and hands, out of the earth, so as to suggest the immensity of what lies underground. The head and limbs protrude from the surface as if the surface were made of water rather than earth, thus making us see the massiveness and pressure of the realm from which the bog bodies emerge.

68. Please note that I am *not* making a distinction here between visual art as spatial and verbal art as temporal, nor am I suggesting that we find watertight barriers between the written and the visual examples I examine—far from it. Such dualisms are always impoverished. What I *am* saying is that while both visual and verbal art are necessarily removed from the "real" body, visual art—or at least some visual art—can emulate the *tangible corporeality* of the bodies to a higher degree than can words.

69. José Ortega y Gasset, "On Point of View in the Arts," in *Writers on Artists,* ed. Daniel Halpern, trans. Poul Snodgrass and Joseph Frank (San Francisco: North Point Press, 1989), 366.

70. For more on "the act of displacement," see Stephen Greenblatt, "Resonance and Wonder," in *Exhibiting Cultures: The Poetics and Politics of Museum Display,* ed. Ivan Karp and Steven Lavine (Washington: Smithsonian Institution Press, 1991), 44.

71. The newly discovered Irish bog body fragment from 2003, known as Oldcroghan Man, could well fool the spectator if placed in a gallery of modern art. The National Museum of Ireland did not permit reproduction of the image. The reader is therefore referred to see it at http://www.bbc.co.uk/history/programmes/timewatch/diary_bog_07 .shtml.

72. But Foster also sees the corpse as central to the fascination with trauma: "If there is a subject of the history for the cult of abjection at all, it is not the Worker, the Woman, or the Person of Color, but the Corpse." Hal Foster, 196–97.

73. José Ortega y Gasset, *The Dehumanization of Art and Notes on the Novel,* trans. Helene Weyl (Princeton, NJ: Princeton University Press, 1948), 44.

74. Ibid., 29.

75. Ibid., 28–29.

76. Here quoted from Steiner, xv.

77. Ernst Van Alphen, *Caught by History: Holocaust Effects in Contemporary Art, Literature, and Theory* (Stanford, California: Stanford University Press, 1998), 11.

78. To see a danger in a likeness too close to corporeal reality was an issue for Kant and for Herder (painted statues and statues with hair broke with aesthetic ideals). If we think for a moment of the categories provided to us by Kant about beauty and the sublime: beauty offers a convergence between the subject and object; in the sublime the subject is overcome by the object, disturbed by it—unless, on the other hand, the subject is in charge of the object, not overwhelmed but in a position to decide whether the object pleases or not. Said differently, beauty is contingent, historically inscribed, always subjectively perceived (Baudrillard). See Jeremy Gilbert-Rolfe, *Beauty and the Contemporary Sublime* (New York: Allworth Press, 1999), 41; See also Steiner, *Venus in Exile.*

79. This renegotiation would seem to allow Anthony-Noel Kelly's "corpses" a place of pride. Kelly's fiberglass and plaster casts of actual cadavers and the development of art production have some critics, such as Godfrey Barker and Laura Stewart, viewing the "new" obsession with decay and death as "post-religious." Riding quotes them as saying: "If one believes there is no greater meaning to life than a 70-year blink in the blackness of eternity, then every moment of breakdown of the body is an existential nightmare."Alan Riding: "Dead but Not Forgotten: Body Art" *New York Times,* April 19, 1997. Other critics lament that "Those who treat the bodies of the recently deceased as though they were canvas and oils or a butcher's off cuts are reducing reverence for the human," or object that there is a double-standard in the use of dead bodies: "If it is done under the cloak of science it becomes palatable . . . . Scientists get a cultural free pass. They say they are not interested in macabre stuff but there they are in the lab sawing skulls apart." See Eleanor Heartney, "Is the Body More Beautiful When it is Dead?" *The New York Times,* June 1,. 2003, AR 37. The article lists and discusses a number of other cases in which the use of dead bodies has been

criminalized. See Mary Ore, "Anatomy as Art, Unsettling but Drawing Crowds," *New York Times,* July 9, 2002, B2.

80. Thomas Laqueur, "Clio Looks at Corporal Politics," in *Corporal Politics,* (Boston: Beacon Press, 1992), 14.

81. David Getz, *Frozen Man* (New York: H. Holt and Co, 1994), 6.

82. Ortega y Gasset, *Dehumanization of Art,* 29.

83. Verlyn Klingenborg, "Some Thoughts on Seeing the Polymerized Remains of Human Cadavers," *New York Times.* April 6, 2005. Klingenborg goes on to observe that the exhibit reduces the corpses to mere "curios, reminders, at best, that the foundation for our knowledge of the human body was revealed through autopsy—a word that means seeing for oneself."

84. Ibid.

85. Mary Ore, "Anatomy as Art, Unsettling but Drawing Crowds." *New York Times,* July 9, 2002, B2. Let me add that in the new world of plastinated bodies, we have come a long way away from the eighteenth- and nineteenth-century Promethean fantasies of virtual life: a long way from romanticized metamorphoses of cold, hard, and dead bodies coming alive in the erotic imagination as in Ovid's description of Galatea, or from equally romanticized visions of the grotesqueness and danger in *false life,* such as in Mary Shelley's *Frankenstein,* where Promethean fire turns cataclysmic. Yet these two different imagined ways of making the inanimate animate—one positive, the other negative—loom implicitly behind aesthetic judgments of false life.

86. See also Hillel Schwartz, 106.

87. Mary Ore, "Anatomy as Art, Unsettling but Drawing Crowds." *New York Times,* July 9, 2002, B2. Ore also makes an observation about Gunther von Hagen's own persona: "He has a cadaverous pallor, a broad smile and an Old World formality, and he is rarely seen without his signature fashion accessory, a broad-rimmed hat not dissimilar to the one worn by Dr. Tulp in Rembrandt's great painting of the anatomy lesson."

CHAPTER 6

1. Notes from visit to Tate New Modern, summer 2004.

2. See Deyan Sudjic, "Engineering Conflict," *The New York Times Magazine,* May 21, 2006, 28. Sudjic notes: ""Shellfire, tanks, bombs, mortars and the effect of 50 years of neglect have turned the Neues Museum into a physical reminder of the horror of the Second World War, the desperate fight for the city at the end of it and the division of Europe into two camps that followed it. [The architect David Chipperfield] is pursuing a strategy of reconstruction that refuses to pretend that nothing has happened here."

3. Hans Ulrich Gumbrecht, *Production of Presence: What Meaning Cannot Convey* (Stanford, CA: Stanford University Press, 2004), 123.

4. Hal Foster, 181–82.

5. Mark Rosenthal, *Anselm Kiefer* (Chicago: Art Institute of Chicago, 1987), 32.

6. Ibid., 32, 36.

7. Tony Bennett, *The Birth of the Museum: History, Theory, Politics* (London: Routledge, 1995), 146–47.

8. Ibid.

9. Susan Stewart, *Poetry and the Fate of the Senses* (Chicago: University of Chicago Press, 2002), 174.

10. Hillel Schwartz makes a similar point: "Left to the cold minutiae of correctness, wax figures today beg for our embrace but forbid our touch." Hillel Schwartz, 104.

11. Mary Wollstonecraft, *Letters Written During a Short Residence in Sweden, Norway, and Denmark* (Lincoln, NE: University of Nebraska Press, 1976), 71.

12. She continues: "—Is this all the distinction of the rich in the grave?—They had better quietly allow the scythe of equality to mow them down with the common mass, than struggle to become a monument of the instability of human greatness. / The teeth, nails and skin were whole, without appearing black like the Egyptian mummies; and some silk, in which they had been wrapt, still preserved its colour, pink, with tolerable freshness. / I could not learn how long the bodies had been in this state, in which they bid fair to remain till the day of judgment, if there is to be such a day; and therefore that time, it will require some trouble to make them fit to appear in company with angels, without disgracing humanity." Ibid., 71–72. I wish to thank Stephanie Buus for calling my attention to this passage.

13. Ibid.

14. Bakhtin goes on to describe the grotesque body as "the open mouth, the genital organs, the breasts, the phallus, the potbelly, the nose." Such an open body "discloses its essence as a principle of growth which exceeds its own limits only in copulation, pregnancy, childbirth, the throes of death, eating, drinking, or defecation [....] The age of the body is most frequently represented in immediate proximity to birth or death, [...] to the womb or the grave." Mikhail Bakhtin, *Rabelais and His World,* trans. Helene Iswolsky (Bloomington, IN: Indiana University Press, 1984), 321.

15. "Every book about disgust is not least a book about the rotting corpse." Winfried Menninghaus, *Disgust: Theory and History of a Strong Sensation.* (Albany, NY: State University of New York Press, 2003), 1.

16. See Barbara Kirshenblatt-Gimblett, "Objects of Ethnography," in *Exhibiting Cultures: The Poetics and Politics of Museum Display,* ed. Ivan Karp and Steven Lavine (Washington: Smithsonian Institution Press, 1991), 416.

17. Susan Stewart, *Poetry and Fate of Senses,* 172.

18. Ibid.

19. She continues: "The semiotic complexity of exhibits of people, particularly those of an ethnographic character, may be seen in reciprocities between exhibiting the dead as if they are alive and the living as if they are dead." See Kirshenblatt-Gimblett, 398.

20. Ibid., 397.

21. "Mødet med Grauballemanden—en udstillingsvandring," unpublished curation notes, kindly provided by archaeologist Pauline Asign. Since I completed my study, Asign has edited (with Niels Lynnerup) the book *Grauballe Man. An Iron Age Body Revisited.* (Aarhus: Aarhus University Press: 2007). Here the reader can follow the process of the reexamination of Grauballe Man.

22. Also in the Danish National Museum's exhibit *The Spoils of Victory* (in 2003), the curators opted to float the bog woman known as Huldremose Woman inside a glass case. In the history and practice of bog body displays, aesthetic concerns with the appearance of the cor-

poreal remains have been part and parcel of decisions concerning which bodies are to be displayed for the public and which not. If bog bodies are too horrid, too frightening, the ethics of display are seen by some curators as threatened by allowing the bodies to be exposed (and dehumanized) without proper respect for human dignity. Conversations with archaeologists Jørgen Jensen (Danish National Museum) and Christian Fischer (Silkeborg Museum). Since 2008 Huldremose Woman has been displayed fully clothed, as she was found, to respect her particular provenance and authenticity. Only her head and feet are visible. Conversation with archaeologist and museum curator Flemming Kaul (Danish National Museum).

23. Svetlana Alpers, "The Museum as a Way of Seeing," in Karp and Lavine.

24. The wonder and power of displaced objects in museums is, as Stephen Greenblatt has pointed out, a way to "stop the viewer in his or her tracks, to convey an arresting sense of uniqueness, to evoke an exalted attention." Looking, he posits, becomes enchanted "when the act of attention draws a circle [. . .] around itself from which everything but the object is excluded, when intensity of regard blocks out all circumambient images, stills all murmuring voices." Greenblatt in Karp and Levine, 42, 49. Such intensity of regard is prioritized differently in the various bog body displays.

25. Although Walter Benjamin claimed that no natural object is vulnerable to the interference of reproduction, bog bodies, as I have shown, are sometimes strangely usurped by reproducibility.

26. See Kjersti Nielsen's "Etikk og godt bevarte menneskelige levninger: etiske aspekter ved utstilling av og forskning på grønlandske mumier og danske moselik," (Tromsø, Norway: unpublished thesis, University of Tromsø, 2002).

27. For a full presentation and interpretation of Grauballe Man's early reception, the concerns regarding exhibition of a dead body, and various misidentifications of the body, see my article "Anachronistic Encounters."

28. I am grateful to my sister and brother-in-law, Inge and Peter Konge, for accompanying me to the museum and offering thoughtful comments that helped me articulate my ideas about the display strategy.

29. Renfrew, 20–21.

30. Hal Foster's concept of the artist as ethnographer similarly suggests a way to overcome the exclusionary nature of established art forms and institutions—"a shift from a subject defined in terms of economic relation to one defined in terms of cultural identity," as he puts it. See Hal Foster, 173. In the context of the present study, we may think of the artist-as-archaeologist along similar lines, as an "artistic interpreter of the cultural text."

31. John H. Jameson, John E. Ehrenhard, and Christine A. Finn, "Introduction: Archaeology as Inspiration—Invoking the Ancient Muses" in *Ancient Muses: Archaeology and the Arts,* ed. John Jameson et.al. (Tuscaloosa, AL: University of Alabama Press, 2003), 2.

32. This ability to recognize and combine found objects harks back not only to the earliest humans but also to our relatives the Neanderthals. Douglas Palmer, for example, tells us: "The careful inclusion of a fossil as part of a tool [made by Neanderthals] shows a further symbolic level of concern and interest. Whilst the fossil is a 'found' object, the stone tool is a conscious selection and the final product becomes equivalent to a 'found' work of art. Although normally associated with the advent of 'modern' 20th-century art, it would seem that the technique is perhaps one of the earliest in the development of art." Douglas Palmer,

Fossil Revolution: The Finds That Changed Our View of the Past (London: HarperCollins, 2003), 10.

33. Wallace, 55.

34. The following description of the Gottorf Castle exhibit is based on notes from a visit in the summer of 2003.

35. Some of the responses were seen only on the bulletin board and were collected from there by me; others are printed in the museum catalogue. Michael Gebühr, *Moorleichen in Schleswig-Holstein* (Archäologisches Landesmuseum der Stiftung Schleswig-Holsteinische Landesmuseen. Schloss Gottorf, 2002) 57.

36. Although recent forensic examinations have cast doubt as to the gender of the body, the reception material with which I am concerned in this study has persistently addressed the body as female. Since the reception material is the object of this study more than are the actual bog bodies, I therefore refer to the body as female.

37. Glob, *The Bog People,* 114.

38. "Kneblen (henholdvis øjenbindet) kan også have tjent irrationelle motiver: man har villet hindre den dødsdømte i at udslynge sine særligt virkningsfulde forbandelser. Den døendes sidste ord og blik var særligt farlige of skadevoldende." Lund, *Moselig,* 30.

39. See Nielsen, 51.

40. There are other issues relating to respectful treatment of mummies that should be mentioned. Kevin Krajick tells us, for example, how when the paleopathologists Arthur and Mary Aufderheide "went shopping for souvenirs at a fancy Bogotá shop, they saw three mannequins draped with replicas of pre-Columbian jewelry; as they stepped closer, they realized that the mannequins were mummies." Kevin Krajick, "The Mummy Doctor," *The New Yorker,* May 16, 2005, 72. Heather Pringle expresses concern in her report that the Aufderheides destroyed ancient bodies that had survived for centuries. Heather Pringle, *The Mummy Congress: Science, Obsession, and the Everlasting Dead* (New York: Hyperion, 2002).

41. Wolfgang Ernst, "Archi(ve)textures of Museology," in *Museums and Memory,* ed. Susan Crane (Stanford, CA: Stanford University Press, 2000), 25–26.

42. According to archaeologist and curator Vincent van Vilsteren, the overlapping of voices at the Drent Museum exhibit was not intentional, but was the result of space restrictions. I wish to thank van Vilsteren for informative guidance through the exhibit in the fall of 2005, and for translating the words of the Yde Girl that appear with the face reconstruction.

43. The many popular television shows on the secrets of mummies (most notably series on the Discovery and National Geographic channels) demonstrate that archaeological programs have turned into entertainment resembling crime shows—both the factual and the fictional variants. One example is PBS's *Return of the Dead* series.

44. In modern forensic archaeology the study of the genetic makeup of ancient human remains can give precise maps of the fabric of humans in the past and link them to humans in the present. Lindow Man, for example, has been linked to a predecessor in the area of his find in England by the use of DNA testing.

45. I am grateful to Toril Moi for accompanying me on one of my visits to Lindow Man and for helping me clarify my project's negotiation of the relationship between human and thing.

46. Lindow Man's fate of being interned in the museum has been used by the band the Triffids in their lyric to "Jerdacuttup Man," words by David McComb and music by David McComb with James Paterson, on the album *Calenture* (Island Records, 1987). I wish to thank Tynan Peterson for calling my attention to this song and for providing me with a copy of the lyrics. We hear, from the viewpoint of the bog man, how it is to "live under glass in the British Museum." Wrinkled and black, old and lonely, Lindow Man takes a ten-minute nap and "never woke up." His sad fate is, however, that both in the bog and in the museum he is frozen in place: "They soaked me in brine and they stewed me in juice / They took out my eyes and replaced them with glass / And with skin made of leather, and teeth made of dice / I slept in the peat under ten feet of ice."

47. R.C. Turner "Discovery and Excavation of the Lindow Bodies," in *Lindow Man: The Body in the Bog,* ed. I. M. Stead, J. B. Bourke, and Don Brothwell (London: British Museum Publications, 1986), 10–11.

48. Wolfgang Ernst: goes on, "The synchronic standard operations of machine memory based on reversible time differ radically from the individual human memory based on the ideas of irreversible time; in computing, archaeological data have nothing to do with remembrance but refer to the ultimate presence. Data are generally stored in files that are temporally but topographically defined, as organized collections of data in space." Ernst, 26.

49. Ibid., 30.

50. The expression is Mortimer Wheeler's from *Archaeology from the Earth* (Oxford: Clarendon Press, 1954). Here cited from Michael Wood's foreword to *World Atlas of Archaeology,* 8.

51. Elizabeth Hallam and Jenny Hockey, *Death, Memory and Material Culture* (Providence, RI, Berg, 2001), 109.

52. Ludwig Wittgenstein, *Philosophical Investigations: 50th Anniversary Commemorative Edition,* trans. G.E.M. Anscombe (Boston: Blackwell Publishing, 2001), 83.

53. Similarly, Stanley Cavell follows up on this question: "To speak sensibly of seeing or treating or taking persons as persons—or of seeing or treating or taking a (human) body as giving expression to a (human) soul—will similarly presuppose that there is some competing way in which persons—or bodies—may be seen or treated or taken." Stanley Cavell, *The Claim of Reason: Wittgenstein, Skepticism, Morality and Tragedy* (Oxford: Oxford University Press, 1979), 372. Once again, I am grateful to Toril Moi for pointing out the relevance of this argument for my book.

54. Bill Brown, "Thing Theory," *Critical Inquiry* (Autumn 2001), 4–5.

55. Barbara Kirschenblatt-Gimblett uses the term "poetics of detachment" in regard to the ethnographic object. It can also be applied, I think, to bog bodies. See her "Objects of Ethnography" in Karp and Levine's *Exhibiting Cultures.*

56. Seamus Heaney, "The Man in the Bog," 4.

57. Regina Bendix, *In Search of Authenticity: The Formation of Folklore Studies* (Madison: University of Wisconsin Press, 1997), 4.

58. Bendix, 8.

59. It is precisely the dismantling of the role of authenticity, Regina Bendix points out, that "explains the emergence during recent years of reflexivity in the scholarly habitus." Ibid., 6. Her aim is to understand the ideological underpinnings in the search of authenticity, and

· to do away with authenticity as a concept that promises transcendence. She argues for a removal of authenticity and its "allied vocabulary." See also introduction, ibid., 3–23.

60. Spencer R. Crew and James E. Sims, "Locating Authenticity," in Karp and Levine, 163.

61. The matter of authenticity in Grauballe Man is particularly interesting in that his original conservator gave him a facelift in order to make him look less horrible. This was not revealed until after CT scanning revealed the inlaid material. Information provided by Niels Lynnerup.

62. See Greenblatt in Karp and Levine, *Exhibiting Cultures*. See also Susan Sontag, *Regarding the Pain of Others*.

63. Mads Mordhorst has pointed out, for example, that a scientific justification for the primacy of an artifact's authenticity often points to the poverty of the copy; yet the aura of authenticity is often, he argues, a result of projection by museum institutions and curators. See "Kildernes magi," *Fortid og Nutid* (June 2001), 124.

64. David Crowther, "Archaeology, Material Culture, and Museums" in *Museum Studies in Material Culture,* ed. Susan M. Pearce (Leicester, UK: Leicester University Press, 1989), 42. See also Wolfgang Ernst, *Museum and Memory,* 32–33.

CHAPTER 7

1. Although faces have become part of organ donations and can be transplanted from one person to another, the ethical questions surrounding such still rare and experimental operations often center on the same ethical issues that once marked discussions of heart transplants. Because the heart is seen as the location of the soul and the face is seen as our outward manifestation of an inner self, transplants are imagined to transfer character traits from donor to recipient.

2. Fingerprints have their own cultural history. See, for example, Colin Beavan, *Fingerprints: The Origins of Crime Detection and the Murder Case That Launched Forensic Science* (New York: Hyperion, 2001). Gilbert Ryle and his thick description further complicate the facial expression ethnographically. See Clifford Geertz, *Interpretation of Cultures* (New York: Basic Books, 1973), 7. The Gilbert Ryle scenario of the significance and interpretation of gestures (winking, twitching, parodying, etc.) asks us to consider how we know that what we see is indeed correct. Geertz says: "Right down to the factual base, the hard rock, insofar as there is any, of the whole enterprise, we are already explicating: and worse, explicating explications. Winks upon winks upon winks." Ibid., 9. See also Jeffrey H. Schwartz, *What the Bones Tell Us* (Tucson: University of Arizona Press, 1993) on facial reconstructions.

3. On the one hand, as Walter Benjamin claimed, historiographical insight can proceed through the face: "To write history means to give dates to their physiognomy." On the other hand, we must understand that the face is not reducible to any positive certainty of what we know. But in a Benjaminian reading the face, or *a* face, cannot be reduced to some categorical knowledge. His concern and apprehension is immanently tied to the tradition of physiognomic interpretation. Walter Benjamin, *The Origin of the German Tragic Drama,* trans. John Osborne (London, NLB, 1977), 183–84. See also Gerhard Richter, *Walter Benjamin and the Corpus of Autobiography* (Detroit: Wayne State University Press, 2000), 104–5, 106. See also Richter's reference to *Passagen-Werk* (N 67; 5:595).

4. Emmanuel Lévinas, *Entre Nous: On Thinking of the Other,* trans. Michael B. Smith and Barbara Harshay (New York: Columbia University Press, 1998). See also Judith Butler, *Precarious Life: The Powers of Mourning and Violence* (London: Verso, 2004), 140. She points out that Levinas "gives us a way of thinking about the relationship between representation and humanization."

5. See Paul G. Bahn, ed., *Cambridge Illustrated History of Archaeology* (Cambridge: Cambridge University Press, 1996), ix.

6. "For the physical anthropologist," write the archaeologist John Prag and the medical artist Richard Neave in *Making Faces,* "total destruction of the body is a disaster." Although discussion of this disaster is avoided in most bog body material, many bog bodies do not provide readable or decipherable faces. I shall return in more detail to John Prag & Richard Neave and *Making Faces: Using Forensic and Archaeological Evidence* (London: British Museum Press, 1997), 12.

7. Holland Cotter: "Expressions so Ancient, yet Familiar" in the *New York Times,* February 18, 2000, B37. Since the so-called Fayoum portraits in this exhibit are considered some of the most realistic portraits from antiquity, with "a range of expression that one does not get from mummies," these reconstructions can help determine the accuracy of the archaeological-artistic method. Caroline Wilkinson, Bob Brier, Richard Neave and Denise Smith. "The Facial Reconstruction of Egyptian Mummies and Comparison with the Fauum Portraits," in *Mummies in a New Millenium: Proceedings of the 4ᵗʰ World Congress on Mummy Studies,* ed. Niels Lynnerup, Claus Andreasen, and Joel Berglund (Greenland: Greenland National Museum and Archives and Danish Polar Center, 2002), 141.

8. Patrizia Magli "The Face and the Soul," in *Fragments of a History of the Human Body. Vol. II,* ed. Michel Feher, Ramona Naddaff, and Nadia Tazi (Cambridge, MA: Urzone Inc, Zone Books and MIT Press , 1978), 87.

9. Ibid.

10. In a conversation in Assen, Holland in 2002, van der Sanden informed me about a number of occasions in which visitors claimed direct ancestry to, or a sense of affinity with, the Yde Girl after her face reconstruction was made public.

11. The fact that Yde Girl would inspire poetic and literary works *after* and *because* her face was reconstructed indicates a kind of interconnectedness between verbal and visual expressions in archaeological, poetic, and literary imagination. In one instance, Yde Girl was seen as a Lolita-youth, innocent and virginal. Love songs were written to her and recorded, with titles like "In My Mind (the Girl from Yde)" and "Forever Young," in which the lyric celebrates the bog girl's long blond hair and asks if she was ever in love: "In my mind I hear your voice / Feels like I've known you before / And in my dreams you come so close / That's how you come alive again." Jack Simon and Harm Posthumus, "In my Mind (the Girl from Yde)" and "Forever Young," from *Snoaren.* compact disc (Holland: Snoaren Productions and Maura Music, 2000). CD kindly provided by Wijnand van der Sanden.

12. Prag and Neave, 170.

13. Ibid., 171.

14. Ibid.

15. Ibid.

16. I wish to thank John Prag for discussing face reconstructions with me in 2004.

17. Prag and Neave, 17.

18. Ibid.,11.

19. Ibid., 19.

20. Ibid., 14.

21. Ibid., 30.

22. Ibid., 33.

23. Ibid., 21.

24. Ibid., 28.

25. Ibid., 219.

26. The expression is borrowed from Michael Taussig, *Defacement: Public Secrecy and the Labor of the Negative* (Stanford, CA: Stanford University Press, 1999), 233.

27. The Danish archaeologist C.J. Thomsen (1788–1865), for example, asked the zoologist and anatomist Frederik Eschrict for descriptions of cranial remains in order to ascertain how Stone Age and Bronze Age peoples looked. See Jørgen Jensen, *Thomsens Museum: Historien om Nationalmuseet* (Copenhagen: Gyldendal, 1992), 269–71. Please note that throughout this book I use the English orthography of "Neanderthal," and not the more correct German "Neandertal."

28. Erik Trinkaus and Pat Shipman, *The Neandertals: Changing the Image of Mankind* (New York: Knopf, 1993), 408.

29. Ibid., 399.

30. Ibid.

31. Ibid., 408, 410.

32. Ibid., 409.

33. Clive Gamble, foreword to Stephanie Moser, *Ancestral Images: The Iconography of Human Origins* (Ithaca, NY: Cornell University Press, 1998), xi.

34. Ibid., xiii.

35. Ibid., xiv.

36. Stephanie Moser, *Ancestral Images: The Iconography of Human Origins* (Ithaca, NY: Cornell University Press, 1998), 173.

37. Mitchell, *Last Dinosaur Book,* 13.

38. Ibid., 19.

39. Ibid., 45.

40. Ibid., 48.

41. Ibid., 50.

42. Ibid., 51–52.

43. Ibid., 282.

44. Ibid., 205. It is the human form that presses the ethics of aesthetics here. There is an interesting aside to be made here in terms of the question of humanization and dehumanization in the display of human remains. And there are some similarities with what W.J.T. Mitchell has culled in regard to the cultural and artistic interest in dinosaurs. In spite of the differences between his archaeological objects and mine, some of Mitchell's points resonate with what happens to bog bodies in art. He shows, for example, how the image of the dinosaur has been excluded from modernistic art because it has been associated with—indeed "contaminated" by—kitsch. Modernism, he writes, "insists that the artist 'make it new,' creat-

ing an object that is forever fresh and self-renewing," and privileges "the original, unique, authentic object created by the artistic imagination" and not a "mere copy of a fragment of a corpse or skeleton, a fossil imprint produced by natural accident, not by human artifice." If we follow Mitchell's criticism of modernism, it would seem that bog bodies, like dinosaurs, symbolize a "petrified stasis" which is excluded from the realm of pure modernism. If bog bodies, unlike dinosaurs, can only be "contaminated" by kitsch through being very loosely associated with popular culture representations of mummies (particularly in horror movies), Mitchell's point that postmodernism, unlike pure modernism, has offered new possibilities for archaeological remains in art making (paleoart, for example) resonates in principle with the history of bog bodies in art. For the most part, however, creative interest in bog bodies does not avoid the sense of nostalgia that Mitchell describes here in regard to paleoart, nor does it ignite the caustic irony that he points to. Part of the explanation could be that when it comes to bog bodies the deep past is not deep enough, and that the morphology of these human remains is not far enough removed from that of modern bodies to avoid entirely associations with what Mitchell calls "pretensions to human greatness."

45. Fr. C.C. Hansen, *Identifikation og rekonstruktion af historiske personers udseende paa grundlag af skelettet* (Copenhagen: J.H. Schultz, 1921).

46. "De væsentligste Ejendommeligheder i Billederne viste sig forenede i Busten, der m.h.t. Liv og karakterfuldt Udtryk overgik hvert enkelt af Portrætterne." Ibid., 47.

47. "At forme Handel's Hoved over Bach's Kranium; men skønt denne Buste af Handel udvendigt lignede godt, var den en anatomisk Umulighed [. . . .] Saadanne Buster er 'anatomiske Løgne.'" Ibid., 47–48.

48. "Rekonstruktion af Ansigtet paa Kranier af forhistoriske eller nulevende Racer." Ibid., 50.

49. "Kollmann antog—vist ikke uden Grund—at Menneskeracerne var relativt konstante gennem temmelig lange Tidsrum, naar der ikke skete nogen Raceblanding. Han mente derfor at kunne faa Oplysninger om Udseendet af præhistoriske Racer ved den plastiske Rekonstruktion af Portrætbuster efter His's Methode, eventuelt suppleret ved nye Maalinger." Ibid, 51.

50. "Visse Holdepunkter i Fysiognomidannelsen hos den nulevende Befolkning" Ibid., 52.

51. "Hvorimod det usædvanlige i Fysiognomiet hos andre Racer nok saa meget virker ved sin typiske Fremmedartethed, saa at de individuelle Differenser f. Ex. ved Kineser- eller Neger-Ansigter af same Art til at begynde med ikke iagttages nær saa fint af Europæere." Ibid., 54–55.

52. "Forsaavidt er det ikke helt uden Berettigelse at tale om Raceportræter som udtryk for det typiske og karakteristiske Gennemsnit." Ibid., 55.

53. "Tom Videbegærlighed eller formaalsløs Nysgerrighed, men som Led i den lange Kæde af Arbejder, som vil udforske det svundne Liv og dets Spor, inden Udslettelsen." Ibid., 91.

54. I am grateful to Nanna Damsholdt for identifying the passage and providing me with information about its place in the *Vulgata*.

55. See Moshe Greenberg, *Ezekiel 21–37: A New Translation with Introduction and Commentary* (New York: Doubleday, 1997), 742.

56. Prag and Neave, 18. The anthropologist V. Suk argues that "All the fossil remains of man which have come down to us are in the form of skeletal bones and can be investigated only as skeletons which can offer us no clues at all for any reconstruction that is true to life." V. Suk. "Fallacies of Anthropological Identifications and Reconstruction: A Critique Based on Anatomical Dissections," (Prague: Publications of the Faculty of Science, Charles University, 1935), 207. Quoted here from Prag and Neave, 17.

57. Prag and Neave, 18.

58. E-mail exchange with James Chatters.

59. John Noble Wilford, "Archaeology and Ancestry Clash Over Skeleton," *New York Times* D4, November 9, 1999.

60. In a conference in 2000 called "Who Owns the Body?" at the University of California–Berkeley, which I attended, Chatters addressed the audience regarding the demands of the Native American Graves Protection and Repatriation Act (NAGPRA), which calls for a return of all human remains to native tribes in the areas of the finds. In his short summary published in the conference program, he argues that "these early people have a right to their place in human history, a place that can be given them through respectful curation and open scientific study, but will be denied if they are summarily reburied for temporal political reasons." His slide presentation included a showing of the face reconstruction.

61. Email exchange with James Chatters.

62. Press release from Astura Folk Assembly, here quoted from Roger Downey, *Riddle of the Bones: Politics, Science, Race, and the Story of Kennewick Man* (New York: Copernicus, 2000), 119. James Chatters distances himself emphatically from the viewpoint of Astura Folk Assembly.

63. See Draaisma, 79.

64. Otter continues: "The angle formed by the forehead and the jaw; the distance between hairline, eyebrow, and nose; the cranial swells and depressions—to mark these lines on the face and head and to imagine them as measures of an interior state was to make visible a new kind of knowledge. These lines were not neutral [....] These lines were filled with meaning. Knowledge was conceived in visual and spatial terms and the linear was seen as giving access to the depth of human character and human difference. Color, texture, contour, line—these are the elements of visual representation and also the strategies of ethnological investigation of the face, the skin, and the head. The racial body was known and owned through reading the lines engraved on the surface. [...] From a phrenological perspective, the eyes are not the window to the soul. There is no need to peer into the eyes when character is inscribed on the surface of the face and head. Here, phrenology is not exactly learning to draw-by-number; rather, it is learning to know by numbers. Phrenology teaches the observer to abstract, delineate, and enumerate features of the head and to read these features as eloquent markers of interior state.". Samuel Otter, *Melville's Anatomies* (Berkeley and Los Angeles: University of California Press, 1999), 34–35.

65. Taussig, 253.

66. Prag and Neave, 227.

67. Docherty, 212–13.

68. Sandberg, 18–19.

69. Ibid. 22.

70. Ibid. 47.

71. Ondaatje, 161.

POSTSCRIPT

1. R. C. Turner, "Dating the Lindow Moss and Other British Bog Bodies," in *Bog Bodies, Sacred Sites and Wetland Archaeology,* ed. Bryony Coles, John Coles, and Mogen Schou Jørgensen (Exeter, UK: WARP, 1999). 233.

2. Ibid., 2.

3. In fact, popular culture oftentimes offers insightful comments on dead bodies' uncanny "sense" of time. Never at loss for a pithy remark, even after having been brought back to life in the wake of his own funeral, James Bond—in *You Only Live Twice*—dryly observes in response to his boss's complaint that he was "late": "We corpses have simply no sense of time." To have or not have "a sense of time" has its own peculiar zest in the annals and archives of bog bodies, albeit not the pithiness of agent 007.

4. Nigel Dennis, *Cards of Identity,* 1955, here cited by and from Lowenthal, xv.

5. Indeed, during the time it has taken me to complete the research for this book, forensics has become astoundingly "hot and sexy" in a number of prime-time TV series. In *CSI* (CBS) a forensic specialist whispers with eerie (and morbid) tenderness to newfound corpses as if they were her babies; in *Bones* (Fox) the leading female protagonist, a forensic archaeologist, seems to funnel her libido entirely into the remains of the dead. "My most meaningful relationship is with dead people," she exclaims in the pilot program while fending off advances from a male colleague (she is subsequently nicknamed "Bones"). A core moment in the show's forensic process features three-dimensional holographic reconstructions used to conjure the "real" faces of crime victims while their raw skeletal parts rest on a gurney nearby. These futuristic visage-hybrids make most face reconstructions used in museum displays of bog bodies and other mummies look positively antiquated. In *Crossing Jordan* (NBC) an oversexed pathologist culls clues from dead bodies to solve possible crimes or misdemeanors, thus *crossing* the boundary of her jurisdiction in the autopsy room; HBO also had its version of corpse entertainment in *Six Feet Under,* in which embalmment in the basement of a funeral home sets the stage for a dysfunctional family's life. In all of these programs, half-draped dead bodies on gurneys serve as centerpieces around which a great deal of flirting takes place.

6. In the daily newspaper *B.T.* on September 12, 1993, Danish readers were told that their national bog star was to be protagonist in a ballet by the Belgian company Tandem. Incidentally, it had been the Copenhagen-born COBRA artist Serge Vandercam who directed that company's attention to the bog body.

7. "I'm taking your ankles—you're back into the bog / I'm taking your legs—you're back into the bog / I'm taking your knees—you're back into the bog," Lyrics by Jean Luc De Meyer from a song named "The Bog," performed by the band Bigod 20, 1990. See *http://bigod20.info/the-bog-4.*

8. In my note-taking for this book I have managed to wear out three pens with the Danish bog body Grauballe Man floating inside. I wish to thank Merrill Kaplan and Kendra Wilson for bringing me the first such pen. Wijnand van der Sanden sees this kind of merchandise

as "worse than kitsch" and suggests that ethical boundaries have been crossed. Wijnand van der Sanden, "Mummies, Mugs, and Museum Shops," online in *Archaeology: A Publication of the Archaeological Institute of America.* August 30, 2005, *http://www.archaeology.org/online/features/bog/exhibit.html.*

9. See also P.K. Lewin's comment on more recent public unveilings of mummies: "Unfortunately many of the investigations have been undertaken in a circus-like atmosphere, with autopsies of 'mummies' being conducted under the glare of media lights, resulting in the destruction of invaluable human remains, with often very little scientific information, if any, being obtained." "Current Technology in the Examination of Ancient Man," in *Human Mummies: A Global Survey of their Status and the Techniques of Conservation,* ed. K. Spindler et al. (Vienna and New York: Springer, 1996), 9.

10. Anthropology, particularly in the works of Johannes Fabian, regards shared time as an epistemological condition. See Johannes Fabian, *Time and the Other: How Anthropology Makes Its Object* (New York: Columbia University Press, 1983), 31.

11. Archaeology often imbeds inorganic artifacts of all sorts with biographies and life stories as if they were persons with whom we could have some sort of conversation, but when the archaeological specimens are already human it is naturally more tempting to "read past minds" than it is with most other ancient artifacts. With the help of I. Kopytoff, Tilley suggests "a direct analogy between the ways in which societies construct persons and the way they construct things. One provides a metaphorical model for the other." Tilley, *Metaphor and Material Culture,* 76.

12. See Fernand Braudel and his description of temporality as geo-historical time, social time, and individual time in *On History.*

13. "In reality," as Bergson famously says, "the past is preserved by itself automatically." He goes on in existential-psychological terms: "In its entirety, probably, it follows us at every instant; all that we have felt, thought and willed from our earliest infancy is there, leaning over the present which is about to join it, pressing against the portals of consciousness that would fain leave it outside." Henri Bergson, *Creative Evolution,* trans. Arthur Mitchell (Mineola, NY: Dover, 1998).

14. Saint Augustine, *The Confessions of Saint Augustine,* trans. Edward Bouverie (N.p.: Plain Label Books, 1954), Bk. XI, 14:17. An explanation of the long philosophical tradition and ontology of time is outside the scope of the present study.

15. The words are from an interview with astronomer Carl Sagan on the PBS television program *Nova.* See www.pbs.org.wgbh.nova/time/sagan.html. Sagan refers to the so-called grandfather paradox: If you travel back in time and kill your own grandfather, you would also erase your own existence. Who then is doing the killing?

16. The possibility of traveling through time or reversing it in order to reconstruct what has happened in the past is part of ongoing discussions in modern physics. Note, for example, Stephen Hawking's reversal of his previous position that information lost in the black hole cannot be retrieved. See also Joan Didion, *The Year of Magical Thinking* (New York: Alfred A. Knopf, 2005), 181.

17. As such, bog bodies always emerge into shifting moments of actuality. See George Kubler's chapter on "The Nature of Actuality," *The Shape of Time: Remarks on the History of Things,* by George Kubler (New Haven, CT: Yale University Press, 1962), 16–24.

18. For another description of time and metaphor, see George Lakoff and Mark Johnson's description of the "time is a moving object metaphor." George Lakoff and Mark Johnson, *Metaphors We Live By* (Chicago: University of Chicago Press, 1980), 43–45.

19. This particular "slice of time" was made famous by Alfred Hitchcock in his thriller *Vertigo* (1958) as a way to stress one of the film's underlying themes: remembrance and memory loss.

20. The modern eye, as the French historian Pierre Nora has argued, "relies entirely on the materiality of the trace, the immediacy of the recording, the visibility of the image." Nora, "Between Memory and History: Les Lieux de Mémoire," 13.

21. I have already shown the importance of Harris lines to a poem on bog bodies.

22. Jean Baudrillard, 38.

23. Ibid.

24. Marcus Valius Martialis writes, "The bee is enclosed, and shines preserved in amber, so that it seems enshrined in its own nectar." Here cited from Palmer, 137.

25. Ibid.

26. See also George O. Poinar and Roberta Poinar, *The Amber Forest: A Reconstruction of a Vanished World* (Princeton, NJ: Princeton University Press, 1999). The authors ask us to "step back in time" and explore with them the amber fossils from a primeval forest in the Dominican Republic. Amber, they write, provides a "clear picture of a frozen moment in the amber forest [. . . .] these candid snapshots [fit] together like pieces of a jigsaw puzzle [. . .] a kaleidoscope of life in its environs. The elegance of these fossils adds to their general enjoyment and fascination. Portrayed will be ancient life forms from a realm that no longer exists in the Greater Antilles, nor anywhere else in the world today! Such extinct forms can answer questions about paleobiodiversity, evolution, and biogeography. They can also enlighten us about past climates, insect-plant interactions, and parasitic relationships."

27. Alexander Pope, "Epistle to Dr. Arbuthnot," in *The Poems of Alexander Pope: A Reduced Version of the Twickenham Text* (New Haven, CT: Yale University Press, 1966), 597.

28. It is worth noting that bogs, like amber and redwoods, are seen as sites for national nostalgic repositories devoted to bygone days. Amber, for example, is regarded as "gold" in the Baltic region, and in the past was often deposited in the bogs as a votive in religious practice. Redwoods have been seen as sacred and natural temples of America's primal past.

29. The imagining of the future is oftentimes fraught with archaeological imagination, as the film *Jurassic Park* testifies. In the film's plot, amber is recast from being the oldest valuable gem to being the emblem of a make-believe future in which DNA extracted from amber insects that have ingested dinosaur blood opens a window to a world of compressed past, present, and future. According to Douglas Palmer, "amulets of amber have been found dating back as far as 35,000 years." Palmer, 137.

30. See Kevin Krajick, 66.

31. Nor is this vision of bringing dormant or dead material back to life new; for ages it has been a staple in the utopian projects of numerous Pygmalions and the dystopian projects of numerous Frankensteins.

32. Getz, *Frozen Man,* 55. An American anthropologist, John Gurche, reshaped the Ice Man's face. "'He looked like a thug when I first made him," Gurche said. "But his appearance

softened. I even had several women come into my studio and see him and say, 'Well, he's not bad looking.'" Ibid., 34.

33. See David Getz, *Frozen Girl.* (New York: Henry Holt, 1998), jacket text.

34. Michael Wilson, "Military Lab Puts Name on Long-Lost Airman." *The New York Times,* March 24, 2006, A1, A15.

35. *New Lexicon Webster's Dictionary of the English Language,* .Encyclopaedic Edition (New York: Lexicon Publications, 1988), s.v. "anachronism."

36. Bill Bryson, *A Short History of Nearly Everything* (London: Black Swan, 2004), 534.

37. Ibid., 534–35. The germinating couple also implies reproduction, and as such suggests continuation into our time.

38. Renfrew, 109–10.

39. *New York Times.* Tuesday, November 9, 1999.The first article was entitled "Museum to Display Times Capsule on Dec. 4"; the other, entitled "New Answers to an Old Question: Who got here first?," was part of a series of articles about the ongoing archaeological dispute over ancestry and origin.

40. The idea that a time capsule of the twentieth century had been conceived, constructed, displayed, and then buried in the floor of the museum for the excavation-pleasure of future generations demonstrates a particular kind of temporal-archaeological sensibility. But the sensibility is not exclusive to this most recent turn of the century. For example, in Thornton Wilder, *Our Town: A Play in Three Acts* (New York: Coward McCann, 1938), a play fraught with temporal ploys, townspeople at the brink of the twentieth century bury mementoes to inform the future of *their* culture anno 1900. Variations of this time-travel fantasy are to be found in Stanley Kubrick's film *2001: A Space Odyssey,* in which a triumphant ape throws a tool into the air, whereupon it morphs into a spaceship, and in *Planet of the Apes* by Franklin Schaffner, in which a journey to a foreign planet turns out to be a return to our own lost civilization. Both the thrill of seeing an ancient tool morph into a futuristic vessel and the shock of returning to a future in which our present time has been archaeologically buried (barely visible through the tips of the Statue of Liberty) illustrate, each in their way, the elasticity of time in the imagination.

41. In the *New York Times,* September 15, 2001, Sarah Boxer writes: "The notion of the inanimate object suffering pain might make somewhat more sense this week as people repeatedly watch the televised images of two airliners banking and entering World Trade Center towers. After seeing the airplanes pierce the buildings, turning them red and black, until they collapse, the idea that they are suffering pain doesn't seem quite so far-fetched. For those who have looked into the ways that pain is rendered and described, the repetitive film may be a pathos formula for our time, a model destined to be imagined, and reimagined, the newest Laocoon."

42. See Eric Lipton and James Glanz, "From the Rubble, Artifacts of Anguish," *New York Times,* January 27, 2002.

43. Bendix, 14–15.

44. Judith Butler's thoughts in *Precarious Life* on the challenges that the humanities face today resonate with some of the ethical questions I raise in this study. She concludes: "If the humanities have a future as cultural criticism, and cultural criticism has a task at the

present moment, it is no doubt to return us to the human where we do not expect to find it, in its frailty and at the limits of its capacity to make sense. We would have to interrogate the emergence and vanishing of the human at the limits of what we can know, what we can hear, what we can see, what we can sense." Butler, 151.

45. Jack Flam, "Introduction: Reading Robert Smithson," in *Robert Smithson: The Collected Writings,* ed. Jack Flam (Berkeley and Los Angeles: University of California Press, 1996), xix. See also Paul Cummings, "Interview with Robert Smithson for the Archives of American Art/Smithsonian Institution," in *Robert Smithson: The Collected Writings,* 296. Here Smithson discusses his relationship with William Carlos Williams, ibid., 284–85. Smithson had written an article that he imagined as an appendix to Williams's long poem "Paterson." See ibid., 298. On the day of my visit to the Spiral Jetty, an artist paid tribute to Smithson by placing shining DVD disks around the edges of the spiral to enhance the reflection of the salt crystals in the sunlight. While some visitors reacted with dismay toward this interference with the authenticity of the art piece (art as sacred), the insertion of technology seemed perfectly appropriate to me, in that it highlighted the collusion between the archaeological, the archaic, and the modern.

46. Seamus Heaney, "Tollund," from *The Spirit Level,* 1996. Here quoted from *Opened Ground,* 410.

47. Glob, *The Bog People,* 18.

48. Here quoted from Turner and Briggs, 145.

49. Smithson, "The Spiral Jetty," in *Robert Smithson: The Collected Writings,* 147.

Allen, Carl Ferdinand. *Haandbog i Fædrelandets Historie med stadigt Henblik paa Folkets og Statens indre Udvikling.* Copenhagen: C.A. Reitzel, 1845.

Allen, Lewis (Abel Meeropol), 1939. "Strange Fruit." In *Scanning the Century: The Penguin Book of the Twentieth Century in Poetry,* by Peter Forbes. New York: Viking, 1999.

Allen, Michael, ed. *Seamus Heaney.* New York: St. Martin's Press, 1997.

Allert, Beate, ed. *Languages of Visuality: Crossings between Science, Art, Politics, and Literature.* Detroit: Wayne State University Press, 1996.

Alpers, Svetlana. "The Museum as a Way of Seeing." In *Exhibiting Cultures: The Poetics and Politics of Display,* edited by Ivan Karp and Steven Lavine. Washington: Smithsonian Institution Press, 1991.

Alphen, Ernst Van. *Caught by History: Holocaust Effects in Contemporary Art, Literature, and Theory.* Stanford, California: Stanford University Press, 1998.

Andrews, Elmer. "The Spirit's Protest." In *Seamus Heaney: A Collection of Critical Essays,* edited by Elmer Andrews. New York: St. Martin's Press, 1992.

———, ed. *Seamus Heaney: A Collection of Critical Essays.* New York: St. Martin's Press, 1992.

Annaler for Nordisk Oldkyndighed (og Historie). 1836/37–1863. Copenhagen: Kongelige Nordiske Oldskriftselskab, 1836–37.

Anthony, David W. "Nazi and Eco-Feministic Prehistories." In *Nationalism, Politics and the Practice of Archaeology,* edited by Philip K. Kohl and Clare Fawcett. Cambridge: Cambridge University Press, 1995.

Aries, Philippe. *Western Attitudes Toward Death: From the Middle Ages to the Present.* Translated by Patricia M. Ranum. Baltimore: The Johns Hopkins University Press, 1974.

———. *The Hour of Our Death*. Translated by Helen Weaver. New York: Oxford University Press, USA, 1991.

Arnold, Bettina, and Henning Hassmann. "Archaeology in Nazi Germany: The Legacy of the Faustian Bargain." In *Nationalism, Politics, and the Practice of Archaeology,* edited by Philip K. Kohl and Clare Fawcett. Cambridge: Cambridge University Press, 1995.

Asign, Pauline, and Niels Lynnerup, eds. *Grauballe Man: An Iron Age Body Revisited*. Aarhus, Denmark: Aarhus University Press: 2007.

Asign, Pauline. "Mødet med Grauballemanden: en udstillingsvandring." Unpublished curation notes.

Atkinson, Paul. *The Ethnographic Imagination: Textual Constructions of Reality*. London: Routledge, 1990.

Atwood, Margaret. "The Bog Man." *Playboy,* January 1991.

———. *Wilderness Tips*. Toronto: M & S, 1991.

Augustine, Saint. *The Confessions of Saint Augustine*. Translated by Edward Bouverie N.p.: Plain Label Books, 1954.

Bachelard, Gaston. *Water and Dreams: An Essay on the Imagination of Matter*. Translated by Edith R. Farrell. Dallas: Pegasus Foundation, 1983.

———. *Earth and Reveries of Will: An Essay on the Imagination of Matter*. Translated by Kenneth Haltman. Dallas: Dallas Institute of Humanities and Culture, 2002.

Bahn, Paul G., ed. *The Cambridge Illustrated History of Archaeology*. Cambridge: Cambridge University Press, 1996.

Bakhtin, M.M. "Forms of Time and Chronotype in the Novel." In *The Dialogic Imagination: Four Essays,* edited by Michael Holquist. Translated by Caryl Emerson and Michael Holquist. Austin, TX: University of Texas Press, 1981.

———. *Rabelais and His World*. Translated by Helene Iswolsky. Bloomington, IN: Indiana University Press, 1984.

Bal, Mieke. *Quoting Caravaggio: Contemporary Art, Preposterous History*. Chicago: University of Chicago Press, 1999.

———. *Travelling Concepts in the Humanities: A Rough Guide*. Toronto: University of Toronto Press, 2002.

Barthes, Roland. "The Photographic Message." In *Image, Music, Text,* translated by Stephen Heath. New York: Hill and Wang, 1977.

———. *Camera Lucida: Reflections on Photography*. Translated by Richard Howard. New York: Hill and Wang, 1981.

Bataille, Georges. *Erotism: Death and Sensuality*. Translated by Mary Dalwood. San Francisco: City Lights Books, 1986.

Baudrillard, Jean. *The System of Objects*. Translated by James Benedict. London: Verso, 2005.

Bazin, Andre. "The Ontology of the Photographic Image." In *Classic Essays on Photography,* edited by Alan Trachtenberg. New Haven, CT: Leete's Island Books, 1980.

Beam, Gordon C.F. "Wittgenstein and the Uncanny." *Soundings: An Interdisciplinary Journal* 76, no. 1 (1993).

Beavan, Colin. *Fingerprints: The Origins of Crime Detection and the Murder Case That Launched Forensic Science*. New York: Hyperion, 2001.

Bergen, C., M.J.L.Th. Niekus, and V.T. van Vilsteren, eds. *The Mysterious Bog People.* Zwolle: Waanders Publishers, 2002.

Bendix, Regina. *In Search of Authenticity: The Formation of Folklore Studies.* Madison: University of Wisconsin Press, 1997.

Benjamin, Walter. *The Origin of German Tragic Drama.* Translated by John Osborne. London: NLB, 1977.

———. *One-way Street, and Other Writings.* Translated by Edmund Jephcott and Kingsley Shorter. London: NLB, 1979.

———. "The Work of Art in the Age of Mechanical Reproduction." In *Illuminations,* edited and with an introduction by Hannah Arendt. Translated by Harry Zohn. New York: Schocken Books, 1986.

———. "Excavation and Memory." In *Selected Writings,* translated by Marcus Paul Bullock, Michael William Jennings, Howard Eiland, and Gary Smith. Edited by Marcus Paul Bullock and Michael William Jennings. Cambridge, MA: Belknap Press, 1996.

Bennett, Tony. *The Birth of the Museum: History, Theory, Politics.* London: Routledge, 1995.

Bergson, Henri. *Creative Evolution.* Translated by Arthur Mitchell. Mineola, New York: Dover, 1998.

Bernhardsen, Christian. *Den stor gåde: Eventyret om P.V. Glob.* Copenhagen: J. Vinten, 1970.

Beuker, J.R. "The Bog: A Lost Landscape." In *The Mysterious Bog People,* edited by C. Bergen, M.J.L.Th. Niekhus, and V.T. van Vilsteren. Zwolle: Waanders Publishers, 2002.

Bigod 20. *The Bog.* A song. Lyrics by Jean Luc De Meyer. Performed by the band Bigod 20, 1990. See http://bigod20.info/the-bog-4.

Blanchot, Maurice. *The Space of Literature.* Translated by Ann Smock. Lincoln, NE: University of Nebraska Press, 1982.

Blicher, Steen Steensen. "Dronning Gunnild," in *Steen Steensen Blichers Samlede Skrifter.* Copenhagen: Det Danske Sprog og Litteraturselskab, 1928.

Bloch, Ernst. *The Utopian Function of Art and Literature: Selected Essays.* Translated by Jack Zipes and Frank Mecklenburg. Cambridge, MA: MIT Press, 1988.

Bloch, Marc. *The Historian's Craft.* Translated by Peter Putnam. New York: Vintage, 1964.

Boettger, Suzaan. *Earthworks: Art and the Landscape of the Sixties.* Berkeley and Los Angeles: University of California Press, 2002.

Bogland Symposium Exhibition. Ireland: Crescent Arts Center, 1990.

Bowdler, Sandra. "Freud and Archaeology." *Anthropological Forum* 7 (1996).

Boym, Svetlana. *The Future of Nostalgia.* New York: Basic Books, 2001.

Braudel, Fernand. *On History.* Translated by Sarah Mathews. Chicago: University of Chicago Press, 1980.

Bronfen, Elisabeth. *Over Her Dead Body: Death, Femininity and the Aesthetic.* New York: Routledge, 1992.

———. *The Knotted Subject: Hysteria and its Discontents.* Princeton, NJ: Princeton University Press, 1998.

Brothwell, Don. *The Bog Man and the Archaeology of People.* London: British Museum Press, 1986.

Brown, Bill. "Thing Theory." *Critical Inquiry,* Autumn 2001.

Bryson, Bill. *A Short History of Nearly Everything.* London: Black Swan, 2004.

Buck-Morss, Susan. *The Dialectics of Seeing: Walter Benjamin and the Arcades Project.* Cambridge, MA: MIT Press, 1989.

Burke, Kenneth Duva. *A Grammar of Motives.* Berkeley and Los Angeles: University of California Press, 1969.

Butler, Judith. *Precarious Life: The Powers of Mourning and Violence.* London: Verso, 2004.

Cavell, Stanley. *The Claim of Reason: Wittgenstein, Skepticism, Morality and Tragedy.* Oxford: Oxford University Press, 1979.

Charlesworth, Michael. "Fox Talbot and the 'White': Mythology of Photography." *Word and Image.* 11, no. 3 (1995).

Christens, J.F. "Oplysningninger om et i en Mose nær Haraldskjær fundet Kvindeligt Lig," in *Annaler for Nordisk Oldkyndighed,* Denmark: Det Kongelige Nordiske Oldskrift-Selskab, 1836–37.

Christensen, Camilla. *Jorden under Høje Gladsaxe: Roman.* Copenhagen: Samleren, 2002.

Claus, Hugo. *Gedichten.* Amsterdam: De Bezige Bij, 2004.

Coles, Bryony, John Coles, and Mogen Schou Jørgensen, eds. *Bog Bodies, Sacred Sites and Wetland Archaeology.* Denmark: National Museum of Denmark, Silkeborg Museum, WARP, 1999.

Conkey, Margaret. "Context in the Interpretative Process." In *Beyond Art: Pleistocene Image and Symbol,* edited by Margaret Conkey et al. San Francisco: California Academy of Sciences, 1997.

Crane, Susan, ed. *Museums and Memory.* Stanford, CA: Stanford University Press, 2000.

Crew, Spencer R., and James E. Sims. "Locating Authenticity." In *Exhibiting Cultures: The Poetics and Politics of Display,* edited by Ivan Karp and Stephen D. Levine. Washington, DC: Smithsonian Institution Press, 1991.

Crescent Art Center. *Bogland Symposium.* Arranged by the Sculpture Society in Ireland under the patronage of Seamus Heaney, 1990.

Crowther, David. "Archaeology, Material Culure, and Museums." In *Museum Studies in Material Culture,* edited by Susan M. Pearce. Leicester, UK: Leicester University Press, 1989.

Cummings, Paul. Interview with Robert Smithson for the Archives of American Art/Smithsonian Institution. In *Robert Smithson: The Collected Writings,* edited by Jack D. Flam. Berkeley and Los Angeles: University of California Press, 1996.

Davis, Whitney. *Replications: Archaeology, Art History, Psychoanalysis.* Edited by Richard W Quinn. University Park: Pennsylvania State University Press, 1996.

Deem, James M. *Bodies from the Bog.* Boston: Houghton Mifflin, 1998.

Derrida, Jacques. *On Touching-Jean-Luc Nancy.* Translated by Christine Irizarry. Edited by Werner Hammacher. Stanford, CA: Stanford University Press, 2005.

Didion, Joan. *The Year of Magical Thinking.* New York: Alfred A. Knopf, 2005.

Docherty, Thomas. "Ana-; or Postmodernism, Landscape, Seamus Heaney." In *Seamus Heaney,* edited by Michael Allen. New York: St. Martin's Press, 1997.

Dollimore, Jonathan. *Death, Desire and Loss in Western Culture.* New York: Routledge, 1998.

Downey, Roger. *Riddle of the Bones: Politics, Science, Race, and the Story of Kennewick Man.* New York: Copernicus, 2000.

Draaisma, Douwe. *Metaphors of Memory: A History of Ideas About the Mind.* Translated by Paul Vincent. Cambridge: Cambridge University Press, 2000.

Drabble, Margaret. *A Natural Curiosity*. New York: Viking, 1989.

———. "Millennium Lecture: 'Runes and Bones.'" From abstract posted online. www .abroad-crwf.com/abroadwritingworkshop.html.

Dunand, Francoise and Roger Lichtenberg. *Mummies: A Voyage Through Eternity*. Translated by Ruth Sharman. New York and London: Harry N. Abrams, 1994.

Elkins, James. *Our Beautiful, Dry and Distant Texts: Art History as Writing*. New York and London: Routledge, 2000.

Ernst, Wolfgang. "Archi(ve)textures of Museology." In *Museums and Memory*, edited by Susan Crane. Stanford, CA: Stanford University Press, 2000.

Evans, Christopher. "Digging With the Pen: Novel Archaeologies and Literary Tradition." In *Interpretative Archaeology*, edited by Chrisopher Tilley. Providence, RI: Berg, 1993.

Fabian, Johannes. *Time and the Other: How Anthropology Makes its Object*. New York: Columbia University Press, 1983.

Finn, Christine. *Past Poetic: Archaeology in the Poetry of W.B. Yeats and Seamus Heaney*. London: Duckworth, 2004.

Fischer, Christian. *På sporet af Tollundmanden*. Copenhagen: Haase, 1988.

———. "Face to Face with Your Past." In *Bog Bodies, Sacred Sites and Wetland Archeology*, edited by Bryony Coles, John Coles, and Mogens Schou Jørgensen. Jutland, Denmark: National Museum of Denmark, Silkeborg Museum, WARP, 1999.

———. "The Tollund Man and the Elling Woman and Other Bog Bodies from Central Jutland." In *Bog Bodies, Sacred Sites and Wetland Archaeology*, edited by Bryony Coles, John Coles, and Mogens Schou Jørgensen. Jutland, Denmark: National Museum of Denmark, Silkeborg Museum, WARP, 1999.

Fisher, John Hayes, producer and director. *The Perfect Corpse*. DVD produced for Nova by Gary Glassman and Providence Pictures, Inc. A BBC/WGBH/ProSieben co-production, 2006.

Flam, Jack. "Introduction: Reading Robert Smithson." In *Robert Smithson: The Collected Writings*, edited by Jack D. Flam. Berkeley and Los Angeles: University of California Press, 1996.

Flon, Christine, ed. *The World Atlas of Archaelogy*. London: Mitchell Beazley, 1985.

Foster, Hal. *The Return of the Real: The Avant-garde at the End of the Century*. Cambridge, MA: MIT Press, 1996.

Foster, John Wilson. *The Achievement of Seamus Heaney*. Dublin: Lilliput Press, 1995.

Foucault, M. *Archaeology of Knowledge*. Translated by A.M. Sheridan Smith. London: Routledge, 2002.

Freud, Sigmund, *Totem and Taboo: Resemblances between the Psychic Lives of Savages and Neurotics*. Harmondsworth: Penguin Books, 1938.

———. "A Note on the Mystic Writing Pad." In *The Standard Edition of the Complete Psychological Works of Sigmund Freud*. London: Hogarth Press, 1961.

———. *The Uncanny*. Translated by David McLintock. London: Penguin, 2003.

———. *The Ego and the Id*. Translated by Joan Riviere. London: L. & V. Woolf; Institute of Psycho-Analysis, 1927.

———. *Civilization and Its Discontents*. Translated by James Strachey. New York: W. W. Norton & Company, 1961.

———. *The Freud Reader.* Edited by Peter Gay. New York: W. W. Norton & Company, 1995.

Frye, Northrop, 1957. *Anatomy of Criticism: Four Essays.* Harmondsworth: Penguin, 1990.

Gamble, Clive. Foreword to *Ancestral Images: The Iconography of Human Origins,* by Stephanie Moser. Ithaca, NY: Cornell University Press, 1998.

———. *Archaeology: The Basics.* London: Routledge, 2001.

Gay, Peter. *Freud: A Life for Our Time.* New York: W.W. Norton & Company, 1989.

Gässler, Ewald. "Wenn Künstler heute ins Moor gehen: Das Moor in der Kunst nach 1945 bis in die Gegenwart." In *Moor—eine verlorene Landschaft.* Oldenburg: Isensee Verlag, 2001.

Gebühr, Michael. *Moorleichen in Schleswig-Holstein.* Schleswig: Archäologishes Landesmuseum der Stiftung Schleswig-Holstrinische Landesmuseum Schloss Gottorf, 2002.

Geertz, Clifford. *Interpretation of Cultures.* New York: Basic Books, 1973.

Gero, Joan M., and Margaret Conkey, eds. *Engendering Archaeology: Women and Prehistory.* Oxford: B. Blackwell, 1991.

Getz, David. *Frozen Man.* New York: H. Holt and Co., 1994.

———. *Frozen Girl.* New York: H. Holt and Co., 1998.

Gilbert-Rolfe, Jeremy. *Beauty and the Contemporary Sublime.* New York: Allworth Press, 1999.

Gilliam, Terry. *Tidelands.* A feature film directed by Terry Gilliam. Capri Films, 2005.

Girard, René. *Violence and the Sacred.* Translated by Patrick Gregory. Baltimore: Johns Hopkins University Press, 1977.

Glob, P. V. *Mosefolket: Jernalderens mennesker bevaret i 2000 aar.* Copenhagen: Gyldendal, 1965.

———. *The Bog People: Iron Age Man Preserved.* Translated by Rupert Bruce-Mitford. Ithaca, NY: Cornell University Press, 1969.

Goethe, Johann Wolfgang. "Römische Elegien (Roman Elegies)." In *Gedenkausgabe der Werke, Briefe und Gespräche, Bd. 1.* Zurich and Stuttgart: Artemis, 1795.

Green, Miranda Aldhouse. *Dying for the Gods: Human Sacrifice in Iron Age and Roman Europe.* Gloucestershire, UK: Tempus Publishing, Limited, 2001.

———. *An Archaeology of Images: Iconology and Cosmology in Iron Age and Roman Europe.* New York: Routledge, 2004.

Greenberg, Moshe. *Ezekiel 21–37: A New Translation with Introduction and Commentary.* New York: Doubleday, 1997.

Greenblatt, Stephen. "Resonance amd Wonder." In *Exhibiting Cultures: The Poetics and Politics of Display,* edited by Ivan Karp and Stephen D. Levine. Washington, DC: Smithsonian Institution Press, 1991.

Grigson, Geoffrey. *Collected Poems 1963–1980.* London: Allison & Busby, 1982.

Gumbrecht, Hans Ulrich. *Production of Presence: What Meaning Cannot Convey.* Stanford, CA: Stanford University Press, 2004.

Halbwachs, Maurice. *On Collective Memory.* Translated by Lewis A Coser. Chicago: University of Chicago Press, 1992.

Hallam, Elizabeth, and Jenny Hockey. *Death, Memory and Material Culture.* Providence, RI: Berg, 2001.

Halpern, Daniel, ed. *Writers on Artists.* San Francisco: North Point Press, 1988.

Hansen, C.C. *Identifikation og rekonstruktion af historiske personers udseende paa grundlag af skelettet.* Copenhagen: J.H. Schultz, 1921.

Hansen, Martin Alfred. *Orm Og Tyr,* 6th edition. Copenhagen: Gydendal, 1963.

Harmon, Maurice. " 'We Pine for Ceremony': Ritual and Reality in the Poetry of Seamus Heaney, 1965–75." In *Seamus Heaney: A Collection of Critical Essays,* edited by Elmer Andrews. New York: St. Martin's Press, 1992.

Harper, Catherine. *A Beginning.* Derry: Orchard Gallery, 1991.

Heaney, Seamus. *Death of a Naturalist.* London: Faber & Faber, 1966.

———. *Door into the Dark.* London: Faber & Faber, 1969.

———. "Feeling into Words." In *Preoccupations: Selected Prose, 1968–1978.* New York: Farrar, Straus, Giroux, 1980.

———. *Wintering Out.* London: Faber and Faber, 1972.

———. *North.* London: Faber and Faber, 1975.

———. *Preoccupations: Selected Prose, 1968–1978.* New York: Farrar, Straus, Giroux, 1980.

———. *The Government of the Tongue: Selected Prose 1978–1987.* New York: Farrar, Straus and Giroux, 1988.

———. *Sweeney's Flight. Based on the Revised Text of "Sweeney Astray": With the Complete Revised Text of "Sweeney Astray."* London: Faber and Faber, 1992.

———. *Spirit Level.* London: Faber and Faber, 1996.

———. *Opened Ground: Selected Poems, 1966–1996.* New York: Farrar, Straus and Giroux, 1998.

———. "The Man in the Bog." In *Bog Bodies, Sacred Sites and Wetland Archeology,* edited by Bryony Coles, John M Coles, and Mogen Schou Jørgensen. Denmark: National Museum of Denmark, Silkeborg Museum WARP, 1999.

Hedeager, Lotte, and Karen Schusboe, eds. *Brugte historier: ti essays om brug og misbrug af historien.* Copenhagen: Akademisk Forlag, 1989.

Heidegger, Martin. "Building Dwelling Thinking" and "The Thing." In *Poetry, Language, Thought,* translated by Albert Hofstadter. New York: Harper and Row, 2001.

Himmler, Heinrich. *Geheimreden 1933 bis 1945 und andere Ansprachen.* Edited by Bradley F. Smith and Agnes F. Peterson. Berlin and Wien, 1974.

Hodder, Ian. *The Archaeological Process: An Introduction.* Oxford: Blackwell, 1999.

Holly, Michael Ann. *Past Looking: The Historical Imagination and the Rhetoric of the Image.* Ithaca, NY: Cornell University Press, 1996.

Hostrup, Jens Christian. "En Spurv i Tranedans." In *Komedier og Digte.* Copenhagen: H. Topsøe-Jensen. Det Danske Forlag, 1954.

Hunter, Jefferson. *Image and Word: The Interaction of Twentieth-century Photographs and Texts.* Cambridge, MA: Harvard University Press, 1987.

Hvass, Lone. *Dronning Gunhild: et moselig fra jernalderen.* Copenhagen: Sesam, 1998.

Jameson, John, John E. Ehrenhard, and Christine A. Finn. "Introduction: Archaeology as Inspiration—Invoking the Ancient Muses." In *Ancient Muses: Archaeology and the Arts,* edited by John Jameson et.al. Tuscaloosa, AL: University Alabama Press, 2003.

Jenkins, Ian. *Archaeologists & Aesthetes: In the Sculpture Galleries of the British Museum 1800–1939.* London: British Museum Press, 1992.

Jensen, Jørgen. *Thomsens Museum: Historien om Nationalmuseet.* Copenhagen: Gyldendal, 1992.

Jorn, Asger. *Asger Jorn og 10.000 aars Nordisk folkekunst.* Edited by Troels Andersen and Tove Nyholm. Silkeborg, Denmark: Silkeborg Kunstmuseums Forlag, 1995.

Jung, C. G. *Memories, Dreams, Reflections.* Translated by Richard Winston and Clara Winston. Edited by Aniela Jaffé. New York: Vintage Books, 1989.

Kalisch, Shoshana, and Barabara Meister. *Yes, We Sang! Songs of the Ghettos and Concentration Camps.* New York: Harper and Row, 1985.

Kantaris, Sylvia. "Couple, Probably Adulterous (Assen, Holland, circa Roman Times)." In *Dirty Washing: New & Selected Poems.* Newcastle upon Tyne: Bloodaxe, 1989.

Karp, Ivan, and Steven Lavine, eds. *Exhibiting Cultures: The Poetics and Politics of Museum Display.* Washington, DC: Smithsonian Institution Press, 1991.

Katz, Joel. *Strange Fruit.* A documentary film, written, produced and directed by Joel Katz. USA: 2002.

Kaul, Flemming. "The Bog-The Gateway to another World." In *The Spoils of Victory: The North in the Shadow of the Roman Empire,* edited by Lars Jørgensen, Birger Storgaard, and Lone Gebauer Thomsen. Copenhagen: Danish National Museum, 2003.

Kimmelman, Michael. *The Accidental Masterpiece: On the Art of Life and Vice Versa.* New York: Penguin Press, 2006.

Kirshenblatt-Gimblett, Barbara. "Objects of Ethnography." In *Exhibiting Cultures: The Poetics and Politics of Display,* edited by Ivan Karp and Steven Lavine. Washington, DC: Smithsonian Institution Press, 1991.

Klein, Kerwin Lee. "On the Emergence of Memory in Historical Discourse." *Representations* 69 (2000).

Klint-Jensen, Ole. A *History of Scandinavian Archaeology.* London: Thames and Hudson, 1975.

Kofman, Sarah. *Freud and Fiction.* Translated by Sarah Wykes. Cambridge: Polity, 1991.

Kohl, Philip L., and Clare P. Fawcett. "Archaeology in the Service of the State: Theoretical Considerations." In *Nationalism, Politics, and the Practice of Archaeology,* edited by Philip L. Kohl and Clare P. Fawcett. Cambridge: Cambridge University Press, 1995.

———, eds. *Nationalism, Politics and the Practice of Archaeology.* Cambridge: Cambridge University Press, 1995.

Kracauer, Siegfried. "Photography." In *Classic Essays on Photography,* edited by Alan Trachtenberg. New Haven, CT: Leete's Island Books, 1980.

Krajick, Kevin. "The Mummy Doctor." *The New Yorker,* May 16, 2005.

Kristiansen, Kristian, ed. *Archaeological Formation Processes: The Representativity of Archaeological Remains from Danish Prehistory.* Copenhagen: Danish National Museum, 1985.

———. "Fortids kraft og kæmpestyrke: om national og politisk brug af fortiden." In *Brugte historier: ti essays om brug og misbrug af historien,* edited by Lotte Hedeager and Karen Schusboe. Copenhagen: Akademisk Forlag, 1989.

———. *Europe before History.* Cambridge: Cambridge University Press, 1998.

Kristeva, Julia. *Powers of Horror: An Essay on Abjection.* Translated by Leon S. Roudiez. New York: Columbia University Press. 1982.

Kubler, George. "The Nature of Actuality." In *The Shape of Time: Remarks on the History of Things*. New Haven, CT: Yale University Press, 1962.

Lakoff, George, and Mark Johnson. *Metaphors We Live By*. Chicago: University of Chicago Press, 1980.

Laqueur, Thomas. "Clio Looks at Corporal Politics." In *Corporal Politics,* edited by Donald Hall. Boston: Beacon Press, 1992.

Lévinas, Emmanuel. *Entre Nous: On Thinking of the Other.* Translated by Michael B. Smith and Barbara Harshay. New York: Columbia University Press, 1998.

Lewin, P.K. "Current Technology in the Examination of Ancient Man." In *Human Mummies. A Global Survey of their Status and the Techniques of Conservation,* edited by K. Spindler et al. Vienna and New York: Springer, 1996.

Lippard, Lucy R. *Overlay: Contemporary Art and the Art of Prehistory*. New York: New Press, 1983.

Llosa, Mario Vargos. "Botero: A Sumptuous Abundance." In *Making Waves* by Mario Vargas Llosa and John King. New York: Farrar, Straus & Giroux, 1996.

Lloyd, David. " 'Pap for the Dispossessed': Seamus Heaney and the Poetics of Identity." *Boundary* 2, vol. 13, no. 2/3 (1985).

Longley, Edna. " 'Inner Emigré' or 'Artful Voyeur?' Seamus Heaney's North." In *Seamus Heaney,* edited by Michael Allen. New York: St. Martin's Press, 1997.

Lowenthal, David. *The Past Is a Foreign Country*. Cambridge: Cambridge University Press, 1985.

Luckenbach, Julie. *Beuys/Logos*. A hyper-essay. http://www.walkerart.org/beuys/hyper.

Lund, Allan A. *Moselig*. Højbjerg, Denmark: Wormianum, 1976.

———. *Hitlers håndlangere: Heinrich Himmler og den nazistiske raceideologi*. Copenhagen: Samleren, 2001.

———. *Mumificerede moselig*. Copenhagen: Høst & Søn, 2002.

Lyon, Christopher. "Beuys. Thinking is Form. The Drawings of Joseph Beuys." In catalogue to exhibit at Museum of Modern Art in New York, February 21–May 4, 1993.

Magli, Patrizia. "The Face and the Soul." In *Fragments of a History of the Human Body. Vol. II,* edited by Michel Feher, Ramona Naddaff, and Nadia Tazi. Cambridge, MA: Urzone Inc, Zone Books, and MIT Press, 1989.

Malraux, Andre. *The Voices of Silence*. Translated by Stuart Gilbert. Princeton, NJ: Princeton University Press, 1978.

Margalit, Avishai. *The Ethics of Memory*. Cambridge, MA: Harvard University Press, 2002.

McCarthy, Cormac. *The Road*. New York: Vintage Books, 2006.

McEvilley, Thomas: "Was hat der Hase gesagt? Fragen über, für oder von Joseph Beuys." In *Joseph Beuys. Skulpturen und Objekte,* edited by Heiner Bastian. Munich: Schirmer/ Mosel, 1988.

McNeill, Daniel. *The Face*. London: Penguin Books, 1998.

Menninghaus, Winfried. *Disgust: Theory and History of a Strong Sensation*. Albany, NY: State University of New York Press, 2003.

Merleau-Ponty, Maurice. *The Visible and the Invisible: Followed by Working Notes*. Evanston, IL: Northwestern University Press, 1968.

Michaels, Anne. *Fugitive Pieces*. New York: Vintage, 1998.

Miller, William Ian. *The Anatomy of Disgust*. Cambridge, MA: Harvard University Press, 1997.

Mitchell, W.J.T. "Going Too Far with the Sister Arts." In *Space, Time, Image, Sign: Essays on Literature and the Visual Arts*, edited by James A.W. Heffernan. New York: Peter Lang, 1987.

———. *Picture Theory: Essays on Verbal and Visual Representation*. Chicago: University of Chicago Press, 1994.

———. *The Last Dinosaur Book: The Life and Times of a Cultural Icon*. Chicago: University of Chicago Press, 1998.

Mordhorst, Mads. "Kildernes magi." *Fortid og Nutid,* June 2001.

Moretti, Franco. *Atlas of the European Novel, 1800–1900*. London: Verso, 1998.

Moseman, Lori Anderson. *Persona*. Jamaica Plain, MA: Swank Books, 2003.

Moser, Stephanie. *Ancestral Images: The Iconography of Human Origins*. Ithaca, NY: Cornell University Press, 1998.

Mueller, Robert Emmett, "Mnemesthics: Art as the Revivification of Significant Consciousness Events." *Leonardo* 21, no. 2 (1988).

Myhre, Lise Nordenborg. "Nationalism, Politics, and the Practice of Archaeology." In *Myter om det nordiske,* edited by Catherina Raudvere, Anders Andrén, and Kristina Jennbert. Lund: Nordic Academic Press, 2001.

Nancy, Jean-Luc. *The Ground of the Image*. Translated by Jeff Fort. New York: Fordham University Press, 2005.

National Museum of Ireland. Website to see image of Old Croghan Man: http://www.bbc .co.uk/history/programmes/timewatch/diary_bog_07.shtml.

Nielsen, Kjersti. "Etikk og godt bevarte menneskelige levninger: etiske aspekter ved utstilling av og forskning på grønlandske mumier og danske moselik." Tromsø, Norway: Unpublished thesis, University of Tromsø, 2002.

Nora, Pierre. "Between Memory and History: Les Lieux de Mémoire." Translated by Marc Roudebush, *Representations,* Spring 1989.

———. "Between Memory and History." In *Realms of Memory: The Construction of the French Past,* translated by Arthur Goldhammer, edited by Lawrence D. Kritzman. New York: Columbia University Press, 1996.

O'Brian, Conor Cruise. *The Listener,* September 25, 1975.

Oldsags-Committeen. "Bemærkninger om et fund af et mumieagtigt kvinde-lig i en mose ved Haraldskjær i Jylland" A forensic report in *Annaler for nordisk oldkyndighed,* Copenhagen: Kongelige Nordiske Oldskriftselskab, 1836–37.

Olsen, Bjørnar. *Fra ting til tekst: teoretiske perspektiv i arkeologisk forskning.* Oslo: Universitetsforlaget, 1997.

Ondaatje, Michael. *Anil's Ghost*. New York: Alfred A. Knopf, 2000.

Ortega y Gaset, José. "On Point of View in the Arts." In *Writers on Artists,* translated by Poul Snodgrass and Joseph Frank, edited by Daniel Halpern. San Francisco: North Point Press, 1989.

———. *The Dehumanization of Art and Notes on the Novel*. Translated by Helene Weyl. Princeton, NJ: Princeton University Press, 1948.

Otter, Samuel. *Melville's Anatomies*. Berkeley and Los Angeles: University of California Press, 1999.

Palmer, Douglas. *Fossil Revolution: The Finds that Changed our View of the Past.* London: HarperCollins, 2003.

Pearce, Susan M., ed. *Museum Studies in Material Culture.* Leicester, UK: Leicester University Press, 1989.

Plato. *Phaedo.* Translated by Harold North Fowler. Cambridge, MA: Harvard University Press, 1913.

Pluciennik, Mark. "Archeological Narratives and Other Ways of Telling." *Current Anthropology* 40, no. 5 (December 1999).

———. "Art, Artefact, Metaphor." In *Thinking Through the Body: Archaeologies of Corporeality,* edited by Yannis Hamilakis, Mark Pluciennik, and Sarah Tarlow. London: Kluwer/Academic Press, 2002.

Poinar, George O., and Roberta Poinar. *The Amber Forest: A Reconstruction of a Vanished World.* Princeton, NJ: Princeton University Press, 1999.

Pope, Alexander, 1734. "Epistle to Dr. Arbuthnot." In *The Poems of Alexander Pope: A Reduced Version of the Twickenham Text.* New Haven, CT: Yale University Press, 1966.

Prag, John, and Richard Neave. *Making Faces: Using Forensic and Archaeological Evidence.* London: British Museum Press, 1997.

Pratt, Mary Louise. *Imperial Eyes: Travel Writing and Transculturation.* London: Routledge, 1992.

Pringle, Heather. *The Mummy Congress: Science, Obsession, and the Everlasting Dead.* New York: Hyperion, 2002.

Purdy, Anthony. "The Bog Body as Mnemotype: Nationalist Archaeologies in Heaney and Tournier." *Style,* Spring 2002.

———. "Unearthing the Past: The Archaeology of Bog Bodies in Glob, Atwood, Hébert and Drabble." *Textual Practice* 16, no. 3 (2002).

Randall, James. "An Interview with Seamus Heaney." *Ploughshares* 5, no. 3 (1979).

Reich, Ebbe Kløvedal. *Fæ og frænde: syv en halv nats fortællinger om vejene til Rom og Danmark.* Copenhagen: Gyldendal, 1977.

Renfrew, Colin. *Figuring It Out: What Are We? Where Do We Come From? The Parallel Vision of Artists and Archaeologists.* London: Thames & Hudson, 2003.

Richter, Gerhard. *Walter Benjamin and the Corpus of Autobiography.* Detroit: Wayne State University Press, 2000.

Ricoeur, Paul. *Interpretation Theory: Discourse and the Surplus of Meaning.* Fort Worth: Texas Christian University Press, 1976.

———. *Oneself As Another.* Translated by Kathleen Blamey. Chicago: University of Chicago Press, 1992.

———. *Memory, History, Forgetting.* Translated by Kathleen Blamey and David Pellauer. Chicago: University of Chicago Press, 2004.

Riegl, Aloïs. *Historical Grammar of the Visual Arts.* Translated by Jacqueline E. Jung. New York: Zone Books, 2004.

Roach, Mary. *Stiff: The Curious Lives of Human Cadavers.* New York: W.W. Norton, 2003.

Rosenthal, Mark. *Anselm Kiefer.* Chicago: Art Institute of Chicago, 1987.

Ross, Anne, and Don Robins. *Life and Death of a Druid Prince: How the Discovery of Lindow Man Revealed the Secrets of a Lost Civilization.* New York: Touchstone, 1991.

Rougemont, Denis de. *Love in the Western World.* Translated by Montgomery Belgion. Princeton, NJ: Princeton University Press, 1983.

Royle, Nicholas. *The Uncanny: An Introduction.* New York: Routledge, 2003.

Rugg, Linda Haverty. *Picturing Ourselves: Photography & Autobiography.* Chicago: University of Chicago Press, 1997.

Sagan, Carl. "Carl Sagan on Time Travel." *Nova* online. www.pbs.org.wgbh.nova/time/sagan.html

Sandberg, Mark B. *Living Pictures, Missing Persons.* Princeton, NJ: Princeton University Press, 2003.

Sanden, Wijnand van der. *Through Nature to Eternity: The Bog Bodies of Northwest Europe.* Translated by Susan J Mellor. Amsterdam: Batavian Lion International, 1996.

———. "Mummies, Mugs, and Museum Shops." Online in *Archaeology: A Publication of the Archaeological Institute of America.* http://www.archaeology.org/online/features/bog/exhibit.html, August 30, 2005.

Sanders, Karin. *Konturer: skulptur- og dødsbilleder fra Guldalderlitteraturen.* Copenhagen: Museum Tusculanum Press, 1997.

———. "The Archaeological Object in Word and Image." *Edda /03. Scandinavian Journal of Literary Research.* Edited by Unni Langås, Andreas G. Lombnæs, Jahn Thon. Oslo: Norwegian University Press (2002).

———. "Bad Bog Babes." *Passage* #50. Edited by Tore Rye Andersen et al. Aarhus, Denmark: Aarhus University Press (2004).

———. "Bodies in Process: Bog Bodies in Contemporary Art and Poetry." *Edda. 4/04. Scandinavian Journal of Literary Research.* Edited by Unni Langås. Oslo: Norwegian University Press (2004).

———. "Fra 'sölen' og 'pölen': et moseligs fortælling i material- og litteraturhistorien." In *Kampen om litteraturhistorien,* edited by Marianne Alenius, Thomas Bredstorff, and Gert Emborg. Copenhagen: Dansklærerforeningens Forlag, 2004.

———. " 'Upon the Bedrock of Material Things': The Journey to the Past in Danish Archaeological Imagination." In *Northbound,* edited by Karen Klitgaard Povlsen. Denmark: Aarhus University Press, 2007.

———. "Anachronistic Encounters: A Reception Story." In *Nordic Naturecultures: Ecocritical Approaches to Film, Art and Literature,* edited by C. Claire Thomson and Christopher Oscarson (forthcoming).

———. "A Portal through Time: Queen Gunhild 81.1" *Scandinavian Studies 81.1* (2009).

———. Imagining Origin: Bog Bodies in the Discourse of Archaeological Ambivalence." In *Culture and Media Studies: European Perspectives,* edited by Maria Pilar Rodrigues. Reno, NV: University of Nevada Press, 2009.

Scarry, Elaine. *The Body in Pain: The Making and Unmaking of the World.* New York: Oxford University Press, 1987.

Schama, Simon. *Landscape and Memory.* New York: Vintage, 1996.

Schjeldahl, Peter. "What on Earth." *The New Yorker,* September 5, 2005.

Schnapp, Alain. *Discovery of the Past.* London: British Museum Press, 1997.

Schwartz, Hillel. *The Culture of the Copy: Striking Likenesses, Unreasonable Facsimiles.* New York: Zone Books, 1996.

Schwartz, Jeffrey H. *What the Bones Tell Us.* Tucson, AZ: University of Arizona Press, 1993.

Scott, Charles E. *The Lives of Things.* Bloomington, IN: Indiana University Press, 2002.

Shelley, Mary Wollstonecraft. *Frankenstein; or, The Modern Prometheus.* Edited by Maurice Hindle. London: Penguin, 2003.

Silberman, Niel Asher. "The Politics and Poetics of Archaeological Narrative." In *Nationalism, Politics and the Practice of Archaeology,* edited by Philip L. Kohl and Clare Fawcett. Cambridge: Cambridge University Press, 1995.

———. "Promised Lands and Chosen Peoples: The Politics and Poetics of Archaeological Narrative." In *Nationalism, Politics, and the Practice of Archaeology,* edited by Philip L. Kohl and Clare Fawcett. Cambridge: Cambridge University Press, 1995.

Simmons, James. "The Trouble with Seamus." In *Seamus Heaney: A Collection of Critical Essays,* edited by Elmer Andrews. New York: St. Martin's Press, 1992.

Simon, Jack and Harm Posthumus. "In my Mind (the Girl from Yde.)" *Snoaren.* Compact Disc, Snoaren Productions and Maura Music, Holland, 2000.

———. "Forever Young." In *Snoaren.* Compact disc, Snoaren Productions and Maura Music, Holland, 2000.

Singer, Charles Joseph, trans. *Galen on Anatomical Procedures: Translation of the Surviving Books with Introduction and Notes.* London: Oxford University Press, 1956.

Smiles, Sam, and Stephanie Moser, eds. *Envisioning the Past: Archaeology and the Image.* Malden, MA: Blackwell, 2005.

Smithson, Robert. *Robert Smithson: The Collected Writings.* Edited by Jack D. Flam. Berkeley and Los Angeles: University of California Press, 1996.

———. "The Spiral Jetty." In *Robert Smithson: The Collected Writings,* edited by Jack D. Flam. Berkeley and Los Angeles: University of California Press, 1996.

Sobchack, Vivian. *Carnal Thoughts: Embodiment and Moving Image Culture.* Berkeley and Los Angeles: University of California Press, 2004.

Sontag, Susan. *On Photography.* New York: Anchor Books, 1990.

———. *Regarding the Pain of Others.* New York: Farrar, Straus and Giroux, 2003.

Spaid, Sue. *Ecovention: Current Art to Transform Ecologies.* Published by Ecoartspace on the occasion of the exhibition "Ecovention," held at the Contemporary Arts Center, Cincinnati, Ohio, June 9–Aug. 18, 2002.

Spindler, Konrad. *The Man in the Ice: The Discovery of a 5,000-Year-Old Body Reveals the Secrets of the Stone Age.* New York: Random House, 1996.

Spindler, Konrad, H. Wilfing, E. Rastbicheer-Zissernig, D. zur Nedden, and H. Nothdurfer, eds. *Human Mummies: A Global Survey of Their Status and the Techniques of Conservation.* Vienna: Springer, 1996.

Stead, I.M., J.B. Bourke, and Don Brothwell, eds. *Lindow Man: The Body in the Bog.* London: British Museum Publications, 1986.

Stegner, Wallace. *The Spectator Bird.* London: Penguin, 1976.

Steiner, Wendy. *Venus in Exile: The Rejection of Beauty in Twentieth-Century Art.* Chicago: University of Chicago Press, 2001.

Stewart, Susan. *On Longing: Narratives of the Miniature, the Gigantic, the Souvenir, the Collection.* Durham, NC: Duke University Press, 1993.

———. *Poetry and the Fate of the Senses.* Chicago: University of Chicago Press, 2002.

Stimilli, Davide. *The Face of Immortality: Physiognomy and Criticism.* Albany, NY: State University of New York Press, 2005.

Stokvis, Willemijn. *Cobra: spontanitetens veje.* Copenhagen: Aschehoug, 2003.

Strieder, Barbara. "'Künstlerpost'—zu einigen plastischen Bildern von Joseph Beuys." In *Joseph Beuys. Plastische Bilder 1947–1970.* Stuttgart: Verlag Gerd Hatje. 1990.

Sudjic, Deyan. "Engineering Conflict." *The New York Times Magazine.* Section 6. May 21, 2006.

Suk, V. "Fallacies of Anthropological Identifications and Reconstruction: A Critique Based on Anatomical Dissections." Prague: Publications of the Faculty of Science, Charles University, 1935.

Tacitus, Cornelius. *The Agricola and the Germania.* Translated by Harold B. Mattingly and S.A. Handford. London: Penguin, 1970.

Talbot, Michael. *The Bog.* New York: Jove Books, 1986.

Taussig, Michael T. *Defacement: Public Secrecy and the Labor of the Negative.* Stanford, CA: Stanford University Press, 1999.

Taylor, John H, and British Museum. *Mummy: The Inside Story.* London: British Museum Press, 2004.

Taylor, Timothy. *The Buried Soul: How Humans Invented Death.* London: Fourth Estate, 2002.

Temkin, Ann, and Bernice Rose. *Beuys. Thinking is Form. The Drawings of Joseph Beuys.* With a contribution by Dieter Koepplin. New York: Museum of Modern Art, 1993.

Theweleit, Klaus. *Männerphantasien: Vol 1. Frauen. Fluten, Körper, Geschichte:* Berlin: Rowohlt, 1977.

Thomas, David Hurst. *Skull Wars: Kennewick Man, Archaeology, and the Battle for Native American Identity.* New York: Basic Books, 2000.

Thomas, Julian. *Time, Culture, and Identity: An Interpretative Archaeology.* London: Routledge, 1996.

———. *Archaeology and Modernity.* London: Routledge, 2004.

Tilley, Christopher, ed. *Interpretative Archaeology.* Providence, RI: Berg, 1993.

———. *Metaphor and Material Culture.* Oxford, UK: Blackwell, 1999.

Tisdall, Caroline. *Joseph Beuys.* London: Thames and Hudson, 1979.

Tonnaer, Désireé. *Leven in het licht van het verstrijken van de tijd. De beelden van Désirée Tonnaer.* Video. Directed by Hans Wynants. N.d.

Tournier, Michel. *The Ogre.* Translated by Barbara Bray. New York: Doubleday & Company, Inc., 1972.

Triffids, The "Jerdacuttup Man." From the album *Calenture.* Words by David McComb and music by David McComb with James Paterson. Island Records, 1987.

Trigger, Bruce. *Alternative Archeologies: Nationist, Colonialist, Imperialist.* Edited by Lotte Headager and Karen Schusboe. Copenhagen: Akademisk Forlag, 1989.

Trilling, Lionel. *Sincerity and Authenticity.* Cambridge, MA: Harvard University Press, 1972.

Trinkaus, Erik and Pat Shipman. *Neandertals, The Changing the Image of Mankind.* New York. Knopf, 1993.

Turner, R.C. "Boggarts, Bogles and Sir Gawain and the Green Knight: Lindow Man and the

Oral Tradition." In *Lindow Man: The Body in the Bog*, edited by I.M. Stead, J.B. Bourke, and Don Brothwell. London: British Museum Publications, 1986.

———. "Dating Lindow Moss and the Other British Bog Bodies." In *Bog Bodies, Sacred Sites and Wetland Archaeology*, edited by Bryony Coles, John Coles, and Mogen Schou Jørgensen. Denmark: National Museum of Denmark, Silkeborg Museum, WARP, 1999.

———. "Discovery and Excavation of the Lindow Bodies." In *Lindow Man: The Body in the Bog*, edited by I.M. Stead, J.B. Bourke, and Don Brothwell. London: British Museum Publications, 1986.

Turner, R.C., and C.S. Briggs. "The Bog Burial of Britain and Ireland." In *Lindow Man: The Body in the Bog*. London: British Museum Publications, 1986.

Vaughan, Kathleen. "Artist Statement: The Bog Series." Unpublished. N.p., 1995–96.

———. "Modes of Knowing and Artistic Practice: Beauty, Bog Bodies, and Brain Science." Unpublished article. N.p., n.d.

———. Website for works in color. http://www.akaredhanded.com/kv3cbogseries.html

Vendler, Helen Hennessy. *Seamus Heaney*. Cambridge, MA: Harvard University Press, 1998.

Vidler, Anthony. *The Architectural Uncanny: Essays in the Modern Unhomely*. Cambridge, MA: MIT Press, 1992.

Vilsteren, V.T. van. "Discoveries in the Bog: History and Interpretation." In *The Mysterious Bog People*, edited by C. Bergen, M.J.L.T. Niekhus, and V.T. van Vilsteren. Zwolle: Waanders Publishers, 2002.

Wailes, Bernard, and Amy L. Zoll. "Civilization, Barbarism, and Nationalism in European Archaeology." In *Nationalism, Politics, and the Practice of Archaeology*, edited by Philip L. Kohl and Clare Fawcett. Cambridge: Cambridge University Press, 1995.

Wallace, Jennifer. *Digging the Dirt: The Archaeological Imagination*. London: Duckworth, 2004.

Wells, H. G. *The Time Machine*. London: Penguin Classics, 2005.

Wheeler, Robert Eric Mortimer. *Archaeology from the Earth*. Oxford, UK: Clarendon Press, 1954.

White, Hayden. *Metahistory: The Historical Imagination in Nineteenth-Century Europe*. Baltimore: Johns Hopkins University Press, 1975.

Wilder, Thornton. *Our Town, a Play in Three Acts*. New York: Coward McCann, 1938.

Wilkinson, Caroline, Bob Brier, Richard Neave, and Denise Smith. "The Facial Reconstruction of Egyptian Mummies and Comparison with the Fauum Portraits." In *Mummies in a New Millenium: Proceedings of the 4th World Congress on Mummy Studies*, edited by Niels Lynnerup, Claus Andreasen, and Joel Berglund. Nuuk, Greenland: Greenland National Museum and Archives and Danish Polar Center, 2002.

Williams, William Carlos. "The Smiling Dane." In *Journey to Love*. New York: Random House, 1955.

Wittgenstein, Ludwig. *Philosophical Investigations: 50th Anniversary Commemorative Edition*. Translated by G.E.M. Anscombe. Boston: Blackwell Publishing, 2001.

Wollstonecraft, Mary. *Letters Written during a Short Residence in Sweden, Norway, and Denmark*. Lincoln, NE: University of Nebraska Press, 1976.

Wood, Michael. Foreword in *The World Atlas of Archaeology*, edited by Christine Flon. London: Mitchell Beazley, 1985.

World Congress on Mummy Studies. *Mummies in a New Millenium: Proceedings of the 4th World Congress on Mummy Studies.* Edited by Niels Lynnerup, Claus Andreasen, and Joel Berglund. Copenhagen: Greenland National Museum and Archives and Danish Polar Center, 2002.

Worsaae, J.J.A. *En oldgranskers erindringer.* Copenhagen: Gyldendal. 1935.

———. *Af en oldgranskers breve.* Copenhagen: Gyldendal, 1938.

Zerlang, Martin. "I historiens pariserhjul: Ebbe Kløvedal Reich." In *Dansk litteraturs historie: 1960–2000,* edited by Klaus P. Mortensen and May Schack. Copenhagen: Gyldendal, 2006.